LAGRANGIAN AND HAMILTONIAN METHODS FOR NONLINEAR CONTROL

A Proceedings volume from the IFAC Workshop,
Princeton, New Jersey, USA, 16 – 18 March 2000

Edited by

N.E. LEONARD
Mechanical and Aerospace Engineering Department,
Princeton University, Princeton, USA

and

R. ORTEGA
SUPELEC-LSS, Gif-sur-Yvette, France

Published for the

INTERNATIONAL FEDERATION OF AUTOMATIC CONTROL

by

PERGAMON
An Imprint of Elsevier Science

UK Elsevier Science Ltd, The Boulevard, Langford Lane, Kidlington, Oxford, OX5 1GB, UK
USA Elsevier Science Inc., 660 White Plains Road, Tarrytown, New York 10591-5153, USA
JAPAN Elsevier Science Japan, Tsunashima Building Annex, 3-20-12 Yushima, Bunkyo-ku, Tokyo 113, Japan

First edition 2000

Library of Congress Cataloging in Publication Data

IFAC Workshop (2000 : Princeton, N.J.)
Lagrangian and Hamiltonian methods for nonlinear control : a proceedings volume from the IFAC Workshop, Princeton, New Jersey, USA, 16-18 March 2000 / edited by N.E. Leonard and R. Ortega.
p. cm.
ISBN 0-08-043658-7
1. Nonlinear control theory--Congresses. 2. Lagrange equations--Congresses. 3. Hamiltonian systems--Congresses. I. Leonard, Naomi Ehrich. II. Ortega, Romeo, 1954- III. Title.

QA402.35 .I43 2000
629.8'36--dc21
00-060640

British Library Cataloguing in Publication Data

A catalogue record for this book is available from the British Library

ISBN 0-08-043658 7

Transferred to digital printing 2005

IFAC WORKSHOP ON LAGRANGIAN AND HAMILTONIAN METHODS FOR NONLINEAR CONTROL

Sponsored by
International Federation of Automatic Control (IFAC)
Technical Committee on Nonlinear Systems
American Automatic Control Council (AACC)
National Science Foundation (NSF)
Army Research Office (ARO)

International Programme Committee (IPC)
Leonard, N.E. (USA) (Chair)
Ortega, R. (F) (Chair)

Bloch, A.M. (USA)
Fradkov, A.L. (RU)
Glad, T. (SE) *(IFAC Representative)*
McClamroch, N.H. (USA)
Samson, C. (F)
Schlacher, K. (A)
Sugie, T. (JP)
van der Schaft, A.J. (NL)

National Organizing Committee (NOC)
Leonard, N.E.
Ortega, R.
Bullo, F.
Ostrowski, J.
Matarese, S.

CONTENTS

OPTIMAL CONTROL FOR HALO ORBIT MISSIONS

Radu Serban * Wang Sang Koon * Martin Lo ****
Jerrold E. Marsden * Linda R. Petzold ***
Shane D. Ross * Roby S. Wilson ****

* *Department of Mechanical and Environmental Engineering, University of California, Santa Barbara, CA 93106.*
** *Navigation and Flight Mechanics, Jet Propulsion Laboratory M/S: 301-140L, 4800 Oak Grove Drive, Pasadena, CA 91109.*
*** *Control and Dynamical Systems, California Institute of Technology 107-81, Pasadena, CA 91125*

Abstract: This paper addresses the computation of the required trajectory correction maneuvers (TCM) for a halo orbit space mission to compensate for the launch velocity errors introduced by inaccuracies of the launch vehicle. By combining dynamical systems theory with optimal control techniques, we produce a portrait of the complex landscape of the trajectory design space. This approach enables parametric studies not available to mission designers a few years ago, such as how the magnitude of the errors and the timing of the first TCM affect the correction ΔV. The impetus for combining dynamical systems theory and optimal control in this problem arises from design issues for the Genesis Discovery mission being developed for NASA by the Jet Propulsion Laboratory. *Copyright ©2000 IFAC*

Keywords: Optimal Control, Mission design, Dynamical systems.

1. INTRODUCTION AND BACKGROUND

The Genesis Mission Genesis is a solar wind sample return mission (see Lo et al [1998]). It is one of NASA's first robotic sample return missions and is scheduled for launch in January 2001 to a halo orbit in the vicinity of the L_1 Lagrange point, one of the five equilibrium points in the three body problem. L_1 is unstable and lies between the Sun and the Earth at roughly 1.5 million km from the Earth in the direction of the Sun. Once there, the spacecraft will remain for two years to collect solar wind samples before returning them to the Earth for study. Figure 1 shows the Genesis halo orbit and the transfer and return trajectories in a rotating frame. This rotating frame is defined by fixing the X-axis along the Sun-Earth line, the Z-axis in the ecliptic normal direction, and with the Y-axis completing a right-handed coordinate system. The Genesis trajectory

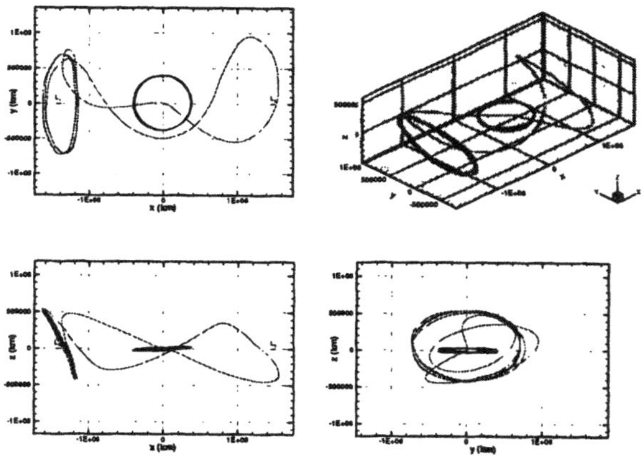

was designed using dynamical systems theory (see Howell et al [1997]). The three year mission, from launch all the way to Earth return, requires only a single small deterministic maneuver (less than 6 m/s) when injecting onto the halo orbit!

Halo Orbit Halo orbits are large three dimensional orbits shaped like the edges of a potato chip. The computation of halo orbits follows standard nonlinear trajectory computation algorithms based on parallel shooting.

The halo orbit, like the L_1 equilibrium point, is unstable. There is a family of asymptotic trajectories that departs from the halo orbit called the unstable manifold; similarly, there is a family of asymptotic trajectories which wind onto the halo orbit called the stable manifold. Each of these families form a two dimensional surface that is a twisted tubular surface eminating from the halo orbit. For Genesis, these manifolds are crucial for the mission design. The stable manifold, which winds onto the halo orbit, is used to design the transfer trajectory which delivers the Genesis spacecraft from launch to insertion onto the halo orbit (HOI). The unstable manifold, which winds off of the halo orbit, is used to design the return trajectory which brings the spacecraft and its precious samples back to Earth via a heteroclinic connection with L_2. See Koon et al [1999] for the current state of the computation of homoclinic and heteroclinic orbits in this problem.

Transfer Trajectory The transfer trajectory is designed using the following procedure. A halo orbit $H(t)$ is first selected, where t represents time. The stable manifold (W^S) of H consists of a family of asymptotic trajectories which take infinite time to wind onto H. Clearly, the exact asymptotic solutions cannot be found numerically and are impractical for space missions where the transfer time needs to be just a few months. Practically, there is a family of trajectories that lie arbitrarily close to W^S and that require just a few months to transfer between Earth and the halo orbit. A simple way to compute an approximation of W^S is based on Floquet theory.

In this paper, we will assume that the halo orbit, H(t), and the stable manifold M(t) are fixed and provided. We will not dwell further on their computation which is well covered in the references. Instead, let us turn our attention to the trajectory correction maneuver (TCM) problem.

TCM Problem The most important error in the launch of Genesis is the launch velocity error. The one sigma expected error is 7 m/s for a boost of approximately 3200 m/s from a circular 200 km altitude Earth orbit. Such an error is rather large because halo orbit missions are extremely sensitive to launch errors. Typical planetary launches can correct launch vehicle errors 7 to 14 days after the launch. In contrast, halo orbit missions must generally correct the launch error within the first day after launch. This correction maneuver is called TCM1, being the first TCM of any mission.

For orbits such as the Genesis transfer trajectory, the correction maneuver, ΔV for change in velocity, grows sharply in inverse proportion to the time from launch. For a large launch vehicle error, which is possible in Genesis' case, the TCM1 can quickly grow beyond the capability of the spacecraft's propulsion system.

The Genesis spacecraft, built in the spirit of NASA's new low cost mission approach, is very basic. This makes the performance of an early TCM1 difficult and risky. It is desirable to delay TCM1 as long as possible, even at the expense of expenditure of the ΔV budget. In fact, Genesis would prefer TCM1 be performed at 2 to 7 days after launch, or even later. However, beyond launch + 24 hours, the correction ΔV based on traditional linear analysis can become prohibitively high.

The desire to increase the time between launch and TCM1 drives one to use a nonlinear approach, based on combining dynamical systems theory with optimal control techniques. We explore two similar but slightly different approaches and are able to obtain in both cases an optimal maneuver strategy that fits within the Genesis ΔV budget of 450 m/s. (1) HOI technique: use optimal control techniques to retarget the halo orbit with the original nominal trajectory as the initial guess. (2) MOI technique: we target the stable manifold. Both methods yield good results.

2. OPTIMAL CONTROL FOR TRAJECTORY CORRECTION MANEUVERS

We now introduce the general problem of optimal control for dynamical systems. We start by recasting the TCM problem as a spacecraft trajectory planning problem. Mathematically they are exactly the same. Then we discuss the spacecraft trajectory planning problem as an optimization problem and highlight the formulation characteristics and particular solution requirements. Then the fuel efficiency caused by possible perturbation in the launch velocity and by different delay in TCM1 is exactly the sensitivity analysis of the optimal solution. The software we use is an excellent tool in solving this type of problem, both in providing a solution for the trajectory planning problem with optimal control and in studying the sensitivity of different parameters. COOPT is developed by the Computational Science and Engineering Group at University of California Santa Barbara (see Users' Guide [1999]).

We emphasize that the objective in this work is not to design the transfer trajectory, but rather to investigate recovery issues related to possible launch velocity errors. We therefore assume that a nominal transfer trajectory (corresponding to zero errors in launch velocity) is available. For the nominal trajectory in our numerical experiments in this paper, we do not use the actual Genesis mission transfer trajectory, but rather an approximation obtained with a restricted model.

Recast TCM as Trajectory Planning Problem We treat two distinct problems: (1) the halo orbit insertion (HOI) problem, in which we target the halo orbit, and (2) the stable manifold insertion (MOI) problem, in which we target the stable manifold associated with the halo orbit. Although different from a dynamical systems' perspective, the two problems are very similar once cast as optimization problems. In the HOI problem, a final maneuver (jump in velocity) is allowed at $T_{HOI} = t_{max}$, while in the MOI problem, the final maneuver takes place on the stable manifold at $T_{MOI} < t_{max}$ and no maneuver occurs at $T_{HOI} = t_{max}$. A halo orbit insertion trajectory design problem can be simply posed as:

> *Find the maneuver times and sizes to minimize fuel consumption (ΔV) for a trajectory starting near Earth and ending on the specified L_1 halo orbit at a position and with a velocity consistent with the HOI time.*

We assume that the evolution of the spacecraft is described by a generic set of six ODEs

$$\mathbf{x}' = \mathbf{f}(t, \mathbf{x}), \tag{2.1}$$

where $\mathbf{x} = [\mathbf{x}^p; \mathbf{x}^v] \in R^6$ contains both positions ($\mathbf{x}^p$) and velocities ($\mathbf{x}^v$). Eq. (2.1) can be either the Circular Restricted Three Body Problem (CR3BP) or a more complex model that incorporates the influence of the Moon and other planets. In this paper, we use the CR3BP model; other models will be investigated in future work.

In order to resolve the discontinuous nature of the resulting optimal control problem, the equations of motion (e.o.m.) are solved simultaneously on each interval between two maneuvers. Let the maneuvers $M_1, M_2, ..., M_n$ take place at times T_i, $i = 1, 2, ..., n$ and let $\mathbf{x}_i(t)$, $t \in [T_{i-1}, T_i]$ be the solution of Eq. (2.1) on the interval $[T_{i-1}, T_i]$ (see Figure 2). Continuity constraints at the position level are imposed at each maneuver, that is,

$$\mathbf{x}_i^p(T_i) = \mathbf{x}_{i+1}^p(T_i), \qquad i = 1, 2, ..., n-1. \tag{2.2}$$

In addition, the final position is forced to lie on the halo orbit, that is, $\mathbf{x}_n^p(T_n) = \mathbf{x}_H^p(T_n)$, where the halo orbit is parameterized by the HOI time T_n. Additional constraints dictate that the first maneuver (TCM1) is delayed by at least a

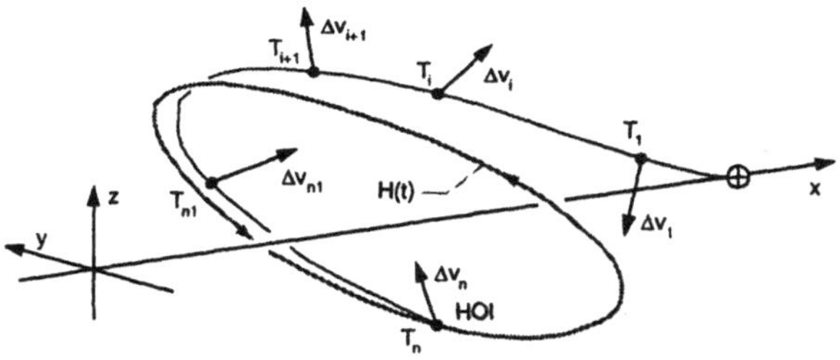

prescribed amount $TCM1_{min}$ after launch, that is,

$$T_1 \geq TCM1_{min}, \tag{2.3}$$

and that the order of maneuvers is respected,

$$T_{i-1} < T_i < T_{i+1}, \qquad i = 1, 2, ..., n-1. \tag{2.4}$$

With a cost function defined as some measure of the velocity discontinuities

$$\begin{aligned} \Delta\mathbf{v}_i &= \mathbf{x}_{i+1}^v(T_i) - \mathbf{x}_i^v(T_i), \qquad i = 1, 2, ..., n-1, \\ \Delta\mathbf{v}_n &= \mathbf{x}_H^v(T_n) - \mathbf{x}_n^v(T_n), \end{aligned} \tag{2.5}$$

the optimization problem becomes

$$\min_{T_i, \mathbf{x}_i, \Delta\mathbf{v}_i} f(\Delta\mathbf{v}_i), \tag{2.6}$$

subject to the constraints in Eqns. (2.2)-(2.5). More details on selecting the form of the cost function are given in Section 3.

Launch Errors and Sensitivity Analysis In many optimal control problems, obtaining an optimal solution is not the only goal. The influence of problem parameters on the optimal solution (the so called sensitivity of the optimal solution) is also needed. In this paper, we are interested in estimating the changes in fuel efficiency (ΔV) caused by possible perturbations in the launch velocity (ϵ_0^v) and by different delays in the first maneuver (TCM1). As we show in Section 3, the cost function is very close to being linear in these parameters ($TCM1_{min}$ and ϵ_0^v). Therefore, evaluating the sensitivity of the optimal cost is a very inexpensive and accurate (especially in our problem) method of assesing the influence of different parameters on the optimal trajectory.

In COOPT, we make use of the Sensitivity Theorem (Bertsekas [1995]) for nonlinear programming problems with equality and/or inequality constraints. The influence of delaying the maneuver TCM1 is directly computed from the Lagrange multiplier associated with the constraint of Eq. (2.3). To evaluate sensitivities of the cost function with respect to perturbations in the launch velocity (ϵ_0^v), we must include this perturbation explicitly as an optimization parameter and fix it to some prescribed value through an equality constraint. That is, the launch velocity is set to

$$\mathbf{v}(0) = \mathbf{v}_0^{\text{nom}} \left(1 + \frac{\epsilon_0^v}{\|\mathbf{v}_0^{\text{nom}}\|}\right), \tag{2.7}$$

where $\mathbf{v}_0^{nom}$ is the nominal launch velocity and

$$\epsilon_0^v = \epsilon, \qquad \epsilon \text{ given.} \tag{2.8}$$

The Lagrange multiplier associated with the constraint in Eq. (2.8) yields the desired sensitivity.

3. NUMERICAL RESULTS

Circular Restricted Three-Body Model As mentioned earlier, we use the equations of motion derived under the CR3BP assumption as the underlying dynamical model in Eq. (2.1). In this model, it is assumed that the primaries (Earth and Sun in our case) move on circular orbits around the center of mass of the system and that the third body (the spacecraft) does not influence the motion of the primaries. In a rotating frame and using nondimensional units, the equations of motion in the CR3BP model are

$$\ddot{x} = 2y + \frac{\partial U}{\partial x}; \; \ddot{y} = -2x + \frac{\partial U}{\partial y}; \; \ddot{z} = \frac{\partial U}{\partial z} \tag{3.1}$$

where $U = \frac{1}{2}(x^2 + y^2) + \frac{1-\mu}{d_\odot} + \frac{\mu}{d_\oplus}$, $d_\odot, d_\oplus$ are the distances between the spacecraft and the two primaries, and μ is the ratio between the mass of the Earth and the mass of the Sun-Earth system. In the above equations, time is scaled by the period of the primaries orbits ($T/2\pi$, where $T = 1$ year), positions are scaled by the Sun-Earth distance ($L = 1.49597927 \cdot 10^8$km), and velocities are scaled by the Earth's average orbital speed around the Sun ($2\pi L/T = 29.80567$km/s).

Choice of Cost Function. A physically meaningful cost function is

$$f_1(\Delta \mathbf{v}) = \sum_{i=1}^{n} \|\mathbf{v}_i\|. \tag{3.2}$$

This function is nondifferentiable when one of the maneuvers vanishes. This problem occurs at the first optimization iteration, as the intial guess transfer trajectory has a single nonzero maneuver at halo insertion. A differentiable cost function is

$$f_2(\Delta \mathbf{v}) = \sum_{i=1}^{n} \|\mathbf{v}_i\|^2. \tag{3.3}$$

Although this second cost function is more appropriate for the optimizer, it raises two new problems. Not only it is not as physically meaningful as the cost function of Eq. (3.2), but, in some particular cases, decreasing f_2 may actually lead to increases in f_1.

To resolve these issues, we use the following three-stage staggered optimization procedure:

(1) Starting with the nominal transfer trajectory as intial guess, and allowing initially n maneuvers, we minimize f_2 to obtain a first optimal trajectory, $\mathcal{T}_1^*$.

(2) Using $\mathcal{T}_1^*$ as initial guess, we minimize f_1 to obtain $\mathcal{T}_2^*$. It is possible that during this optimization stage some maneuvers can become very small. After each optimization iteration we monitor the feasibility of the iterate and the sizes of all maneuvers. As soon as at least one maneuver decreases under a prescribed threshold (0.1 m/s) at some fesible configuration, we stop the optimization algorithm.

(3) If necessary, a third optimization stage, using $\mathcal{T}_2^*$ as initial guess and f_1 as cost function is performed with a reduced number of maneuvers $\bar{n}$ (obtained by removing those maneuvers identified as 'zero maneuvers' in step 2).

Merging Optimal Control with Dynamical Systems Theory We present some results for the HOI problem and the MOI problems. A more indepth study will be given in a forthcoming publication. In both cases we investigate the effect of varying times for $TCM1_{min}$ on the optimal trajectory, for given perturbations in the nominal launch velocity. The staggered optimization procedure described above is applied for values of $TCM1_{min}$ ranging from 1 day to 5 days and perturbations in the magnitude of the injection velocity, ϵ_0^v, ranging from -7 m/s to $+7$ m/s. We present typical transfer trajectories, as well as the dependency of the optimal cost on the two parameters of interest. In addition, using the algorithm presented in Section 2, we perform a sensitivity analysis of the optimal solution. For the Genesis TCM problem, sensitivty information of first order is sufficient to characterize the influence of $TCM1_{min}$ and ϵ_0^v on the spacecraft performance.

The merging of optimal control and dynamical systems has been done through either (1) the use of the nominal transfer trajectory as a really accurate initial guess, or (2) the use of the stable invariant manifold.

Halo Orbit Insertion (HOI) Problem. In this problem we directly target the selected halo orbit with the last maneuver taking place at the HOI point. Using the optimization procedure described in the previous section, we compute the optimal cost transfer trajectories for various combinations of $TCM1_{min}$ and ϵ_0^v. In all of our computations, the launch conditions are those corresponding to a given nominal transfer trajectory with the launch velocity perturbed as described in Section 2.

As an example, we present complete results for the case in which the launch velocity is perturbed by -3 m/s and the first maneuver correction is delayed by at least 3 days. Initially, we allow for $n = 4$ maneuvers. In the first optimization stage, the second type of cost function has a value of $f_2^* = 1153.998$ (m/s)^2 after 5 iterations. This corresponds to $f_1^* = 50.9123$ m/s. During the second optimization stage, we monitor the sizes of all four maneuvers, while minimizing the cost function (3.2). After 23 iterations, the optimization was interrupted at a feasible configuration when at least one maneuver decreased below a preset tolerance of 0.1 m/s. The corresponding cost function is $f_1^{**} = 45.1216$ m/s with four maneuvers of sizes 33.8252 m/s, 0.0012 m/s, 0.0003 m/s, and 11.2949 m/s. In the last optimization stage we remove the second and third maneuvers and again minimize the cost function f_1. After 7 optimization iterations an optimal solution with $f_1^{***} = 45.0292$ m/s is obtained. The two maneuvers of the optimal trajectory have sizes of 33.7002 m/s and 11.3289 m/s and take place at 3.0000 and 110.7969 days after launch, respectively. Lagrange multipliers associated with the constraints of Eqs. (2.3) and (2.8) give the sensitivities of the optimal solution with respect to launching velocity perturbation, -10.7341 (m/s)/(m/s), and delay in first maneuver correction, 4.8231 (m/s)/days.

Launch Errors and Sensitivity Analysis. The staggered optimization procedure was applied for all values of $TCM1_{min}$ and ϵ_0^v in the region of interest. In a first experiment, we investigate the possibility of correcting for errors in the launch velocity using at most two maneuvers ($n = 2$). Numerical values of optimal cost as a function of these two parameters are given in Table 3. Except for the cases in which there is no error in the launch velocity (and for which the final optimal transfer trajectories have only one maneuver at HOI), the first correction maneuver is always on the prescribed lower bound $TCM1_{min}$. For all cases investigated, halo orbit insertion takes place at most 18.6 days earlier or 28.3 days later than in the nominal case ($T_{HOI} = 110.2$).

ϵ_0^v (m/s)	TCM1 (days) 1	2	3	4	5
-7	64.8086	76.0845	88.4296	99.6005	109.9305
-6	54.0461	67.0226	77.7832	86.8630	95.8202
-5	47.1839	57.9451	66.6277	74.4544	81.8284
-4	40.2710	48.8619	55.8274	62.0412	67.9439
-3	33.4476	39.8919	45.0290	49.6804	54.1350
-2	26.6811	30.9617	34.3489	37.3922	40.3945
-1	19.9881	22.2715	23.7848	25.2468	26.6662
0	13.4831	13.3530	13.4606	13.3465	13.2919
1	23.1900	21.9242	23.2003	24.4154	25.5136
2	26.2928	30.2773	33.3203	35.9203	38.3337
3	34.6338	38.8496	43.5486	47.7200	51.6085
4	41.4230	47.5266	53.9557	62.3780	65.1411
5	45.9268	56.2245	64.4292	75.0188	81.4325
6	53.9004	64.9741	76.6978	83.8795	95.2313
7	61.4084	75.9169	85.4875	98.4197	106.0411

Several conclusions can be drawn. First, for all cases that we investigated, the optimal costs are well within the ΔV budget allocated for trajectory correction maneuvers (450 m/s for the Genesis mission). The cost function is very close to being linear with respect to both $TCM1_{min}$ time and launch velocity error. Also, the halo orbit insertion time is always close enough to that of the nominal trajectory as not to affect either the collection of the solar wind or the rest of the mission (mainly the duration for which the spacecraft evolves on the halo orbit before initiation of the return trajectory).

Manifold Orbit Insertion (MOI) Problem. In the MOI problem the last nonzero maneuver takes place on the stable manifold and there is no maneuver to insert onto the halo orbit. A much larger parameter space is now investigated (we target an entire surface as opposed to just a curve) thus making the optimization problem much more difficult than the one corresponding to the HOI case. The first problem that arises is that the nominal transfer trajectory is not a good enough initial guess to ensure convergence to an optimal solution. To obtain an appropriate initial guess we use the following procedure: (1) We start by selecting an HOI time, T_{HOI}. This yields the position and velocity on the halo orbit. (2) With the above position and velocity as initial conditions, the equations of motion in Eq. (3.1) are then integrated backwards in time for a selected duration T_S along the stable manifold. This yields an MOI point which is now fixed in time, position, and velocity. (3) For a given value of $TCM1_{min}$ and with $\epsilon_0^v = 0$, and using the nominal transfer trajectory as initial guess, we use COOPT to find a trajectory that targets this MOI point, while minimizing f_1.

With the resulting trajectory as an initial guess and the desired value of ϵ_0^v we proceed with the staggered optimization presented before to obtain the final optimal trajectory for insertion on the stable manifold. During the three stages of the optimization procedure, both the MOI point and the HOI point are free to move (in position, velocity, and time) on the stable manifold surface and on the halo orbit, respectively. Taking the

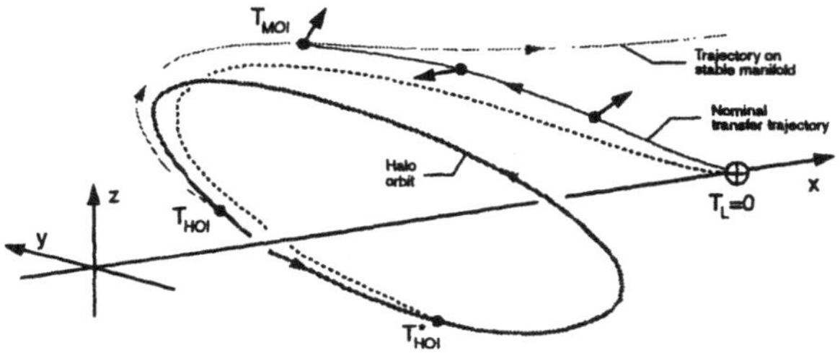

launch time to be $T_L = 0$ and the HOI time (T^*_{HOI}) of the nominal transfer trajectory as a reference point on the halo orbit, we can investigate a given zone of the design space by an appropriate choice of the HOI point of our initial guess trajectory with respect to T^*_{HOI} (step 1 of the above procedure). That is, we select a value T_0 such that $T_{HOI} = T^*_{HOI} + T_0$. The point where the initial guess trajectory inserts onto the stable manifold is then defined by selecting the duration T_S for which the equations of motion are integrated backwards in time (step 2 of the above procedure). This gives a stable manifold insertion time of $T_{MOI} = T_{HOI} - T_S = T^*_{HOI} + T_0 - T_S$. Next, we use COOPT to evaluate these various choices for the initial guess trajectories (step 3 of the above procedure). A schematic representation of this procedure is shown in Figure 3.

Regions Best Suited for MOI Insertion. Using the values of $f_1(\Delta V)$, we can identify regions of the stable manifold that are best suited for MOI insertion. Examples are: (1) (*Region A*) MOI trajectories that insert on the halo orbit in the same region as the nominal transfer trajectory and which therefore correspond to initial guess trajectories with small T_0; (2) (*Region B*) MOI trajectories that have HOI points on the 'far side' of the halo orbit and which correspond to initial guess trajectories with halo insertion time around $T^*_{HOI} + 1.50$ ($T_0 = 1.50 \cdot 365/2\pi = 174.27$ days).

At first glance, trajectories in Region B might appear ill-suited to the Genesis mission as they would drastically decrease the duration for which the spacecraft evolves on the halo orbit (recall that design of the return trajectory dictates the time at which the spacecraft must leave the halo orbit). But for a typical MOI trajectory, all trajectories on the stable manifold asymptotically wind onto the halo orbit and are thus very close to the halo orbit for a significant time. This means that collection of solar wind samples can start much earlier than halo orbit insertion, therefore providing enough time for all scientific experiments before the spacecraft leaves the halo orbit.

After choosing a region of the stable manifold by selecting an initial guess trajectory, we perform a similar analysis as in the HOI problem. Consider correcting for perturbations in launch velocity by seeking optimal MOI trajectories in Region B, that is, on the far side of the halo from the Earth. For non-zero ϵ_0^v and $TCM1_{min}$, we compute an MOI initial guess trajectory with $T_0 = 1.50$ and $T_S = 0.75$ and then use the staggered optimization procedure to find an optimal MOI trajectory in this vicinity.

The results are given in Table 3. Note that the optimal MOI trajectories are close (in terms of the cost function f_1) to the corresponding HOI trajectories. Therefore, either method provides an excellent solution to the TCM problem.

4. CONCLUSIONS AND FUTURE WORK

This paper explored new approaches for autmomated parametric studies of optimal trajectory correction maneuvers for a halo orbit mission. Using the halo orbit insertion approach, for all the launch velocity errors and $TCM1_{min}$ considered we found optimal recovery trajectories. The cost functions (fuel consumption in terms of ΔV) are within the allocated budget even in the worst case

$TCM1_{min}(days)$	ϵ_0^v (m/s)	f_1 (m/s)
3	-3	45.1427
	-4	55.6387
	-5	65.9416
	-6	76.7144
	-7	87.3777
4	-3	49.1817
	-4	61.5221
	-5	73.4862
	-6	85.7667
	-7	99.3405
5	-3	53.9072
	-4	66.8668
	-5	81.1679
	-6	94.3630
	-7	109.2151

(largest $TCM1_{min}$ and largest launch velocity error).

Using the stable manifold insertion approach, we obtained similar results to those found using HOI targeted trajectories. The failure of the MOI approach to reduce the ΔV significantly may be because the optimization procedure (even in the HOI targeted case) naturally finds trajectories 'near' the stable manifold. We will investigate this interesting effect in future work.

For now, the main contribution of dynamical systems theory to the optimal control of recovery trajectories is in the construction of good initial guess trajectories in sensitive regions where optimizers have the greatest flexibility.

Acknowledgements. This research was supported in part by NSF/KDI grant NSF ATM-9873133.

Ascher, U. M. and L. R. Petzold [1998] *Computer Methods for Ordinary Differential Equations and Differential-Algebraic Equations*, SIAM, 1998.

Bertsekas, D.R. [1995] *Nonlinear Programming*, Athena Scientific, Belmont, Ma.

Howell, C., B. Barden and M. Lo [1997] Application of dynamical systems theory to trajectory design for a libration point mission, *The Journal of the Astronautical Sciences* **45(2)**, April 1997, 161-178.

Koon, W.S., M.W. Lo, J.E. Marsden and S.D. Ross [1999] Heteroclinic connections between periodic orbits and resonance transitions in celestial mechanics, *Submitted for publication.*

Lo, M., B.G. Williams, W.E. Bollman, D. Han, Y. Hahn, J.L. Bell, E.A. Hirst, R.A. Corwin, P.E. Hong, K.C. Howell, B. Barden, and R. Wilson [1998] Genesis Mission Design, *Paper No. AIAA 98-4468.*

Li, S. and L.R. Petzold [1999] *Design of New DASPK for Sensitivity Analysis*, UCSB Department of Computer Science Technical Report.

Serban, R. [1999] *COOPT - Control and Optimization of Dynamic Systems - Users' Guide*, Report UCSB-ME-99-1, UCSB Department of Mechanical and Environmental Engineering.

LOW ENERGY TRAJECTORIES FOR SPACE TRAVEL USING STABILITY TRANSITION REGIONS

Edward Belbruno

Princeton University

Abstract: Low energy trajectories for motion for spacecraft, or small bodies, can be constructed by using chaotic dynamics associated with stability transition regions. These regions are generally near planetary bodies, and can be numerically estimated. We present an explicit analytic approximation for a region of this type. A new transfer from the earth to the moon was flown in 1991 demonstrating this approach. A number of applications are discussed, to both spaceflight and astronomy.

Keywords: Chaos, Space Flight, Stability Domains, Stabilization, Low Energy, Differential Equations, Dynamics, Optimization.

1. INTRODUCTION

Recent results in celestial mechanics over the past two decades have shown that regions of very sensitive, or chaotic, motions exist in the three and four-body problems which are not well understood. This is the case, for example, in the motion of a particle of a negligible mass moving about the moon under the gravitational perturbation of the earth.

In 1986 Belbruno found a way numerically estimate a region about the moon where a particle moves in a chaotic fashion [2]. The region defines a transition between the gravitational capture and escape and is termed a ***weak stability boundary*** (WSB). It can be used to construct a new type of low energy transfer between the earth and moon, where a particle is automatically captured into elliptic lunar orbit. This represents a different methodology from using the Hohmann transfer where a large maneuver must be done in order the slow a spacecraft down so it can be elliptically captured. The automatic capture is also less risky for operational purposes. The lack of a capture maneuver implies less energy.

A transfer of this type was operationally demonstrated in October 1991 when the Japanese spacecraft Hiten arrived at the moon [3,4]. Since then, this transfer, termed a WSB transfer, has been refined and better understood. Easier ways have also been developed to generate them [5]. In particular, as is discussed in [5] and Section 3, a standard forward searching Newton's algorithm can be used. These transfers have been studied for use in capture at other bodies such as Europa, which is discussed in Section 3, by Sweetser, et. al. [13]. Japan's Lunar-A mission is planning the use of one of these, and they have been studied for the possible use by the U.S. Air Force Academy [5]. In Section 3 we discuss ballistic ejection from the earth-moon system and the recent work of Penzo [12].

In Section 2 we study the WSB as a special case of a more general object we call a stability transition boundary. An analytic representation for the WSB is derived in Section 2, which can be used to estimate the energy a particle has to have when moving in this region. This result shows that a region R of heteroclinic intersections of invariant manifolds connecting neighborhoods of the Lagrange points L_1, L_2, recently numerically proven to exist by Koon, Lo, Marsden, Ross [9], is a subset of the WSB, and may be equivalent to it for energy values slightly less than the energy of a particle at L_1.

An interesting resonance motion was noticed in 1990 near the WSB where a particle can quickly transition between different resonance states [3]. This was related in 1996 to cometary motions by Belbruno and Marsden [6]. In [6] it was suggested that the comets were following near the invariant manifolds associated to the WSB at Jupiter. This was verified by Koon et. al [9] where a numerically assisted proof was given, showing that comets were following near invariant manifolds associated to the region R. In Section 4, an interesting sequence of resonant motion is described near the moon, [7], which hints at the complexities of motions near the WSB awaiting to be discovered.

2. STABILITY TRANSITION REGIONS AND THE WEAK STABILITY BOUNDARY

In this section stability transition regions are defined both from physical considerations and then mathematically. An explicit analytic representation is given, and related to results in [9].

2.1 *Model*

We begin for the sake of simplicity with the planar circular restricted three-body problem. This is defined by the motion of a particle m in the gravitational field generated by the uniform circular motion of angular frequency ω of two larger mass points which we label m_1, m_2. For example, $m_1 =$ earth, $m_2 =$ moon. It is assumed that m moves in the same plane as m_1, m_2, and has a negligible mass with respect to them.

We assume that m moves in an inertial coordinate system X, Y, centered at the origin which is the centered mass of m_1 and m_2. We then transform to a rotating coordinate system x, y by the transformation

$$\begin{pmatrix} x \\ y \end{pmatrix} = \begin{pmatrix} c & s \\ -s & c \end{pmatrix} \begin{pmatrix} X \\ Y \end{pmatrix}, \qquad (1)$$

$c = \cos\omega t, \sin\omega t, t \in \mathbb{R}$ is time, which rotate with the same frequency ω as m_1 and m_2. In the (x, y) coordinate system m_1, m_2 are fixed. Without loss of generality, we can set $\omega = 1$, and place m_1 at the position $(\mu, 0)$, and m_2 at $(-1 + \mu, c)$, and where the mass of m_1 is normalized to $1 - \mu$ and m_2 is μ. μ is the mass ratio of m_1/m_2. For the earth and moon, $\mu = .012$. The differential equations are given by

$$\begin{aligned} \ddot{x} - 2\dot{y} &= x + \Omega_x \\ \ddot{y} + 2\dot{x} &= y + \Omega_y, \end{aligned} \qquad (2)$$

where $\dot{} \equiv \frac{d}{dt}$,

$$\Omega = \frac{1-\mu}{r_1} + \frac{\mu}{r_2},$$

$r_1 =$ distance of m to $m_1 = [(x + \mu)^2 + y^2]^{\frac{1}{2}}, r_2 =$ distance of m to $m_2 = [(x + 1 - \mu)^2 + y^2]^{\frac{1}{2}}$. The right hand side of (2) represents the sum of the radially directed centrifugal force $\mathbf{F_C} = (x, y)$ and the sum $\mathbf{F_G}$ of the gravitational forces due to m_1, m_2, respectively.

We will assume $0 < \mu \ll 1$, e.g. $\mu = .012$, and consider the motion of m near m_1. Setting

$$\mathbf{F} = \mathbf{F_C} + \mathbf{F_G} = \mathbf{0}$$

together with setting $\dot{x} = 0, \dot{y} = 0$ yields the 5 Lagrange points $L_k, k = 1, 2, 3, 4, 5$ [10]. Physically, this represents a balancing of the sum of the forces on m, under the condition that the velocity of m is *zero*. However, as m moves, the force on m can nearly balance *without* the velocity of m being zero. In this case, $\mathbf{F} \approx \mathbf{0}$.

In such a situation, the motion of m can be sensitive, and as m moves about m_2 its motion may be quite unstable. A way to numerically estimate such a region is described next, which has many applications.

2.2 *Weak Stability Boundaries*

A way to numerically estimate a region of sensitive motion about the moon was done in 1987 [2]. See also [4,7]. We briefly describe it here. We consider a radial line L from m_2 as shown in Figure 1. Trajectories are propagated from L under the assumptions,

(1) At any distance r_2 from m_2 on L a trajectory is propagated for (2) starting at time $t = 0$, assuming the initial velocity vector is normal to L.
(2) The two-body Kepler energy of m with respect to m_2 at $t = 0$ is negative.
(3) The eccentricity $e \in [0, 1)$ is held fixed at any distance along L to the same value e at $t = 0$, which is accomplished by adjusting the initial inertial velocity magnitude $(\dot{X}^2 + \dot{Y}^2)^{\frac{1}{2}}$ with respect to m_2 as the origin to $(\mu(1+e)/r_2)^{\frac{1}{2}}$.

Thus, m starts its motion on an osculating ellipse which we assume is at its periapsis. The inertial Kepler energy $E_2 = \frac{1}{2}(\dot{X}^2 + \dot{Y}^2) - \mu r_2^{-1}$ of m with respect to m_2 at the origin is therefore given by

$$E_2 = \frac{\mu}{2}\left(\frac{e-1}{r_2}\right) < 0. \qquad (3)$$

We define stability in the following way,

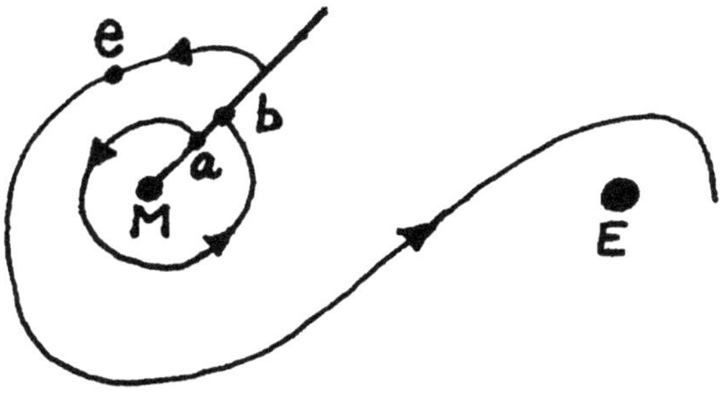

Fig. 1. Definition of Stability about the Moon, M

The motion of m is *stable* about m_2 if, after leaving L at $t = 0$, it makes a full cycle about m_2, without going around m_1, and returns to L with $E_2 < 0$.

This is shown in Figure 1. m leaves a at $t = 0$ and returns to b.

The motion of m is *unstable* if one of the following two things happens:

(1) It performs a full cycle about m_2, without going about m_1, and returns to b, where $E_2 \geq 0$, or
(2) m moves away from m_2 towards m_1 and makes a cycle about m_1. In this case there is a point e on the trajectory, shown in Figure 1, where $E_2 = 0$.

As the initial conditions are varied along L in the manner just described, keeping e fixed, a well defined finite distance r^* is estimated with the properties

(1) If $r_2 < r^*$, the motion is stable
(2) If $r_2 > r^*$, the motion is unstable

r^* depends on two parameters, the polar angle, θ, L makes with the x-axis, and e. Thus, $r^* = r^*(\theta, e)$. We define

$$\mathcal{W} = \{r^*(\theta, e) | \theta \in [0, 2\pi], e \in [0, 1)\}.$$

Thus, the $\mathcal{W}$ is two-dimensional (See [2]).

It is numerically found that motion of m near $\mathcal{W}$ is very sensitive, and m can be abruptly ejected from m_2. Resonant motion of m upon ejection from m_2 is typical, and is discussed in Section 4.

$\mathcal{W}$ defines a stability transition boundary we call the *Weak Stability Boundary* (WSB). $\mathcal{W}$ can be viewed as a surface of two dimensions in the phase space of position and velocity $(x, y, \dot{x}, \dot{y})$. It is noted that if we had only considered a two-body problem between m and m_2, r^* would be infinite. Thus the effect of the gravitational perturbation of m_1 is to allow r^* to be finite where escape can occur. Given r^*, the critical velocity required to be in $\mathcal{W}$ is near escape velocity of magnitude $(2\mu/r^*)^{\frac{1}{2}}$ in inertial coordinates, in many cases.

It is remarked in we are considering the motion of m as defined by (2) which is in rotating coordinates (x, y), and that as specified, E_2 is defined as the Kepler energy in inertial coordinates (X, Y). Thus, when determining r^* by the above method, quantities defined by inertial coordinates need to be transformed to rotating coordinates. In rotating coordinates (x, y) defining (2),

$$E_2 = \frac{1}{2}(\dot{x}^2 + \dot{y}^2) - \frac{\mu}{r_2} + \frac{1}{2}r_2^2 + L, \qquad (4)$$

$L = \dot{x}y - \dot{y}(x + 1 - \mu)$.

2.3 *Analytical Estimate of $\mathcal{W}$*

It is numerically found that for motion near the WSB,

$$E_2 \lesssim 0. \qquad (5)$$

See [7]. To analytically estimate this region, the approximation

$$E_2 = 0$$

is made. This represents a three-dimensional set $\Sigma = \{x, y, \dot{x}, \dot{y} | E_2 = 0\}$. Since the motion of m must satisfy (2), then another expression we have is the energy of m, called the Jacobi energy, J, which is constant along solutions of (2). It is given by

$$J = -(\dot{x}^2 + \dot{y}^2) + (x^2 + y^2) + \mu(1 - \mu) + 2\Omega$$

Thus, we define an invariant three-dimensional surface for solutions of (2),

$$\mathcal{J} = \{x, y, \dot{x}, \dot{y} | J = C, C \in \mathbb{R}\},$$

The set $\Sigma \cap \mathcal{J}$ is two-dimensional and represents an approximation to the WSB.

It can be analytically represented by substituting $E_2 = 0$, where E_2 is determined by (4), into the equation $J = C$, yielding after some simplification,

$$r_2^2 + 2L + A - C = 0, \qquad (6)$$

where

$$A = 2(1 - \mu)r_1^{-1} + r^2 + \mu(1 - \mu),$$

$r^2 = x^2 + y^2$. (6) represents a two-dimensional surface which approximates $\mathcal{W}$. If we transform to m_2-centered coordinates $(\overline{x}, \overline{y})$,

$$\overline{x} = x - (-1 + \mu), \ \overline{y} = y,$$

then (6) becomes

$$r_2^2 + 2\overline{L} + \overline{A} - C = 0,$$

$\overline{L} = \dot{\overline{x}}\overline{y} - \dot{\overline{y}}\overline{x}, \overline{A} = A(\overline{x}, \overline{y})$. In polar coordinates $\overline{x} = r_2 \cos\theta, \overline{y} = r_2 \sin\theta$, this takes the form, after simplification, since $\overline{L} = -r_2^2\dot{\theta}$,

$$r_2^2(1 - 2\dot{\theta}) + \tilde{A} - C = 0, \tag{7}$$

$\tilde{A} = \overline{A}(r_2, \theta)$. Thus we have

Proposition 1. The WSB can be approximated by (7), under the above assumptions, which implicitly yields $r_2 = r_2(\dot{\theta}, \theta)$.

The constant C can be estimated from (7) if it is assumed $r_2 \gtrsim 0$, i.e. near but not equal to zero. In this case $r \approx 1 - \mu, r_1 \approx 1$ implying

$$A \approx 2(1 - \mu) + (1 - \mu)^2 + \mu(1 - \mu) = 3 - 3\mu$$

Thus, since $\dot{\theta}$ may be large in magnitude, we have

Proposition 2. For motion on the WSB, where $r_2 \gtrsim 0$, then

$$C \approx 3 - 3\mu - 2\dot{\theta}r_2^2 \tag{8}$$

(8) has good agreement with numerically observed results, which will be referred back to in Sections 3, 4.

The preceding results are valid for (2), and can be generalized for more complicated modeling of the earth, moon, and other bodies using the planetary ephemeris DE403, and where m moves in three dimensions. It is remarked that for many WSB motions, r_2 need not be small, and thus (8) is not valid.

3. BALLISTIC CAPTURE TRANSFERS

For sake of argument, we consider the case m_1 = earth, m_2 = moon and m = spacecraft. For the moment, consider that m moves in three-dimensional space, and m_1, m_2 are modeled with DB403. As is described in [4,5], the classical transfer from the earth to the moon is the Hohmann transfer. A disadvantage of this route is that although it arrives in approximately 3 days from a low circular earth, at an altitude of say 200 km, to a low lunar periapsis, at say 100 km, it arrives with a hyperbolic excess velocity of approximately 1 km/s, so that a large maneuver, Δv_c, must be done to change the velocity to be captured into lunar orbit with a given eccentricity $e \in [0,1)$. If $e = 0$, then $\Delta v_c \approx .8$ km/s. Δv_c could be reduced significantly if it arrived at lunar periapsis with $\Delta v_c = 0$, so that it was already captured with a value of $e \in [0,1)$. Thus, we have the following definition,

A capture of a trajectory for m at lunar periapsis is called ***ballistic*** if $E_2 < 0$ implying $\Delta v_c = 0$. A trajectory from the earth to the moon is called a ***ballistic capture trajectory*** if m arrives at lunar periapsis in ballistic capture.

A ballistic capture transfer was first numerically demonstrated in 1987 [2]. A more useful version of such a transfer was operationally demonstrated by the Japanese spacecraft Hiten which arrived at the moon in October of 1991 [4,5,3]. This type of transfer has a trip time on the order of 3 months, for example, it arrives at lunar capture at 100 km altitude at lunar periapsis with $e \approx .94$. This implies that .6 km/s is required for lunar capture instead of .8 km/s for a Hohmann transfer thus yielding an approximate 25% improvement in Δv_c. The Hohmann transfer and this new ballistic capture transfer both require about the same Δv to leave low earth orbit. This savings over the Hohmann transfer can be used to substantially increase the payload that can be delivered into lunar orbit [5]. This so-called WSB ballistic capture transfer is shown in Figure 2. It is documented in [5]. It flies out to 1.5 million km from the earth, and then falls back to lunar capture at the WSB at the lunar far side. Solar perturbations play a key role, and at the apoapsis A 1.5 million km, m is on the WSB of the earth due to the sun. Its trip time is 91 days. This transfer has a negligible maneuver of 4 m/s at apoapsis. It's capture is over the south lunar pole.

3.1 *Construction*

Ballistic capture transfers can be numerically generated by starting at the desired capture point on the lunar WSB and performing a backwards integration which can bring it back to low earth periapsis by variation of lunar parameters such as slightly varying e. Dynamically, the orbit is following an invariant manifold from the WSB, in backwards time, from the lunar WSB, to earth apoapsis at 1.5 million km from the earth.

A new forward search algorithm was developed in 1997 which allows an easy and robust forward search to lunar capture by varying two parameters at the earth and targeting to two lunar parameters. This is described in detail in [5]. Thus, the WSB-transfer can be calculated in a straight forward manner.

Recent results in [9] for (2) have shown more fully the global invariant manifold structure near the moon near where the Lagrange point L_1 is located. This provides more insight into the

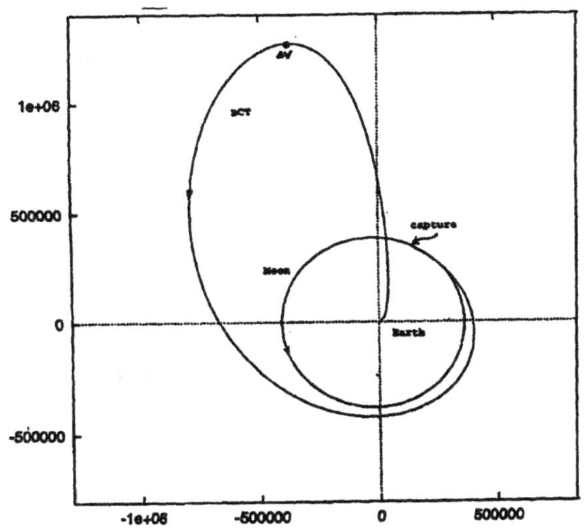

Fig. 2. A WSB Lunar Transfer, ecliptic projection

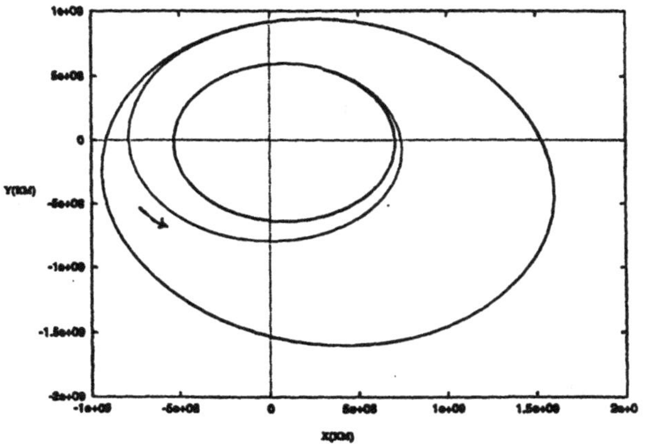

Fig. 3. The Comet 74P/Smirnova-Chernykh transitioning from a 6 : 13 to a 7 : 5 ellipse

dynamics of ballistic capture for suitable energies. They considered values where $C \lesssim C_1$, where C_1 is the value of C when J is evaluated at the Lagrange point L_1. This implies that to achieve capture for these energies, m passes *near* the location of L_1, where its velocity is nonzero.

3.2 *Consistency of Results*

In [9], where $C \lesssim C_1$, and hence where m must pass near L_1 for capture, is consistent with those values of C predicted by Proposition 2.

This is the case since for m approaching ballistic capture for those energies, it passes *near* L_1. This implies that $\mathbf{F} \approx \mathbf{0}$ and $\dot{\theta} \neq 0$, and so the particle is moving near a stability transition region defined above, approximated by the WSB, as numerically and analytically defined. This implies that Proposition 2 should also then yield an estimate of C near C_1. Estimates for such transfers near L_1 imply $\dot{\theta} = .05, r_2 = .184$. This implies from Proposition 2 that $C \approx 2.961$. Since C_1 =3.184, then C is also slightly smaller than C_1.

This also shows that for motions near L_1 and, more generally, motions with $C \lesssim C_1$ near the region $R(C)$ of heteroclinic intersection of invariant manifolds associated to Lyapunov orbits from L_1 and L_2 described in [9], that these motions also satisfy Proposition 2. Thus, we have demonstrated,

Proposition 3. For $C \lesssim C_1$, $R(C)$ is a stability transition boundary satisfying the conditions of the WSB. Thus, $R(C) \subset$WSB.

The lunar WSB as defined by Proposition 2 is not limited to energies $C \lesssim C_1$ and thus describes a more general situation.

The WSB transfer from [5] is generated for modeling using DE403 for the motion of the earth an moon. It implies $C \approx$2.965 near lunar capture.

It is demonstrated in an interesting paper by Sweetser, et. al. [13], that ballistic capture transfers can be constructed at Europa using the WSB methodology.

3.3 *Ballistic Ejection*

It was demonstrated in [3] that dynamics near the WSB at the moon gave rise to trajectories that led to ejection from the earth-moon system with a significant hyperbolic excess velocity. Penzo at JPL has developed a new type of ejection transfers using both WSB ideas and classical gravity assist [12], by replacing lunar WSB dynamics with classical gravity assist, and by flying out near the earth WSB, due to solar gravitational perturbations. This class of ejection transfers are being planned for future Mars missions by NASA.

4. RESONANCE TRANSITIONS

It was noticed using numerical simulations in 1990 that motion near the lunar WSB gave rise to periodic orbits about the earth in resonance with the moon [3]. A resonant orbit, in general, is said to be of type $i : j, i = 1, 2, \ldots, j = 1, 2, \ldots$, if m makes i revolutions about the earth in j lunar periods. It was suggested from [3] that, when a trajectory for m enters the WSB on an $i : j$ resonant orbit, it is ejected onto an $l : k$ resonant orbit, where l, k are generally different integers than i, j. This defines a ***resonance transition*** or ***hop*** from $i : j$ to $l : k$. Utilizing WSB ideas, Belbruno and Marsden in 1996 [6] studied a number of comets making resonance transitions 1996, and showed that they were occurring at Jupiter's WSB, due to Jupiter-Sun perturbations. (See Figure 3.)

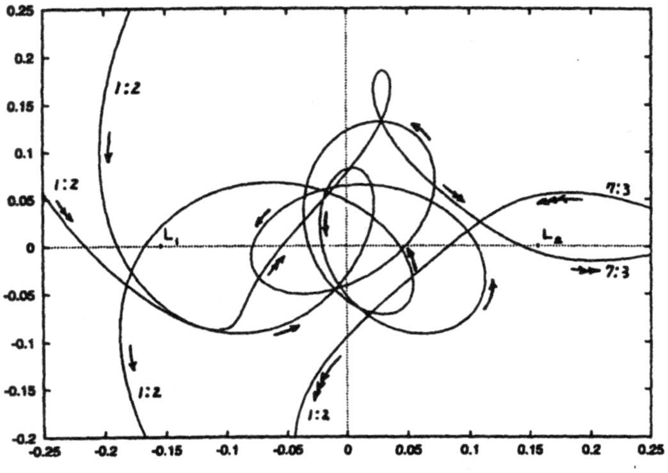

Fig. 4. A Sequence of Hops near the Moon

It was predicted, and not proven, in [6] that the comet orbits were moving near invariant manifolds associated to resonant regions within the WSB, by alluding to more general results of Mather [11]. This was numerically proven in [9] that for one of the comets studied in [6], where $C \lesssim C_1$, the hop was associated to region of resonance dynamics lying within $R(C)$. This is a substantial result. For C much less than C_1, this type of proof as it is is constructed may not be applicable. In this case, Proposition 2, or more generally, Equ.(7), may be useful in locating regions which have sensitive motion.

The interchange of resonance hops near the moon can be complicated, as was studied in [7]. Figure 4, from [7], shows a neighborhood of the moon and a set of different resonance hops taking place. This only hints at the complexity of such motions near the WSB and also $R(C)$.

Applications of these ideas to Kuiper belt objects, many of which move in 2 : 3 resonances with Neptune, are discussed in [8].

5. REFERENCES

[1] V.I. Arnold, *Mathematical Methods of Classical Mechanics* (GTM 60, Springer Verlag, 1989).

[2] E.A. Belbruno, Lunar Capture Orbits, A Method of Constructing Earth-Moon Trajectories and the Lunar GAS Mission, in: *Proceedings of AIAA/DGLR/JSASS Inter. Propl. Conf.* AIAA Paper No. 97–1054, (May 1987).

[3] E.A. Belbruno, Examples of the Nonlinear Dynamics of Ballistic Capture and Escape in the Earth-Moon System, in: *Proceedings of the Annual AIAA Astrodynamics Conf.* AIAA Paper No. 90–2896, (August 1990).

[4] E.A. Belbruno, J. Miller, Sun-Perturbed Earth-to-Moon Transfers with Ballistic Capture, *J. Guid., Control and Dynamics* **16** No. 4, (July–August 1993) 770–775.

[5] E.A. Belbruno, R. Humble, J. Coil, Ballistic Capture Lunar Transfer Determination for the U.S. Air Force Academy Blue Moon Mission, *Advances in Astronautical Science, Spaceflight Mechanics* **95** AAS Publ., Paper No. AAS 97-171, (1997).

[6] E.A. Belbruno, B. Marsden, Resonance Hopping in Comets, *The Astronomical Journal* **113** No. 4, (April 1997) 1433–1444.

[7] E.A. Belbruno, Fast Resonance Shifting as a Mechanism of Dynamic Instability Illustrated by Comets and CHE Trajectories, in: J. Remo, ed., *Annals of the New York Academy of Sciences, Near Earth Objects: The United Nations International Conf.* **822** (1997) 195–226.

[8] E.A. Belbruno, Resonance Hopping in the Kuiper Belt, in: V. Steves and A. Roy, eds., *The Dynamics of Small Bodies in the Solar Systems* Series C, Mathematical and Physical Sciences, **522** (Kluwer Academic Publishers, 1999) 37–49.

[9] W.S. Koon, M.W. Lo, J.E. Marsden, S.D. Ross, *Heteroclinic Connections between Periodic Orbits and Resonance Transitions in Celestial Mechanics* (Preprint, 1999).

[10] J.B. Marion, *Classical Dynamics of Particles and Systems* (Academic Press, 1970).

[11] J.N. Mather, A Criterion for the Non-Existence of Invariant Circles, *Publ. Math. IHES* **63**, (1982) 155–204.

[12] P.A. Penzo, A Survey and Recent Developments of Lunar Gravity Assist, in: *Space Manufacturing 12, Proceedings of the SSI Princeton Conference on Space Manufacturing* (2000).

[13] T. Sweetser, et. al., Trajectory Design for an Europa Orbiter Mission: A Plethora of Astrodynamic Challenges, in: *Proceedings of AAS/AIAA Space Flight Mechanics Meeting* Paper No. AAS 97–174, (February 1997).

[14] H. Yamakawa, J. Kawaguchi, N. Ishii, H. Matsuo, On Earth-Moon Transfer Trajectory with Gravitational Capture, in: *Proceedings AAS/AIAA Astrodynamics Specialists Conf.* Paper No. AAS93–633, (August 1993).

HAMILTONIAN STRUCTURE OF GENERALIZED CUBIC POLYNOMIALS

Peter Crouch* Fátima Silva Leite,[1] Margarida Camarinha**,[1]**

* *Systems Science and Engineering Research Center, Arizona State University*
Tempe, AZ 85287–USA
Phone: +1-480-965 1722 Fax: +1-480-965 2267

** *Departamento de Matemática, Universidade de Coimbra*
3000 Coimbra–Portugal
Phone: +351-239 791150 Fax: +351-239 832568

Abstract. We present a Hamiltonian formulation of a second order variational problem on a Riemannian manifold $(Q, < \cdot,\cdot >)$, which gives rise to generalized cubic polynomials on Q, and explore the possibility of writing down the extremal solutions of that problem as a flow in the space $\cup_{q\in Q} T_qQ \oplus T_q^*Q \oplus T_q^*Q$. For that we utilize the connection ∇ on Q, corresponding to the metric $< \cdot,\cdot >$. We exhibit the extremal equations in Hamiltonian form and identify the correct symplectic form. In general the results depend upon a choice of frame for TQ, but for the special situation when Q is a Lie group G with Lie algebra $\mathcal{G}$, our results are global and the flow reduces to a flow on $G \times \mathcal{G} \times \mathcal{G}^* \times \mathcal{G}^*$.

Keywords. Optimal control, Differential geometry, Nonlinear systems.

1. INTRODUCTION

Modelling complex mechanical systems is often accomplished using the notational convenience of differential geometry and in particular Riemannian geometry and symplectic geometry. For systems without dissipation one can choose a Lagrangian approach or a Hamiltonian approach. The basic phase space is usually taken as TQ in the Lagrangian approach, and T^*Q in the Hamiltonian approach, where Q is the configuration space. Indeed, in the Hamiltonian approach, the flow is specified by a Hamiltonian vector field on T^*Q, with Hamiltonian H, and using the canonical symplectic form Ω on T^*Q. In the Lagrangian case, the flow is Hamiltonian on TQ, but in this case, one must use a suitable symplectic structure on TQ, which is obtained by pulling back Ω to TQ via a suitable bundle mapping $\Sigma : TQ \to T^*Q$. For a Lagrangian defined by kinetic plus potential energies and specified by a Riemannian metric, $< \cdot,\cdot >$ on Q, the map Σ is just the linear mapping associated to the metric and defined by

$$(\Sigma X)(Y) = < X, Y >, \quad X, Y \in \Gamma(TQ), \quad (1)$$

where $\Gamma(TQ)$ denotes the set of smooth vector fields on Q. However, there is another formulation available to us in the Lagrangian case above in which one uses the Levi-Civita connection ∇ on Q, which is compatible with the Riemannian metric $< \cdot,\cdot >$, to write the system in terms of

[1] Work supported in part by ISR and research network contract ERB FMRXCT-970137.

a higher order differential equation on Q. Indeed, if $L = \frac{1}{2} < \dot{q}, \dot{q} > +V(q)$, $\dot{q} \in T_qQ$, is such a Lagrangian function and Σ is defined as above we write $\Sigma\dot{q} = p \in T_q^*Q$ for the momentum of the system and $\Delta_q V \in T_q^*Q$ for the gradient of the potential function V at the point $q \in Q$, $< \Delta_q V, Z >= dV_q(Z)$. Also, if $E \to Q$ is a vector bundle over Q, and F is a function on E, we define the direccional derivative of F by setting

$$(D_qF)(\bar{e})(e) = \lim_{h \to 0} \frac{1}{h}(F(q, e + h\bar{e}) - F(q, e)),$$

where $e, \bar{e} \in E_q$, the fiber in E over q. It follows that $D_qL = p$.

We also write $\frac{D}{dt}$, to denote the covariant derivative corresponding to ∇. It follows that the Euler Lagrange equation corresponding to L is just

$$\frac{Dp}{dt} = \Delta_q V, \tag{2}$$

or

$$\frac{D^2q}{dt^2} = \Sigma^{-1}\Delta_q V. \tag{3}$$

The corresponding Hamiltonian is of course

$$\begin{aligned} H(q,p) &= p(\dot{q}) - L(\dot{q}, q)\,|_{p=\Sigma\dot{q}} \\ &= < \dot{q}, \dot{q} > -L(\dot{q}, q)\,|_{p=\Sigma\dot{q}} \\ &= 1/2 < \dot{q}, \dot{q} > -V(q)\,|_{p=\Sigma\dot{q}} \\ &= 1/2\, p(\Sigma^{-1}p) - V(q). \end{aligned} \tag{4}$$

Now the Hamiltonian equations for H, corresponding to the Euler-Lagrange equations (2) or (3), are of course given by a vector field on T^*Q, that is a system of equations in TT^*Q. Writing down these equations depends upon fixing a system of coordinates in which to express Ω, etc. whereas the same effect has been accomplished in the Lagrangian case, reducing a system of equations in TTQ to a system of equations, either in TQ or T^*Q, through the connection ∇.

In studying control problems for such mechanical systems, and in particular optimal control problems, one encounters even higher order bundles. For example, the following optimal control problem is a typical situation.

$$\min_u \int_0^T \frac{1}{2} < u, u > dt \tag{5}$$

subject to:

$$\dot{q} = V, \quad \frac{DV}{dt} = u, \tag{6}$$

$$\begin{aligned} q(0) = q_0, &\quad \dot{q}(0) = V_0, \\ q(T) = q_T, &\quad \dot{q}(T) = V_T, \end{aligned} \tag{7}$$

where q_0 and q_T are given points in Q, V_0 and V_t are tangent vectors to Q, at q_0 and q_T respectively.

As observed above, the system of equations (6) may already be viewed as the reduction to TQ, of a system which is viewed in the Hamiltonian setting as one in TT^*Q. To solve the optimal control problem, however, the maximum principle instructs us that extremal solutions are projections of a Hamiltonian flow in TT^*TQ. This situation is clearly very cumbersome, utilizing the canonical symplectic form on T^*TQ. In this paper we explore the possibility of writing down the extremal solutions of the problem (5) - (6) - (7) as a flow on the space $E = \cup_{q\in Q} T_qQ \oplus T_q^*Q \oplus T_q^*Q$. We exhibit the extremal equations in Hamiltonian form and identify the correct symplectic form. The idea is to utilize the connection ∇ as before, but unlike the case of flow (2) our result is dependent upon a choice of frame for TQ. Thus we obtain global results, only in the case that Q is parallelizable. For the particular case when $Q = G$, a Lie group, our results are global and the flow reduces to a flow on $G \times \mathcal{G} \times \mathcal{G}^* \times \mathcal{G}^*$ where $\mathcal{G}$ is the Lie algebra of G.

Our solution to the above problem will be sought by treating it as a constrained variational problem, and utilizing the Lagrange multipliers as costates, in a typical fashion. Indeed, extremals for the problem (5) - (6) - (7), are characterized as projections of the flow resulting from the following equation.

$$\frac{D^4q}{dt^4} + R(\frac{D^2q}{dt}, \frac{Dq}{dt})\frac{Dq}{dt} = 0. \tag{8}$$

(See (Crouch and Silva Leite, 1995) and (Crouch and Silva Leite, 1991)).

This result is obtained by treating the problem as the unconstrained variational problem of minimizing the following functional subject to (7)

$$\int_0^T < \frac{D^2q}{dt^2}, \frac{D^2q}{dt^2} > dt. \tag{9}$$

Our approach follows this one with some minor modifications.

Now let $\{X_1, \cdots, X_n\}$ be a frame of vector fields on Q and $\{\omega_1, \cdots, \omega_n\}$ a co-frame of covector fields such that $\omega_k(X_j) = \delta_{kl}$. This selection must be local, unless Q is parallelizable. In terms of these frames we may write any vector field Y and covector field η along a curve $t \to q(t)$ in the following way

$$Y(q(t)) = Y(t) = \sum_{i=1}^n y_i(t)X_i(q(t)) \in T_{q(t)}Q,$$

$$\eta(q(t)) = \eta(t) = \sum_{i=1}^n \eta_i(t)w_i(q(t)) \in T_{q(t)}^*Q.$$

Thus, although the X_i's and the w_i's are defined on some open set in Q, $Y(t)$ and $\eta(t)$ are only defined along the curve $q(t)$. Setting

$$\dot{q}(t) = \sum_{i=1}^n v_i(t)X_i(q(t)) = V(t) \in T_{q(t)}Q,$$

it follows that the covariant derivatives of Y and η, along the curve $t \to q(t)$ with velocity vector field V, are given by:

$$\frac{DY}{dt} = \sum_{i=1}^{n} \dot{y}_i(t)X_i(q(t)) + \sum_{i=1}^{n} y_i(t)(\nabla_V X_i)(q(t)),$$

$$\frac{D\eta}{dt} = \sum_{i=1}^{n} \dot{\eta}_i(t)w_i(q(t)) + \sum_{i=1}^{n} \eta_i(t)(\nabla_V w_i)(q(t)).$$

We denote these expressions by the contracted forms

$$\frac{DY}{dt} = \dot{Y} + \nabla_V Y, \quad \frac{D\eta}{dt} = \dot{\eta} + \nabla_V \eta. \quad (10)$$

2. A VARIATIONAL APPROACH

We consider solving the optimal control problem (5) - (6) - (7) through the following variational problem:

$$\min_u J(q, V, p_1, p_2, u) = \int_0^T (p_1(\dot{q} - V) + p_2(\frac{DV}{dt} - u) + \frac{1}{2} < u, u >)dt, \quad (11)$$

subject to the boundary conditions (7) *and the dynamics* (6).

Here $p_1(t)$, $p_2(t)$ belong to $T^*_{q(t)}Q$, $u(t)$, $V(t)$ belong to $T_{q(t)}Q$.

Using the fundamental theorem of the calculus of variations, one may derive the corresponding Euler-Lagrange equations. For details on the proof of the following result and results hereafter, see (Crouch *et al.*, 1999).

Theorem 1. The extremals of the optimal control problem (5) - (6) - (7) may be expressed as solutions of the following system of equations, relative to the local choice of frame and co-frame for TQ and T^*Q:

$$\begin{cases} \dot{q} = V \\ \dot{V} = \Sigma^{-1}p_2 - \nabla_V V \\ \dot{p}_1 = -dp_1(V, .) + p_2(\nabla(\Sigma^{-1}p_2)) \\ \dot{p}_2 = -p_1 + p_2(\nabla V) - \nabla_V p_2 \end{cases} . \quad (12)$$

Notice that the optimal control u^* is given by $u^* = \Sigma^{-1}p_2$ and that, since $V \in T_{q(t)}Q$, $p_1, p_2 \in T^*_{q(t)}Q$, the system evolves on $\cup_{q\in Q} T_qQ \oplus T^*_qQ \oplus T^*_qQ$.

3. A HAMILTONIAN APPROACH

The Hamiltonian function associated with the optimal control problem (5) - (6) - (7) is given by

$$H(q, V, p_1, p_2) = -\frac{1}{2} < \Sigma^{-1}p_2, \Sigma^{-1}p_2 > + p_1(V) + p_2(\dot{V}) \quad (13)$$

or, equivalently, by

$$H(q, V, p_1, p_2) = -\frac{1}{2} < \Sigma^{-1}p_2, \Sigma^{-1}p_2 > + p_1(V) - p_2(\nabla_V V). \quad (14)$$

Our next objective is to exhibit the extremal equations in Hamiltonian form, for this Hamiltonian, and identify the correct symplectic form. We start with some preliminary results.

Lemma 2.

$$p(\nabla_X Y) = -(\nabla_X p)(Y),$$

for any vector fields X, Y and covector field p along a curve $t \to q(t)$.

Corollary 3. An alternative formula for the Hamiltonian is the following:

$$H(q, V, p_1, p_2) = 1/2\, p_2(\Sigma^{-1}p_2) + (\nabla_V p_2)(V) + p_1(V). \quad (15)$$

Theorem 4. The extremals of the optimal control problem (5) - (6) - (7) satisfy:

$$\begin{cases} \dot{q} = D_{p_1}H \\ \dot{V} = D_{p_2}H \\ \dot{p}_1 + dp_1(\dot{q}, .) = -D_q H \\ \dot{p}_2 = -D_V H \end{cases} , \quad (16)$$

where D_{p_1}, D_{p_2}, D_V are fiber derivatives in the bundle $E = \cup_{q\in Q} T_qQ \oplus T^*_qQ \oplus T^*_qQ$.

Proof: From (14) it is clear that $D_{p_1}H = V$ and so the first equation in (12) implies that $D_{p_1}H = \dot{q}$.

From (14) and the second equation in (12) we have

$$D_{p_2}H = \Sigma^{-1}p_2 - \nabla_V V = \dot{V}.$$

We now proceed to get the fourth equation. From (12) and (10) respectively, one gets $\frac{Dp_2}{dt} = -p_1 + p_2(\nabla V)$ and $\frac{Dp_2}{dt} = \dot{p}_2 + \nabla_V p_2$. From these identities and lemma 2 it follows that

$$\dot{p}_2 = -p_1 - (\nabla p_2)(V) - \nabla_V p_2. \quad (17)$$

On the other hand, it follows from corollary 3 that

$$D_V H = p_1 + (\nabla p_2)(V) + \nabla_V p_2, \quad (18)$$

and so, according to (17), $D_V H = -\dot{p}_2$ as required. Finally, to obtain the equation for $\dot{p}_1$ we consider H in the form given in (13),

$$H = -1/2\, p_2(\Sigma^{-1}p_2) + p_2(\dot{V}) + p_1(V).$$

Thus, for any vector field X along q we have

$$dH(X) = X(H) = -p_2\left(\nabla_X(\Sigma^{-1}p_2)\right),$$

since $p_2(\dot{V}) + p_1(V)$ does not depend on q. But, from the third equation in (12), we may write

$$dH(X) = -\dot{p}_1 - dp_1(\dot{q}, .)$$

as required. □

Having derived equations (16), we now explain in what sense they are Hamiltonian. We first define a map

$$\sigma : \begin{matrix} E & \longrightarrow & \hat{E} \\ (q, V, p_1, p_2) & \to & (q, v^i, p_1^i, p_2^i) \end{matrix} , \quad (19)$$

where $E = \cup_{q\in Q} T_qQ \oplus T^*_qQ \oplus T^*_qQ$, $\hat{E} = Q \oplus \mathbb{R}^n \oplus \mathbb{R}^{n*} \oplus \mathbb{R}^{n*}$ and v^i, p_1^i, p_2^i are respectively the coordinates of the vector field V and covector fields

p_1, p_2 in Q, with respect to the global frames $\{X_1, \cdots, X_n\}$ and $\{w_1, \cdots, w_n\}$. That is,

$$V = \sum_i v^i X_i, \; p_1 = \sum_i p_1^i w_i, \; p_2 = \sum_i p_2^i w_i.$$

σ is a diffeomorphism between the differentiable manifolds E and $\hat{E}$.

Lemma 5. $\hat{E}$ is a symplectic manifold with symplectic form

$$\hat{\Omega} = \sum_i (w_i \wedge dp_1^i - p_1^i dw_i + dv^i \wedge dp_2^i). \quad (20)$$

Proof : Since $\sum_i dv^i \wedge dp_2^i$ is the natural symplectic structure on $\mathbb{R}^n \oplus \mathbb{R}^{n*}$, it is enough to show that $\sum_i (w_i \wedge dp_i - p_1^i dw_i) = -d(\sum_i p_1^i w_i)$ is a symplectic form on $T^*Q \simeq Q \oplus \mathbb{R}^{n*}$. Assume that $q_1, \cdots, q_n$ are local coordinates on Q, so that $\{dq_1, \cdots, dq_n\}$ is a basis of T_q^*Q. So, any 1-form on Q, in particular the w_i's, may be expressed in terms of that basis and we may write, $\sum_i p_1^i w_i = \sum_i \hat{p}_1^i dq_i = \theta$, where $p_1^i = \sum_j \hat{p}_1^j dq_j(X_i)$. Since the matrix $[dq_j(X_i)]_{i,j}$ is nondegenerate everywhere, $\{\hat{p}_1^i\}_i$ is another set of local coordinates for $\mathbb{R}^{n*}$. It is therefore sufficient to show that $d\theta$ is nondegenerate, but this is a simple exercise in local coordinates. So, $\sum_i (w_i \wedge dp_i - p_1^i dw_i)$ is the symplectic form $-d\theta$ on T^*Q. □

Since $\sigma : E \to \hat{E}$ is a diffeomorphism and $\hat{\Omega}$ is a symplectic form on $\hat{E}$, we may define the pull back $\sigma^*\hat{\Omega}$, of $\hat{\Omega}$ by σ, by

$$(\sigma^*\hat{\Omega})_x(X_1, X_2) = \hat{\Omega}_{\sigma(x)}(\sigma_*(X_1), \sigma_*(X_2)),$$

where $x \in E$, $X_1, X_2 \in T_x E$ and σ_* is the derivative of σ at x. Ω is clearly a symplectic form on E. However Ω is complicated to explicitly write down.

Now let $\hat{H} = (\sigma^{-1})^* H$ be the pull back of H by σ^{-1}. That is,

$$\begin{aligned} \hat{H}(q, v^i, p_1^i, p_2^i) &= H \circ \sigma^{-1}(q, v^i, p_1^i, p_2^i) \\ &= H(q, V, p_1, p_2). \end{aligned}$$

Theorem 6. On the symplectic space $(\hat{E}, \hat{\Omega})$, the equations

$$\left\{ \begin{aligned} \dot{q} &= \frac{\partial \hat{H}}{\partial p_1^i} X_i \\ \dot{v}^i &= \frac{\partial \hat{H}}{\partial p_2^i} \\ \dot{p}_1^i + dp_1(\dot{q}, X_i) &= -d\hat{H}(X_i) \\ \dot{p}_2^i &= -\frac{\partial \hat{H}}{\partial v^i} \end{aligned} \right. \quad (21)$$

are Hamiltonian, with Hamiltonian function $\hat{H}$.

Proof : Simply expand the equation $\hat{\Omega}(X_{\hat{H}}, .) = d\hat{H}$, in the coordinates (q, v^i, p_1^i, p_2^i), with $X_{\hat{H}} = (q, v^i, p_1^i, p_2^i)$. □

Theorem 7. The system (16) is a Hamiltonian system with Hamiltonian H on the symplectic space (E, Ω).

Proof : The map σ maps the dynamics (16) onto the dynamics (21) and by construction σ is a symplectic morphism of (E, Ω) onto $(\hat{E}, \hat{\Omega})$. □

4. APPLICATIONS TO EARLIER FORMULATIONS

In this section we briefly apply the equations (12) to obtain a new interpretation of existing results.

We start with the following lemma.

Lemma 8. For $Z, V \in \Gamma(TM)$, $\eta \in \Gamma(T^*M)$, one has

$$\begin{aligned} \frac{D}{dt}\eta(\nabla V)|_Z &= \frac{D\eta}{dt}(\nabla_Z V) \\ &+ \eta\left(\frac{D}{dt}(\nabla_Z V) - \nabla_{\frac{DZ}{dt}} V\right). \end{aligned}$$

Theorem 9. The system of equations (12) solves the extremal flow (8).

Proof : From (12) $\frac{Dp_2}{dt} = p_2(\nabla V) - p_1$ so, from lemma 8,

$$\begin{aligned} \frac{D^2 p_2}{dt^2}(Z) &= -\frac{Dp_1}{dt} + \frac{Dp_2}{dt}(\nabla_Z V) \\ &+ p_2\left(\frac{D}{dt}(\nabla_Z V) - \nabla_{\frac{DZ}{dt}} V\right). \end{aligned}$$

Again, substituting from (12) and using $\frac{DZ}{dt} = \dot{Z} + \nabla_V Z$ and the symmetry of ∇ we obtain

$$\begin{aligned} &\frac{D^2 p_2}{dt^2}(Z) = p_1(\nabla_Z V) - p_2(\nabla_Z \frac{DV}{dt}) + p_2(\nabla_{\nabla_Z V} V) \\ &-p_1(\nabla_Z V) + p_2\left(\frac{D}{dt}(\nabla_Z V) - \nabla_{\frac{DZ}{dt}} V\right) \\ &= p_2(-\nabla_Z \dot{V} - \nabla_Z \nabla_V V + \nabla_{\nabla_Z V} V + \nabla_{\dot{Z}} V \\ &+\nabla_Z \dot{V} + \nabla_V \nabla_Z V - \nabla_{\dot{Z}} V - \nabla_{\nabla_V Z} V) \\ &= p_2(\nabla_{\nabla_Z V - \nabla_V Z} V + \nabla_V \nabla_Z V - \nabla_Z \nabla_V V) \\ &= p_2(\nabla_{[Z,V]} V + \nabla_V \nabla_Z V - \nabla_Z \nabla_V V) \\ &= p_2(R(V, Z)V). \end{aligned}$$

Now, since from (12) $\frac{DV}{dt} = \Sigma^{-1} p_2$, it follows that $p_2(X) = (\Sigma \frac{DV}{dt})(X) = < \frac{DV}{dt}, X >$. If X is chosen to be parallel along q, that is $\frac{DX}{dt} = 0$, we have $\frac{Dp_2}{dt}(X) = < \frac{D^2 V}{dt^2}, X >$ and $\frac{D^2 p_2}{dt^2}(X) = < \frac{D^3 V}{dt^3}, X >$. As a consequence

$$\begin{aligned} < \frac{D^3 V}{dt^3}, X > &= p_2(R(V, X)V) \\ &= < \frac{DV}{dt}, R(V, X)V > \end{aligned}$$

and using the symmetries of the curvature tensor R (Milnor, 1969), we have

$$\frac{D^3 V}{dt^3} + R(\frac{DV}{dt}, V)V \equiv 0, \quad (22)$$

which is the equation (8) . □

We also have an expression (14) for the Hamiltonian function corresponding to the Hamiltonian system (16). In (Camarinha, 1996), Camarinha found an invariant for the flow (22) or (8), namely the function I in the next lemma. But I is indeed H given by (14), as shown in the following result.

Lemma 10. The function

$$I = \frac{1}{2} < \frac{DV}{dt}, \frac{DV}{dt} > - < \frac{D^2V}{dt^2}, V >$$

is an invariant of the flow (22) and $I = H$, where H is the Hamiltonian function (14).

5. THE LIE GROUP CASE

We now specialize to the case where $Q = G$, is a compact or semi-simple Lie group, with Lie algebra $\mathcal{G}$. In this case Q is parallelizable and the equations (12), and indeed the equations (16), may be given a global interpretation. In this case we also have an explicit expression for the connection corresponding to the unique bi-invariant metric on G, $\nabla_X Y = \frac{1}{2}[X,Y]$, (see, for instance, (Milnor, 1969)). This corresponds to the choice where $\Sigma : \mathcal{G} \to \mathcal{G}^*$ is defined by $(\Sigma X)(Y) =< Y, X >$.

We may assume that $\{X_1, \ldots, X_n\}$ is a basis of left-invariant vector fields and $\{w_1, \ldots, w_n\}$ is a dual basis of left-invariant one-forms. It follows that the equations (12) are indeed globally defined and we may identify V, p_1, p_2, as elements of $\mathcal{G}$, $\mathcal{G}^*$ and $\mathcal{G}^*$ respectively.

At this point it is important to recall a few formulas when Q is a Lie group G. If for $X \in \mathcal{G}$, ad_X denotes the adjoint map

$$\begin{array}{rcl} ad_X : \mathcal{G} & \longrightarrow & \mathcal{G} \\ Y & \rightarrow & [X,Y] \end{array},$$

the co-adjoint map of ad_X is defined by:

$$ad^*_X\eta(Y) = -\eta(ad_X Y) = -\eta([X,Y]), \quad (23)$$

for $\eta \in \mathcal{G}^*, Y \in \mathcal{G}$.

Now, if X and Y are left-invariant vector fields on G and η is a left-invariant one-form on G, then $Y(\eta(X)) = 0, \ \forall X, Y,$ and consequently

$$\eta(\nabla_X Y) = -(\nabla_X \eta)Y, \quad (24)$$
$$d\eta(X,Y) = -\eta([X,Y]). \quad (25)$$

Also taking into account that $\nabla_Y X = \frac{1}{2}[Y,X]$ it follows from (25) and (24) that

$$d\eta(X,Y) = -2\eta(\nabla_X Y) = 2(\nabla_X\eta)Y,$$

and, also, using (23)

$$\nabla_X \eta = \frac{1}{2} ad^*_X \eta. \quad (26)$$

We now turn our attention to the problem of identifying the correct Hamiltonian and symplectic structure, which will be a generalization of that from T^*G to T^*TG, identified in Abraham and Marsden (Abraham and Marsden, 1978).

Lemma 11. In the case of a compact or semisimple Lie group G with Lie algebra $\mathcal{G}$, the extremal equations (12) may be written in the form

$$\begin{cases} \dot{q} = L_{q_*}(V) \\ \dot{V} = \Sigma^{-1}p_2 \\ \dot{p}_1 = -ad^*_V p_1 \\ \dot{p}_2 = -p_1 \end{cases}, \quad (27)$$

where $V \in \mathcal{G}, p_1, p_2 \in \mathcal{G}^*$, L_q is left translation in the Lie group G and L_{q_*} is the derivative of L_q.

Proof : As a consequence of (25) and(26) we may write the equations (12), with $V \in \mathcal{G}, p_1, p_2 \in \mathcal{G}^*$, in the form

$$\begin{cases} \dot{q} = L_{q_*}(V) \\ \dot{V} = \Sigma^{-1}p_2 \\ \dot{p}_1 = -ad^*_V p_1 + \frac{1}{2} ad^*_{\Sigma^{-1}p_2} p_2 \\ \dot{p}_2 = -p_1 \end{cases}.$$

But from (25) and the definition of Σ,

$$\begin{aligned} ad^*_{\Sigma^{-1}p_2} p_2(X) &= -p_2([\Sigma^{-1}p_2, X]) \\ &= - < \Sigma^{-1}p_2, [\Sigma^{-1}p_2, X] >= 0. \end{aligned}$$

□

Note that the system (27) is Hamiltonian on $G \times \mathcal{G} \times \mathcal{G}^* \times \mathcal{G}^* = E$ with Hamiltonian function $H = \frac{1}{2}p_2(\Sigma^{-1}p_2) + p_1(V)$ and symplectic form

$$\Omega(\dot{q}, \dot{V}, \dot{p}_1, \dot{p}_2, \dot{\bar{q}}, \dot{\bar{V}}, \dot{\bar{p}}_1, \dot{\bar{p}}_2) = \omega_i(\dot{q})\, \dot{\bar{p}}^i_1 - \omega_i(\dot{\bar{q}})\, \dot{p}^i_1$$

$$-(ad^*{}_{L_q{}^{-1}{}_*(\dot{q})} p_1)(L_q{}^{-1}{}_*(\dot{\bar{q}})) + \dot{v}^i\, \dot{\bar{p}}^i_2 - \dot{\bar{v}}^i\, \dot{p}^i_2.$$

Corollary 12. The system of equations (27) satisfies

$$\dddot{V} + [V, \ddot{V}] = 0. \quad (28)$$

Equations (28) were first written down in this generality, as a specialization of the extremal flow (22) to the Lie group case, in (Crouch and Silva Leite, 1995), but see also (Noakes *et al.* , 1989).

We can also write the extremal equations in the lemma 11 using the natural identification of elements of $\mathcal{G}$ and $\mathcal{G}^*$. More precisely, define $p_1 =< A_1, . >, p_2 =< A_2, . >, A_1, A_2 \in \mathcal{G}$, and $\hat{E} = G \times \mathcal{G} \times \mathcal{G} \times \mathcal{G}$. Then,

Theorem 13. The extremal equations on $\hat{E}$ have the form

$$\begin{cases} \dot{q} = L_{q_*}(V) \\ \dot{V} = A_2 \\ \dot{A}_1 - [A_1, V] = 0 \\ \dot{A}_2 = -A_1 \end{cases}. \quad (29)$$

Proof : We only need to prove that the third equation is satisfied. It follows from identity (23) and $< X, [Y,Z] >=< [X,Y], Z >$ that

$$\begin{aligned} \dot{p}_1 = -ad^*_V p_1 &\Leftrightarrow < \dot{A}_1, . >=< A_1, [V, .] > \\ &\Leftrightarrow < \dot{A}_1, . >= - < [V, A_1], . >, \end{aligned}$$

which completes the proof. □

Theorem 14. $\hat{E} = G \times \mathcal{G} \times \mathcal{G} \times \mathcal{G}$ is a symplectic manifold with symplectic form

$$\hat{\Omega}(\dot{q}, \dot{V}, \dot{A}_1, \dot{A}_2, \dot{\overline{q}}, \dot{\overline{V}}, \dot{\overline{A}}_1, \dot{\overline{A}}_2) = < \dot{\overline{A}}_1, {L_q^{-1}}_*(\dot{q}) >$$

$$- < \dot{A}_1, {L_q^{-1}}_*(\dot{\overline{q}}) > + < \dot{V}, \dot{\overline{A}}_2 > - < \dot{\overline{V}}, \dot{A}_2 >$$

$$- < ad_{{L_q^{-1}}_*(\dot{q})} A_1, {L_q^{-1}}_*(\dot{\overline{q}}) > .$$

Moreover, the equations (29) are Hamiltonian with Hamiltonian function

$$\hat{H} = \frac{1}{2} < A_2, A_2 > + < A_1, V > .$$

5.1 *Example*

A specific problem in optimal control of the form (5)-(6)-(7) was treated in (Bloch and Crouch, 1994) where an analysis was made between the Hamiltonian and the Lagrangian formulation of higher order optimal control problems. We treat the example again here in a slightly different setting.

Consider the problem:

$$\min_u \int_0^T \frac{1}{2} < u, u > dt, \ \textit{subject to:}$$

$$\begin{cases} \dot{Q} = \Omega_1 Q \\ \dot{\Omega}_1 = u \end{cases} \quad Q \in SO(n), \ u, \Omega_1 \in so(n) .$$

and boundary conditions

$$Q(0) = Q_0, \quad Q(T) = Q_T,$$
$$\dot{Q}(0) = \dot{Q}_0, \quad \dot{Q}(T) = \dot{Q}_T.$$

Here $< A, B >= trace(A^T B)$. To solve the problem we construct the Hamiltonian

$$H(u, Q, \Omega_1, p_1, p_2) = < p_2, u > + < p_1, \Omega_1 Q > -1/2 < u, u > . \tag{30}$$

Thus the optimal control is $u^* = p_2 \in so(n)$, from which we get

$$H = 1/2 < p_2, p_2 > + < p_1, \Omega_1 Q > .$$

Using properties of the trace of a matrix we obtain

$$\dot{p}_2 = -1/2(p_1 Q^T - Q p_1^T), \quad \dot{p}_1 = -\Omega_1^T p_1. \tag{31}$$

We hypothesize a solution where $p_1 = \Omega_2 Q$, with $\Omega_2 \in so(n)$. If we make this assumption it follows from (31) that

$$\dot{p}_2 = -\Omega_2,$$
$$\dot{\Omega}_2 = [\Omega_1, \Omega_2]$$

and so, the full extremal equations may be written as

$$\begin{cases} \dot{Q} = \Omega_1 Q \\ \dot{\Omega}_1 = p_2 \\ \dot{p}_2 = -\Omega_2 \\ \dot{\Omega}_2 = [\Omega_1, \Omega_2] \end{cases} \tag{32}$$

The equations (32) are precisely the equations of theorem 13 and the corresponding Hamiltonian function (30) is

$$\begin{aligned} H &= 1/2 < p_2, p_2 > + < p_1, \Omega_1 Q > \\ &= 1/2 < p_2, p_2 > + < \Omega_2 Q, \Omega_1 Q > \\ &= 1/2 < p_2, p_2 > + < \Omega_2, \Omega_1 > . \end{aligned}$$

Note that the symplectic structure on $SO(n) \times so(n) \times so(n) \times so(n)$, is that in theorem 14. We also note that the equations

$$\begin{cases} \dot{Q} = \Omega_1 Q \\ \dot{\Omega}_1 = p_2 \\ \dot{p}_2 = -1/2(p_1 Q^T - Q p_1^T) \\ \dot{p}_1 = -\Omega_1^T p_1 \end{cases}$$

are Hamiltonian with respect to the Hamiltonian function

$$H = 1/2 < p_2, p_2 > +1/2 < p_1 Q^T - Q p_1^T, \Omega_1 > .$$

6. REFERENCES

Abraham R. and Marsden J. E. (1978). *Foundations of Mechanics*, 2nd ed., Springer-Verlag, New York, USA.

Bloch A. and Crouch P. (1994). Reduction of Euler Lagrange problems for constrained variational problems and relation with optimal control problems. In: *Proceedings of 33th IEEE CDC.* pp. 2584-2590.

Camarinha M. (1996). *The Geometry of Cubic Polynomials on Riemannian manifolds.* Ph. D. thesis (English translation). Department of Mathematics, Coimbra University, Portugal.

Crouch P. and Silva Leite F. (1995). The Dynamic Interpolation Problem on Riemannian Manifolds, Lie Groups and Symmetric Spaces. *Journal of Dynamical and Control Systems* **1**, N.2, pp. 177–202.

Crouch P. and Silva Leite F. (1991). Geometry and the Dynamic Interpolation Problem. In: *Proceedings of the American Control Conference.* Boston. pp. 26–29.

Crouch P., Silva Leite F. and Camarinha M. (1998). A second order Riemannian variational problem from a Hamiltonian perspective. *Pré Publicações* **98-17**. Departamento de Matemática, Universidade de Coimbra, Portugal.

Milnor J. (1969). *Morse Theory.* Ann. Math. Stud., **51**. Princeton University Press.

Noakes L., Heinzinger G. and Paden B. (1989). Cubic splines on curved spaces. *IMA Journal of Mathematical Control & Information* **6**. pp. 465-473.

Nomizu K. (1969). *Foundations of Differential Geometry,* Vol I, II. Interscience Tracts in Pure and Applied Mathematics, John Willey & Sons, New York, USA.

QUANTIZED CONTROL SYSTEMS AND DISCRETE NONHOLONOMY

A. Bicchi * A. Marigo ** B. Piccoli ***

* *Centro "E. Piaggio, Università di Pisa, Via Diotisalvi 2, 56100 Pisa, Italy.* bicchi@ing.unipi.it
** *SISSA - ISAS, Int. School Advanced Studies, 34014 Trieste, Italy.* marigo@sissa.it
*** *DIIMA, University of Salerno, 84084 Fisciano (SA), Italy.* piccoli@sissa.it

Abstract: In this paper we study control systems whose input sets are quantized, and in particular finite or countable but nowhere dense. We specifically focus on problems relating to the structure of the reachable set of such systems, which may turn out to be either dense or discrete. We report results on the rechable set of linear quantized systems, and on a particular but interesting class of nonlinear systems, forming the discrete counterpart of driftless nonholonomic continuous systems. Implications and open problems in the analysis and synthesis of quantized control systems are addressed.

Keywords: Quantized control systems, Nonholonomic systems, Hybrid systems.

1. INTRODUCTION

In this paper we consider systems of the type

$$x^+ = g(x,u),\ x \in \mathbb{R}^n,\ u \in U \subset \mathbb{R}^m \quad (1)$$

where the input set, U, is quantized, i.e. finite or numerable but nowhere dense in $\mathbb{R}^m$. Quantized control systems arise in a number of applications because of many physical phenomena or technological constraints. In the control literature, quantization of inputs has been considered mainly as due to D/A conversion, and mostly regarded as a disturbance to be rejected (Bertram (1958); Slaughter (1964); Delchamps (1990)). Typical results in this spirit are those provided by Hou *et al.* (1997), who show how a nonlinear system with quantized feedback, whose linear approximation (without quantization) has an asymptotically stable solution, has uniformly ultimately bounded solutions; and how such bounds can be made small at will by refinig quantization sufficiently.

More recently, some attention has been focused on quantized control systems as specific models of hierarchically organized systems with interaction between continuous dynamics and logic (Wong and Brockett (1999); Elia and Mitter (1999)). In these papers, quantization of inputs is regarded as a fundamental characteristic of systems where the resources for implementing the control scheme are limited, such as e.g. when communications between the plant and the controller can only happen through a finite capacity channel. As a consequence of taking such viewpoint, the focal point of research is to understand how to quantize the control system best (in some suitable sense), rather than assessing robustness of design with respect to quantization. In their papers, both Wong and Brockett (1999) and Elia and Mitter (1999) focus on the stabilization problem. Authors of the latter paper provide a result on the optimal (coarsest) quantization for asymptotically stabi-

[1] Work partially supported through grants ASI ARS-99-170 and MURST "RAMSETE"

lizing a linear discrete–time system, that turns out to require a countable symmetric set of logarithmically decreasing inputs, namely $U = \{\pm u_i : u_{i+1} = \rho u_i, -\infty \leq i \leq +\infty\} \cup \{0\}$. Although this choice (and the corresponding partition induced in the state space) captures the intuitive notion that coarser control is necessary when far from the goal, it still needs input values that are arbitrarily close to each other near the equilibrium.

An observation common to many papers on stabilization with quantized control is that, if the available quantized control set is finite, or countable but nowhere dense (in the natural topology of $\mathbb{R}^m$) then stability can only be achieved in a weak sense — be it ultimate boundedness (Hou *et al.* (1997)), containability (Wong and Brockett (1999)), or practical stability (Elia and Mitter (1999)).

The focus of our paper is on the study of particular phenomena that may appear in quantized control systems, which have no counterpart in classical systems theory, and that deeply influence the qualitative properties and performance of the control system. These concern the structure of the set of points that are reachable by system (1), and particularly its density. We will address two instances of the general system (1), namely linear systems, and driftless nonlinear systems. In particular, among the latter, we will focus our attention on the (discrete counterpart of) nonholonomic systems. We report on conditions under which the rechable set for these systems is dense in $\mathbb{R}^n$, or otherwise when it possesses a lattice structure. We will discuss applications to problems in steering nonholonomic systems, and discuss possible implications and open problems in the analysis and synthesis of quantized control systems.

2. FIRST DEFINITIONS AND EXAMPLES

We will consider systems defined as follows

Definition 1. Let a system be defined by a quintuple $(\mathcal{X}, \mathcal{T}, \mathcal{U}, \Omega, A)$, *where* $\mathcal{X}$ *denotes the configuration set,* $\mathcal{T}$ *an ordered time set,* $\mathcal{U}$ *a set of acceptable input symbols (possibly depending on the configuration),* Ω *a set of acceptable input words, and* $\mathcal{A}$ *is a state-transition map* $\mathcal{A} : \mathcal{T} \times \Omega \times \mathcal{X} \to \mathcal{X}$. *Denote* $\mathcal{A}_{t,\omega}(x) = \mathcal{A}(t, \omega, x)$, *with composition by concatenation* $\mathcal{A}(x_1, a_2, t_1) \circ \mathcal{A}(x_0, a_1, t_0) = \mathcal{A}(\mathcal{A}(\mathbf{x}_0, a_1, t_0), a_2, t_1)$.

In particular, we will focus here on $\mathcal{T} = \mathbb{N}$, as most interesting phenomena relating with quantization appear as linked to discrete time. A system as in definition 1 with both $\mathcal{X}$ and $\mathcal{U}$ discrete sets essentially represents a sequential machine or an automaton, while for $\mathcal{X}$ and $\mathcal{U}$ continuous sets, a discrete–time, nonlinear control system is obtained. We are interested in studying reachability problems that arise when $\mathcal{X}$ has the cardinality of a continuum, but $\mathcal{U}$ is discrete (i.e., finite or countable, but nowhere dense), i.e. when inputs are *quantized*. The following example motivates the generality of the definition above with a specific robotics application.

Example1. We will consider the discrete analogue of a well known continuous nonholonomic system, which is the plate–ball system (see e.g. Brockett and Dai (1993); Jurdjevic (1993); Levi (1993)). A ball rolls without slipping between two parallel plates, of which one is fixed and the other one translates. If the moving plate is driven along a closed trajectory, in particular e.g. it is translated to the right by some amount, then forward, left, and backward by the same amount, the same will happen to the ball centre, which will end up in the same initial position. However, the final orientation of the sphere will be changed by a net amount. Indeed, it can be shown (Marigo and Bicchi (2000)) that an arbitary orientation in $SO(3)$ can be reached by rolling arbitrary pairs of non–isomorphic surfaces, which fact was used as a basis for building simplified dextrous robot hands.

Consider now a similar experiment with a polyhedron replacing the ball. For practical reasons, possible actions on this system (studied in detail in Ceccarelli *et al.* (2000)), are only rotations about one of the edges of the face lying on the plate, by exactly the amount that brings an adjacent face on the plate. A generic configuration of the polyhedron can be described by giving the index of the face sitting on the plate, the position of the projection on the plate of the centroid, and the orientation of the projection of an inner diagonal of the cube. Hence, the configuration set is represented by the stratified manifold $\mathcal{X} = \mathbb{R}^2 \times S^1 \times \mathcal{F}$, where $\mathcal{F}$ denotes the set of faces of the polyhedron. Given the discrete nature of input actions, we take $\mathcal{T} = \mathbb{N}_+$, $\mathcal{U} = \mathcal{F}$, Ω the (configuration-dependent) set of all sequences of adjacent faces starting with the face of the present configuration, and $\mathcal{A}_{t,\omega}(x)$ the configuration reached at the end of a t–long sequence of tumbles $\omega \in \Omega$ allowed at x. $\triangleleft$

Definition 2. A configuration x_f *is reachable from* x_0 *if there exists a time* $t \in \mathcal{T}$ *and an acceptable input string* $\omega \in \Omega$ *that steers the system from* x_0 *to* $x_f = \mathcal{A}_{t,\omega}(x_0)$.

In the following we shall denote by R_x the reachable set from x, i.e. the set of configurations that can be reached from x. For differentiable systems, the notion of *reachability from* x is introduced when $R_x = \mathcal{X}$. For discrete-time systems with quantized inputs, however, Ω is a subset of all possible finite sequences ω of symbols in the discrete set $\mathcal{U}$, hence R_x is a countable set and, in the general case that the configuration set $\mathcal{X}$ has the cardinality of a continuum, it will not make sense checking whether R_x equals $\mathcal{X}$.

Example1–b. The set of configurations that can be reached starting from a given configuration of the polyhedron of Example 1, in a large but finite number of steps N, may have different characteristics. Consider for instance (intutively, or by simulation) positions reached by the centroid of different polyhedra after N steps: only points lying on a regular grid can be reached by rolling a cube, while for a generic parallelepiped or pyramid they tend to fill the plane as N grows. Also, orientations obtained by rolling the cube or the parallelepiped are only multiples of $\pi/2$, while orientations reached by the generic pyramid tend to fill the unit circle as N grows. Conditions under which the reachable set is dense, and a description of the lattice structure in discrete cases, have been studied by Y.Chitour *et al.* (1996); Ceccarelli *et al.* (2000). $\triangleleft$

We introduce a concept of *approachability*, which is stated in the further assumption that the state space is a metric space with distance $d(\mathbf{x}_1, \mathbf{x}_2)$:

Definition 3. A configuration x_f can be approached *from x_0 if $\forall \epsilon$, $\exists t \in \mathcal{T}$, $\exists \omega \in \Omega$ such that $d(\mathcal{A}_{t,\omega}(x_0), x_f) < \epsilon$. We say that the system is* approachable *from x_0 if the reachable set R_{x_0} is dense in $\mathcal{X}$, and is* locally approachable *from x_0 if the closure of the reachable set R_{x_0} contains a neighborhood of x_0. Finally, the system is* approachable *if*

$$\text{closure}\ (R_{\mathbf{x}}) = \mathcal{X},\ \forall x \in \mathcal{X}.$$

Lack of density of R_x will be referred to as *discreteness* of the reachable set. The term *dense in a subset* $\mathcal{X}' \subset \mathcal{X}$ will be used to indicate that

$$\text{closure}\ (R_x) \cap \mathcal{X}' = \mathcal{X}',\ \forall x \in \mathcal{X},$$

Notice that the possibility that the reachable set of a quantized control system is discrete, separates such systems from differentiable systems; on the other hand, the possibility of having a dense reachable set distinguishes quantized control systems from classical finite–state machines.

In practical aplications, it may be important to measure the coarseness of discrete reachable sets. We will then say that a configuration x_f is ϵ–*approachable* from x_0 if $\exists t \in \mathcal{T}, \omega \in \Omega$, such that $d(\mathcal{A}_{t,\omega}(x_0), x_f) < \epsilon$. The set of configurations that are ϵ–approachable from x is denoted by R^ϵ_x. The system will be said ϵ–approachable if $R^\epsilon_x = \mathcal{X}$, $\forall x \in \mathcal{X}$.

Let us consider a quantized control system in discrete time in the form

$$x^+ = g(x, u),\ u \in U, \qquad (2)$$

where $x \in \mathcal{X} = M$, a manifold, and U a finite set. For simplicity, also let Ω be comprised of all strings of symbols in U. Obviously, such definition is equivalent to assigning a finite number of maps $g_u : M \to M$.

In this case the reachable set from a point $x \in M$ is $R_x = \{g_{u_1} \cdots g_{u_n}(x) : n \in \mathbb{N}_0, u_i \in U\}$ ($\mathbb{N}_0$ includes the number 0 so that $x \in R_x$). Moreover, we introduce the relation $\sim$ over the elements of M by setting $x \sim y$, $x, y \in M$, if $y \in R_x$. We want to focus on a special class of systems that we call invertible systems.

Definition 4. The system (2) is said to be invertible if for every $x \in M$ and $u \in U$ there exists a finite sequence of controls $u_i \in U$, $i = 1, \ldots, n$, such that $g_{u_1} \cdots g_{u_n}(g(x, u)) = x$.

The following proposition is obvious:

Proposition 1. The relation $\sim$ is an equivalence relation if and only if the system is invertible.

If the system is invertible, we can partition the state space into a family of reachable sets. This is equivalent to take the quotient $M/\sim$ with respect to the equivalence relation $\sim$. We call the set $\widetilde{M} = M/\sim$ the reachability set of the system (2) and we endow $\widetilde{M}$ with the quotient topology, that is the largest topology such that $\pi : M \to \widetilde{M}$, the canonical projection, is continuous.

Example 2. Consider the system

$$x^+ = x + u$$

where $x \in \mathbb{R}$ and $u \in \mathcal{U}$, $\mathcal{U}$ finite subset of $\mathbb{R}$. If $\mathcal{U} = \{0, 1/2, -1\}$ then the system is invertible. The reachable set from the origin R_0 is the subgroup of $\mathbb{R}$ generated by 1/2 and the reachability set $\widetilde{M}$ is homeomorphic to S^1. If $\mathcal{U} = \{\sqrt{2}, -1\}$ then the system is not invertible. For example $\sqrt{2} \in R_0$, but, since $\sqrt{2}$ is irrational, $0 \notin R_{\sqrt{2}}$. $\triangleleft$

Example3. Consider the system

$$x^+ = g(x, u)$$

where $x \in \mathbb{R}$, $\mathcal{U} = \{\pm 1/2, \pm 2\}$ and $g(x, u) = u \cdot x$. The system is invertible, $R_0 = \{0\}$ and for every $x \neq 0$ $R_x = \{\pm 2^i x : i \in \mathbb{Z}\}$. The reachability set $\widetilde{M}$ is homeomorphic to the set $S^1 \cup \{\alpha\}$, where on S^1 there is the usual topology while the only neighborhood of α is the whole space. $\triangleleft$

Notice that in example 3, the reachable set R_x for $x \neq 0$ has only one accumulation point, namely 0. If we assume that M is a metric space and the maps g_u are isometries then we have a dicotomy illustrated by next proposition:

Proposition 2. Consider an invertible system (2). Let (M, d) be a metric space and assume that $x \to g(x, u)$ is an isometry for every $u \in \mathcal{U}$ then each reachable set R_x is formed either by accumulation points or by isolated points.

Proof. Assume that the set R_x admits an accumulation point $\bar{x} \in R_x$. Let $x_k \in R_x$ be such that $x_k \to \bar{x}$ and the set $\{x_k : k \in \mathbb{Z}\}$ is infinite. Since the system is invertible, for every k there exists $\tilde{u}_k = (u_k^1, \ldots, u_k^{n_k})$ such that $u_k^i \in U$ and $g_{u_k^1} \cdots g_{u_k^{n_k}}(x_k) = x$. Define $y_k = \lim_m g_{u_k^1} \cdots g_{u_k^{n_k}}(x_m)$. For every k and m we have:

$$\begin{aligned} & d(g_{u_k^1} \cdots g_{u_k^{n_k}}(x_m), x) = \\ & d(g_{u_k^1} \cdots g_{u_k^{n_k}}(x_m), g_{u_k^1} \cdots g_{u_k^{n_k}}(x_k) = \\ & d(x_m, x_k). \end{aligned}$$

Passing to the limit in m, we have $d(y_k, x) = d(\bar{x}, x_k)$. Clearly the sequence y_k converge to x and contains infinitely many distinct points, so x is an accumulation point for R_x. Now it easily follows that all points of R_x are accumulation points for R_x. $\square$

The system:

$$x^+ = x + u \tag{3}$$

with $x \in R^n$ is an interesting special case. It is clear that for every $x_0 \in R^n$ the reachable set $X(x_0)$ from x_0 is equal to $x_0 + X_0$ where X_0 is the reachable set from the origin. The hypothesis of the above Proposition are satisfied. Notice that if $n = 1$ and U is simmetric then the set X_0 is either everywhere dense or nowhere dense in $\mathbb{R}$ (since it is a subgroup of $\mathbb{R}$), hence presenting a stonger dicotomy of the one illustrated by the above Proposition. For $n > 1$ we may have directions along which the reachable set X_0 is dense and directions along which is discrete. This is precisely the case of $n = 2$ and $U = \{(\pm 1, 0), (\pm\sqrt{2}, 0), (0, \pm 1)\}$. Notice that if we define $\pi_v : \mathbb{R}^n \to \mathbb{R}$ to be the ortogonal projection on the direction of the vector v, then $\pi_v(X_0)$ is dense in $\mathbb{R}$ for every v not parallel to $(0, 1)$ (and this corresponds to the fact that the projection of the reachable set is precisely the reachable set of the projection of the system). On the other side, $X_0 \cap \{\lambda v : \lambda \in \mathbb{R}\}$ is discrete for every v not parallel to $(1, 0)$.

Another key aspect of reachability analysis for nonlinear control systems is nonholonomy. Definitions for continuous systems are typically formulated in terms of well-known integrability conditions on the constraint codistribution. For a system such as that in 1, a more general definition is necessary:

Definition 5. A system is said to be nonholonomic at x_0 if it is possible to decompose $\mathcal{X}$ in a projection or base *space $\mathcal{B} = \Pi(\mathcal{X})$ and a* fiber *$\mathcal{F}$, such that $\mathcal{B} \times \mathcal{F} = \mathcal{X}$ (that is $\mathcal{X}$ is a trivial bundle) and there exists $\omega \in \Omega$ and $t \in \mathcal{T}$ such that $\mathcal{A}_{t,\omega}$ steers the system from x_0 to $x^\star = \mathcal{A}_{t,\omega}(x_0)$ with $\Pi(x_0) = \Pi(x^\star)$ but $x_0 \neq x^\star$.*

Example1–c. An illustration of discrete nonholonomy is obtained by the rolling polyhedron system. When a sequence of rotations on the right, hence forward, left and backward is considered, the center of the die returns to its initial position, while the orientation has changed by a finite amount. $\triangleleft$

3. SYNTHESIS OF LINEAR QUANTIZED CONTROL SYSTEMS

In this section, we report some results on systems of the form

$$x^+ = Ax + Bu, \; u \in U \tag{4}$$

with U a quantized set as usual, and (A, B) a controllable pair. Reachability questions that may be asked about such system can be divided in two types:

Definition 6.

Q1 given a pair (A, B), find conditions under which a quantized control set U exists such that the reachable set $R(0, U)$, from 0 and corresponding to the set U, is dense in $\mathbb{R}^n$. If possible, find such a U.

Q2 given a pair (A, B), a quantized set U, and initial conditions $x(0)$, determine whether or not the corresponding reachable set is dense.

We will refer to question **Q1** as to a synthesis problem, and to **Q2** as to an analysis problem.

The synthesis problem has been extensivley studied in Chitour and Piccoli (2000). Main results are reported below.

Theorem 1. Necessary and sufficient conditions for a quantized control set U to exist such that the reachable set $R(0, U)$ from 0 of (4) is dense in $\mathbb{R}^n$ are that

(1) (A, B) is controllable;
(2) if λ is an eigenvalue of A, then $|\lambda| \geq 1$.

The necessity of the first condition is obvious. If the second condition does not hold, the reachable set is bounded in some component. However, a similar density result can still be obtained (provided that no eigenvalue of A is zero) if local approachability at the origin is considered instead.

Remark. Conditions for a positive answer to the synthesis problem are very weak. Proofs given in Chitour and Piccoli (2000), though far from trivial, are constructive, as they provide explicitly a *standard* control set $\mathcal{U} = \{0, \pm u_1, \pm u_2, \ldots\}$ that achieves density for a fixed system. Furthermore, results are shown to be uniform with respect to both initial conditions and eigenvalues changes.

A further twist to the synthesis problem results from restricting control values to belong to a subset of $\mathbb{R}^n$. In particular, in applications involving D/A conversions, inputs will be restricted as $U \in \mathbb{Q}^n$. The case $U \in \mathbb{Z}^n$ is also relevant to many applications. For this case we immediately have the following:

Theorem 2. Consider the system (4) and assume that the matrices A and B have integer entries. Let $U = \{i\,\alpha : i \in \mathbb{Z}\}$ for some $\alpha \in \mathbb{R}$. Then the reachable set $R(0, U)$ is discrete.

In general if we allow the control set U to be discrete but infinite then unless we are in the situation of the above theorem we expect density of $R(0, U)$ to be generic. The situation is profoundly different if we consider finite control set U even without uniform bound on the cardinality. There is a special class of algebraic numbers that play a key role. We recall that an algebraic number λ is a real number that is root of a polynomial P with integer coefficients. If, moreover, the leading coefficient of P is 1 then λ is called an algebraic integer. For an algebraic number λ we can determine the minimal polynomial P_λ that is the polynomial of minimal degree such that $P_\lambda(\lambda) = 0$, moreover if λ is an algebraic integer P_λ can be chosen with leading coefficient 1. Given an algebraic number λ we call the other roots of P_λ the Galois conjugates of λ (obviousy they cannot be real).

Definition 7. An algebraic integer $\lambda > 1$ is a Pisot number if all its Galois conjugates have modulus strictly less than one.

The following theorem follows form the analysis of Chitour and Piccoli (2000).

Theorem 3. Consider a system (4) satisfying the assumptions of Theorem 1 (necessary for density) and assume that A is in Jordan form, $B = I$ (the identity matrix). The reachable set $R(0, U)$ is not dense in $\mathbb{R}^n$ for every finite set $U \subset \mathbb{Q}^n$ if and only if there exists an eigenvalue of A whose modulus is a Pisot number.

Notice the strength of the Theorem implying that in the case in which an eigenvalue (or its modulus) is a Pisot number, then whatever choice of a finite set $U \subset \mathbb{Q}^n$ with arbitrarily large finite cardinality gives a nondense reachable set $R(0, U)$. The set of Pisot number is obviously countable but the surprising fact is that it is close. Hence, it is not dense in $\mathbb{R}$ and indeed is "small" in topological sense. Many facts are indeed known about the set T of Pisot numbers. For example T admits a minimum value $\lambda \sim 1.33$, that is the unique positive root of $x^3 - x - 1$. The smallest accumulation point of T is the well known golden number $(1 + \sqrt{5})/2$ that is root of $x^2 - x - 1$. We refer the reader to Chitour and Piccoli (2000) and references therein for information about Pisot numbers.

On the other side, if all eigenvalues are not Pisot then it is possible to obtain density of $R(0, U)$ choosing a large enough number M (of the order of the modulus of the biggest eigenvalue) and all controls with integer coordinates in $[-M, M]$. See Erdös *et al.* (1998) and Chitour and Piccoli (n.d.).

We want also to point out that sampled systems with D/A conversions and usage of computers naturally lead to system of type (4) with U finite subset or $\mathbb{Q}^n$. It is then clear the importance of the above result.

4. ANALYSIS PROBLEMS

The analysis question is indeed much more difficult to answer. To understand the difficulty we refer the reader to Keane *et al.* (1995) where the so called $\{0, 1, 3\}$-problem is studied. This corresponds exactly to the analysis of the Hausdorff measure of the reachable set for the system $x^+ = \lambda x + u$, $x \in \mathbb{R}$, $\lambda < 1$, $u \in U = \{0, 1, 3\}$, if we allow infinite sequences of controls. The analysis problem has some partial answer in the cited paper and references therein.

Another strictly linked number theory problem is the one considered in Erdös *et al.* (1998). We refer the reader to Chitour and Piccoli (2000) for a deeper discussion of the links between these hard mathematical problems. From the results of Erdös *et al.* (1998) it is even more clear the role played by Pisot numbers.

In this section, we provide some results on the analysis question concerning some simple examples of driftless systems of the type

$$x^+ = x + g(x)u, \; u \in U \tag{5}$$

Given two real numbers $r_1, r_2 \in \mathbb{R}$ we write $r_1 \sim r_2$ to indicate that r_1, r_2 have rational ratio, that is $\frac{r_1}{r_2} \in \mathbb{Q}$. It is easy to check that $\sim$ is an equivalence relation. Consider the control system

$$x^+ = x + u \tag{6}$$

where $x \in \mathbb{R}$ and u takes values in a finite set $U \subset \mathbb{R}$. Our aim is to prove that the following condition is necessary and sufficient in order to have that the reachable set from any initial point is dense in $\mathbb{R}$.

(C) There exist $u, v \in U$ such that $u \not\sim v$ and $u \cdot v < 0$.

First notice that condition (C) is equivalent to the following

(C') There exist $u, v \in U$ such that $u \not\sim v$ and there exist $u', v' \in U$ such that $u' \cdot v' < 0$.

Indeed, obviously (C) implies (C'). On the other hand, assume that (C') is true, then $U^{\pm} = U \cap R^{\pm}$ are nonempty. If for every $u \in U^+$ and $v \in U^-$ we have $u \sim v$ then, since $\sim$ is an equivalence relation we get that all control have rational ratio reaching a contradiction.

We start noticing the following fact (see e.g. Chitour and Piccoli (2000)):

Proposition 3. Let R_0 be a reachable set for the system (6) from the origin. Then R_0 is dense if and only if there exist two sequences $c_k \in R_0$ and $d_k \in R_0$ both converging to zero such that $d_k < 0 < c_k$.

Let us now prove the following

Theorem 4. Let R_0 be a reachable set for the system (6) from the origin. Then R_0 is dense if and only if (C) holds true. Moreover, if R_0 is not dense then is nowhere dense.

Proof. Let us first assume that (C) holds true and let $u, v \in \mathcal{U}$ be as in (C). Since the ratio $\frac{u}{v}$ is not rational we can consider the sequence $\frac{p_k}{q_k} \in Q$, p_k, q_k integers, $q_k > 0$, given by its continued fraction. We have:

$$\frac{u}{v} - \frac{p_k}{q_k} = (-1)^k \varepsilon_k$$

where $0 < \varepsilon_k < \frac{1}{q_k^2}$ and q_k grows to infinity. We get immediately:

$$q_k\, u + (-p_k)v = (-1)^k v \varepsilon_k q_k.$$

From $u \cdot v < 0$ we get $-p_k > 0$, hence $q_k\, u + (-p_k)v \in R_0$. Now the required sequences are obtained setting, if $v > 0$, $c_k = q_k\, u + (-p_k)v$ for k even and $d_k = q_k\, u + (-p_k)v$ for k odd and the opposite if $v < 0$.

Assume now that (C) does not hold. Then either $u \cdot v > 0$ for every $u, v \in \mathcal{U}$ or $u \sim v$ for every $u, v \in \mathcal{U}$. In the first case it is obvious that the set R_0 is contained either in R^+ or in R^-. In the latter case, the proof is as follows. Let $\mathcal{U} = \{u_1, \ldots, u_n\}$ with $u_1 \neq 0$. Then any point of the reachable set R_{x_0} from x_0 can be written as $x_0 + a$, $a = m_1\, u_1 + \ldots + m_n\, u_n$ with u_i positive integers. We have $\frac{u_i}{u_1} = \frac{p_i}{q_i} \in Q$ ($\frac{p_1}{q_1} = 1$), thus:

$$a = m_1\, u_1 + \ldots + m_n\, u_n = u_1 \left(\sum_{i=1}^{n} \frac{m_i\, p_i}{q_i} \right) =$$

$$= u_1 \left(\frac{\sum_{i=1}^{n} m_i\, p_i\, q_1 \cdots q_{i-1} q_{i+1} \cdots q_n}{q_1 \cdots q_n} \right).$$

Now if $a \neq 0$ we have that the numerator of the above expression is different from zero and being an integer is at least of modulus 1. Therefore, if $a \neq 0$ we get

$$|a| > \frac{|u_1|}{|q_1 \cdots q_n|}$$

and obviously R_0 can not be dense. Moreover, from the same expression we have that a is always a multiple of $u_1/(q_1 \cdots q_n)$ hence R_0 is indeed nowhere dense. □

Since the reachable set from a point x_0 is exactly $x_0 + R_0$ we have a dicotomy similar to that of Section 2, even if in this case (due to the possible non simmetry of $\mathcal{U}$) R_0 may fail to be a subgroup of $\mathbb{R}$.

Let us consider the system (6) but now with $x \in \mathbb{R}^n$, that is

$$x^+ = x + u \tag{7}$$

with $x \in \mathbb{R}^n$, $u \in \mathcal{U} \subset \mathbb{R}^n$. From the above analysis we get:

Theorem 5. A necessary condition for the reachable set X from the origin to be dense is that $\mathcal{U}$ contains $n+1$ controls of which n are linealy independent. If $u_1, \ldots, u_n \in \mathcal{U}$ are linearly independent and there exists n irrational negative numbers $\alpha_1, \ldots, \alpha_n$ such that $v_i = \alpha_i u_i \in \mathcal{U}$ for every $i = 1, \ldots, n$ then X is dense in $\mathbb{R}^n$.

5. NONHOLONOMIC SYSTEMS

We are interested in studying the structure of the reachability set for nonlinear system that exhibit nonholonomic behaviours. To do so, we consider the discrete–time analog of a much studied class of continuous–time nonholonomic systems that are written in chained form

$$\begin{aligned} \dot{x}_1 &= u_1 \\ \dot{x}_2 &= u_2 \\ \dot{x}_3 &= x_2 u_1 \\ \vdots &= \vdots \\ \dot{x}_n &= x_{n-1} u_1 \end{aligned} \quad . \tag{8}$$

The chained form was introduced in R. M. Murray (1993) because it allows a rather simple steering method, using sinusoids at integrally related frquencies. The technique consisted in driving system (8) to the desired value of the variables x_1, x_2; then applying a low frequency sinusoidal input to steer x_3 while bringing back x_1, x_2 after a cycle; and so on iteratively with higher frequency sinusoids. At each step, the amplitude of the sinusoids is adjusted so as to reach the desired value of the corresponding coordinate.

A different technique for steering continuous nonholonomic systems that are in strictly triangular form[2] has been proposed in Marigo and Bicchi (1998). The idea there was to purposefully introduce quantization of the input space, by defining a set of fixed input functions on compact time sets. Such control "quanta" can then be concatenated, and form a group acting on the left on the configuration space. The ST form of the system guarantees that the action of the subgroup of the control quanta group that takes the base variables (x_1, x_2) back to their initial value, is abelian on variables x_3. Furthermore, the action of proper subgroups (the derived flag of the control quanta group) is also abelian on corresponding sections of the fiber. Although an infinite number of generators for the control quanta group should in principle be considered, authors proposed to use a finite set generating a discrete reachable set with a lattice structure. These properties allow to steer the system to a desired configuration variable after variable, by simply writing the lattice generators in Hermite normal form, planning on the lattice, then using the generalized inverse Euclid algorithm.

Consider now the discrete system

$$\begin{aligned} x_1^+ &= x_1 + u_1 \\ x_2^+ &= x_2 + u_2 \\ x_3^+ &= x_3 + x_2 u_1 + u_1 u_2 \frac{1}{2} \\ x_4^+ &= x_4 + x_3 u_1 + x_2 u_1^2/2 + u_1^2 u_2 \frac{1}{6} \\ \vdots &= \vdots \\ x_n^+ &= \sum_{i=0}^{n-2} x_{n-i} \frac{u_1^i}{i!} + u_1^{n-2} u_2 \frac{1}{(n-1)!} \end{aligned} \tag{9}$$

[2] A system is in ST form if $\dot{x}_i = g(x_{i+1}, \cdots, x_n)u$. ST systems include, but are not limited to, nilpotent systems Marigo (1999), and are hence much more general than chained form systems.

which can be regarded as system (8) under unit sampling. Notice that this system is invertible (as opposed e.g. to the forward Euler approximation of (8)). Indeed, for any state–independent, symmetric set of input symbols $\mathcal{U}$, the group of input words $\Omega = \{\text{strings of symbols in } \mathcal{U}\}$ with inverse $(w_1 w_2 \cdots w_m)^{-1} = -w_m \cdots - w_b - w_a$, $\pm w_i \in \mathcal{U}, \forall i$, acts on the configuration space through the end–point map such that $\mathcal{A}(\omega^{-1}, \mathcal{A}(\omega, x)) = x$.

We are interested in studying the reachability set of system (9), and in providing a steering method for the system.

One can readily check that the system is nonholonomic in the sense of definition 5, by taking (x_1, x_2) as the base variables. Reachability in the base space can be studied by results reported above for linear driftless systems. We will hence focus on the reachability of the fiber corresponding to a given base point $(\bar{x}_1, \bar{x}_2)$. Simple calculations show that the reachable set in the fiber does not depend on the base variables, hence we may consider $\bar{x}_1 = 0, \bar{x}_2 = 0$ without loss of generality.

Consider the subgroup $\tilde{\Omega} \in \Omega$ of control words that take the base variables back to their initial configuration. These are sequences of inputs such that the sum of the first and second components are zero. The action of this subgroup on the fiber is commutative: namely, $\mathcal{A}(\tilde{\omega}_1, \mathcal{A}(\tilde{\omega}_2, x)) = \mathcal{A}(\tilde{\omega}_2, \mathcal{A}(\tilde{\omega}_1, x))$, $\forall \tilde{\omega}_1, \tilde{\omega}_2 \in \tilde{\Omega}$. Notice that this represents a significant departure from the behaviour of the continuous model (8), where the action of the generic cyclic control is abelian only on the first fiber variable, x_3, and more restricted subgroups should be searched that have the commutative action property on the rest of the fiber.

To be more specific, let us consider the case that $\mathcal{U} = \{\pm(1,0), \pm(0,1), \pm(a,b)\}$. The rechable set for the base variables is described by

$$\begin{aligned} x_1 &= m_1 + a m_3 \\ x_2 &= m_2 + b m_3 \end{aligned} \tag{10}$$

with $m_i \in \mathbb{Z}$. If a and b are rational, and $a = \frac{p_a}{q_a}$, $b = \frac{p_b}{q_b}$, p_a, q_a, p_b, q_b integers and pairwise coprime, the rechable set of base space is clearly a lattice. The subgroup $\tilde{\Omega}$ is given by all control words with $(m_1, m_2, m_3) = \alpha(p_a q_b, p_b q_a, -q_a q_b)$, $\alpha \in \mathbb{Z}$ (this means, words where the symbol $(1,0)$ is used $m m_1$ times, $(0,1)$ is used m_2 times, and $(-a,-b)$ is used m_3 times). For $\alpha = 1$, there are

$$N = \frac{(p_a q_b + p_b q_a + q_a q_b)!}{p_a q_b! p_b q_a! q_a q_b!}$$

possible words. The reachable set as a whole is discrete: the i–th coordinate is an integer multiple of $1/\mu_i$ dove $\mu_i = (i-1)! q^(i-1)$, $q = \max\{q_a, q_b\}$. Coordinates of higher index have a finer resolution.

6. CONCLUSIONS

In this paper, we have considered reachability problems in quantized control systems. We have shown that the reachable set may be dense or discrete depending on the quantized set of inputs, and have provided some results in the analysis and synthesis problems. We have also provided a definition and some characterization of nonholonomic phenomena occurring in nonlinear quantized control systems. Many open problems remain in this field, that is in our opinion among the most important and challenging for applications of embedded control systems and in several other applications. Although some problems have been shown to hard, we believe that a reasonably complete and useful system theory of quantized control system could be built by merging modern discrete mathematics techniques with classical tools of system theory.

References

Bertram, J. E. (1958). The effect of quantization in sampled feedback systems. *Trans. AIEE Appl. Ind.* **77**, 177–181.

Brockett, R. and L. Dai (1993). Non-holonomic kinematics and the role of elliptic functions in constructive controllability. In: *Nonholonomic Motion Planning* (Z. Li and J.F. Canny, Eds.). Kluwer Academic Publ.

Ceccarelli, M., A. Marigo, S. Piccinocchi and A. Bicchi (2000). Planning motions of polyhedral parts by rolling. *Algorithmica.* in press.

Chitour, Y. and B. Piccoli (2000). Controllability for discrete systems with a finite control set. *Math. Control Signals Systems.* to appear.

Chitour, Y. and B. Piccoli (n.d.). Reachability of quantized control systems. work in progress.

Delchamps, D. F. (1990). Stabilizing a linear system with quantized state feedback. *IEEE Trans. Autom. Control* **35**(8), 916–926.

Elia, N. and S. K. Mitter (1999). Quantization of linear systems. In: *Proc. 38th Conf. Decision & Control.* IEEE. pp. 3428–3433.

Erdös, P., I. Joó and V. Komornik (1998). On the sequence of numbers of the form $\varepsilon_0 + \varepsilon_1 q + \ldots + \varepsilon_n q^n, \varepsilon_i \in \{0, 1\}$. *Acta Arith.* **LXXXIII**(3), 201–210.

Hou, L., A. N. Michel and H. Ye (1997). Some qualitative properties of sampled-data control systems. *IEEE Trans. Autom. Control* **42**(12), 1721–1725.

Jurdjevic, V. (1993). The geometry of the plate-ball problem. *Arch. Rational Mech. Anal.* **124**, 305–328.

Keane, M., M. Smorodinsky and B. Solomyak (1995). On the morphology of γ-expansions with deleted digits. *Trans. Amer. Math. Soc.* **347**, 955–966.

Levi, M. (1993). Geometric phases in the motion of rigid bodies. *Arch. Rational Mech. Anal.* **122**, 213–229.

Marigo, A. (1999). Constructive necessary and sufficient conditions for strict triangularizability of driftless nonholonomic systems. In: *Proc. IEEE Int. Conf. on Decision and Control.* pp. 2138–2143.

Marigo, A. and A. Bicchi (1998). Steering driftless nonholonomic systems by control quanta. In: *Proc. IEEE Int. Conf. on Decision and Control.* pp. 4164–4169.

Marigo, A. and A. Bicchi (2000). Rolling bodies with regular surface: Controllability theory and applications. *IEEE Trans. Autom. Control.*

R. M. Murray, S. S. Sastry (1993). Nonholonomic motion planning: Steering using sinusoids. *IEEE Trans. Autom. Control* **38**, 700–716.

Slaughter, J. B. (1964). Quantization errors in digital control systems. *IEEE Trans. Autom. Control* **9**, 70–74.

Wong, W. S. and R. Brockett (1999). Systems with finite communication bandwidth constraints – ii: stabilization with limited information feedback. *IEEE Trans. Autom. Control* **44**(5), 1049–1053.

Y.Chitour, A.Marigo, D.Prattichizzo and A.Bicchi (1996). Rolling polyhedra on a plane, analysis of the reachable set. In: *Workshop on Algorithmic Foundations of Robotics.*

PORT CONTROLLED HAMILTONIAN REPRESENTATION OF DISTRIBUTED PARAMETER SYSTEMS

B.M.J.Maschke*, ** A.J. van der Schaft *

* *Faculty of Mathematical Sciences, Department of Systems, Signals and Control, University of Twente, P.O.Box 217, 7500 AE Enschede, The Netherlands, e-mail: {maschke,a.j.vanderschaft}@math.utwente.nl*
** *Laboratoire d'Automatisme Industriel, Conservatoire National des Arts et Métiers, Paris, France*

Abstract: A port controlled Hamiltonian formulation of the dynamics of distributed parameter systems is presented, which incorporates the energy flow through the boundary of the domain of the system, and which allows to represent the system as a boundary control Hamiltonian system. This port controlled Hamiltonian system is defined with respect to a Dirac structure associated with the exterior derivative and based on Stokes' theorem. The definition is illustrated on the examples of the telegrapher's equations, Maxwell's equations and the vibrating string.

Keywords: distributed parameter systems, Hamiltonian systems, Dirac structures, boundary control.

1. INTRODUCTION

The Hamiltonian formulation of classes of distributed parameter systems has been a challenging and fruitful area of research for quite some time. A nice introduction, especially with respect to systems stemming from fluid dynamics, can be found in Chapter 7 of the book by Olver (Olver, 1993), where also a historical account is provided. The identification of the underlying Hamiltonian structure of sets of p.d.e.'s has been very instrumental in proving all sorts of results on integrability, the existence of soliton solutions, stability, reduction, etc., and in *unifying* existing results, see e.g. (Holm *et al.*, 1985).

Recently, there has been also a surge of interest in the *control* of nonlinear distributed parameter systems, motivated by various applications. At the same time, for *finite-dimensional* nonlinear systems a satisfactory theory has been developed concerning the generalized Hamiltonian modelling of physical systems with external (input and output) variables. This has led to the notions of *port-controlled Hamiltonian (PCH) systems*, and *port-controlled Hamiltonian systems with dissipation* (PCHD systems) (van der Schaft and Maschke, 1995), (Dalsmo and van der Schaft, 1999), (Ortega *et al.*, 1999*b*), (van der Schaft, 2000). This theory is aimed at applications in the consistent modelling and simulation of complex *interconnected* physical systems, and in the design and *control* of such systems, exploiting the Hamiltonian and passivity structure in a crucial way (van der Schaft, 2000), (Ortega *et al.*, 1999*a*), (Ortega *et al.*, 1999*b*).

In the present paper we start to expand the research program on finite-dimensional PCH and PCHD systems to the distributed parameter (or, infinite-dimensional) case. The first idea for doing so is to try to extend the theory as for instance exposed in (Olver, 1993) to distributed parameter systems *with external variables* (inputs and outputs). However, a fundamental difficulty which arises is the treatment of *boundary conditions.* Indeed, from a control and interconnection point of view it is quite essential to describe a distributed parameter system with varying boundary conditions inducing *energy exchange through the boundary,* since in many applications the interaction with the environment (e.g. actuation or measurement) will actually take place through the boundary of the system. Obvious examples are the telegraphers equations (describing the dynamics of a transmission line), where the boundary of the system is described by the voltages and currents at both ends of the transmission line, or the vibrating string (or, more generally, a flexible beam), where it is natural to consider the forces and velocities at one or both ends of the string as the external variables of the system. On the other hand, the treatment of infinite-dimensional Hamiltonian systems in the literature (see again (Olver, 1993)) seems mostly focussed on systems with infinite spatial domain, where the variables go to zero for the spatial variables tending to infinity, or on systems with boundary conditions such that the energy exchange through the boundary is *zero.* Furthermore, it is not obvious how to incorporate non-zero energy flow through the boundary in the existing framework. The problem is already illustrated by the Hamiltonian formulation of e.g. the Korteweg-de Vries equation. Here for zero boundary conditions a Poisson bracket can be formulated with the use of the differential operator $\frac{d}{dx}$, since by integration by parts this operator is obviously skew-symmetric. However, for boundary conditions corresponding to non-zero energy flow the differential operator is not skew-symmetric anymore (since the remainders are not zero when integrating by parts). Also the interesting paper (Lewis *et al.*, 1986) does not really solve the problem, since this latter paper is concerned with the modification of the Poisson bracket in case of a free boundary.

In the present paper we propose a framework to overcome this fundamental problem, by defining a *Dirac structure* on certain spaces of differential forms on the spatial domain and its boundary. This construction of the Dirac structure is based on the use of Stokes' theorem. Then we employ the definition of a port-controlled Hamiltonian system with respect to a Dirac structure, as already given in previous papers (see e.g. (van der Schaft and Maschke, 1995)) for the finite-dimensional case, to describe *implicit* PCH systems, in order to formalize distributed parameter systems with boundary external variables as infinite-dimensional PCH systems. This framework is then applied to the port-controlled Hamiltonian formulation of the telegrapher's equations, the vibrating string, and Maxwell's equations on a bounded domain. Due to space limitations the port-controlled Hamiltonian formulation of systems arising in fluid dynamics (e.g. Euler's equations) or in elastodynamics (e.g. flexible beams or plates) will be deferred to a future paper.

2. DIRAC STRUCTURES AND FINITE-DIMENSIONAL PCHD SYSTEMS

From a network modeling perspective (Paynter, 1961), (Breedveld, 1984), (Maschke *et al.*, 1992), a lumped parameter physical system is naturally described by a set of (possibly multi-dimensional) *energy-storing* elements, a set of *energy-dissipating* or *resistive* elements, and a set of *ports* (by which interaction with the environment can take place), interconnected to each other by a *power-conserving interconnection*, as depicted in Figure 1.

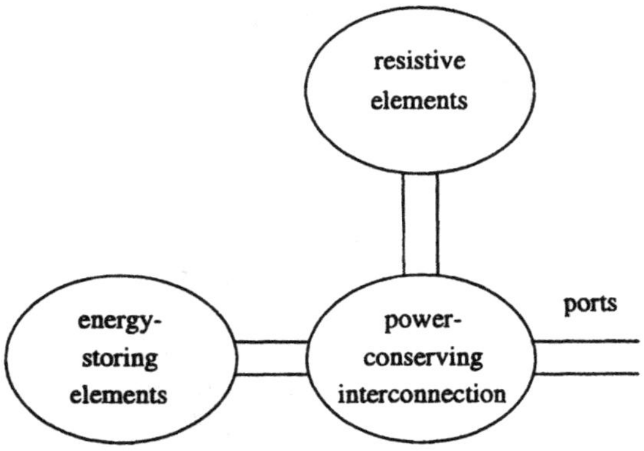

Fig. 1. Implicit port-controlled Hamiltonian system with dissipation

The power-conserving interconnection includes interconnection constraints like Kirchhoff's laws and Newton's third law, as well as power conserving elements like (in the electrical domain) transformers, gyrators, and (in the mechanical domain) transformers, kinematic pairs and kinematic constraints. To every two-fold line (also called *power bond*) there are associated two kind of variables, usually called *flows and efforts*, whose product is *power.* In the mechanical domain the flows and efforts are generalized velocities and forces, while in the electrical domain they are currents and voltages. The power-conserving interconnection relates the flows and efforts corresponding to the energy-storing elements, the resistive elements and the ports to each other in such a way that the total incoming power into the interconnection structure is always zero.

Associated with the energy-storing elements are in the lumped parameter case energy-variables $x_1, \cdots, x_n$, being local coordinates for some n-dimensional state space manifold $\mathcal{X}$, and a total energy $H : \mathcal{X} \to \mathbb{R}$.

The power-conserving interconnection is formalized by a *Dirac structure*, as was introduced in (Courant, 1990), (Dorfman, 1993). For our purposes in this paper we only need *constant* Dirac structures on vector spaces. Thus, let $\mathcal{V}$ be (finite or infinite) dimensional linear space, and denote its dual (the space of linear functions on $\mathcal{V}$) by $\mathcal{V}^*$. The product space $\mathcal{V} \times \mathcal{V}^*$ is considered to be the space of flow and effort variables, with power defined by

$$P = \langle v^* | v \rangle, \quad (v, v^*) \in \mathcal{V} \times \mathcal{V}^*, \qquad (1)$$

where $\langle v^* | v \rangle$ denotes the duality product, that is, the linear function $v^* \in \mathcal{V}^*$ acting on $v \in \mathcal{V}$. Often we call $\mathcal{V}$ the space of *flows* f, and $\mathcal{V}^*$ the space of *efforts* e, with the power of an element $(f, e) \in \mathcal{V} \times \mathcal{V}^*$ denoted as $\langle e | f \rangle$.

Example 1. Let $\mathcal{V}$ be the space of generalized *velocities*, and $\mathcal{V}^*$ be the space of generalized *forces*, then $< e|f >$ is mechanical power. Similarly, let $\mathcal{V}$ be the space of *currents*, and $\mathcal{V}^*$ be the space of *voltages*, then $\langle e|f \rangle$ is electrical power.

There exists on $\mathcal{V} \times \mathcal{V}^*$ a canonically defined symmetric bilinear form

$$\langle\langle (f_1, e_1), (f_2, e_2) \rangle\rangle := \langle e_1 | f_2 \rangle + \langle e_2 | f_1 \rangle \qquad (2)$$

for $f_i \in \mathcal{V}$, $e_i \in \mathcal{V}^*, i = 1, 2$. Now consider a linear subspace

$$S \subset \mathcal{V} \times \mathcal{V}^* \qquad (3)$$

and its orthogonal complement with respect to the bilinear form $\langle\langle , \rangle\rangle$ on $\mathcal{V} \times \mathcal{V}^*$, denoted as

$$S^\perp \subset \mathcal{V} \times \mathcal{V}^*. \qquad (4)$$

Definition 1. (Courant, 1990), (Dorfman, 1993), (Dalsmo and van der Schaft, 1999). A constant Dirac structure on $\mathcal{V}$ is a linear subspace $\mathcal{D} \subset \mathcal{V} \times \mathcal{V}^*$ such that

$$\mathcal{D} = \mathcal{D}^\perp \qquad (5)$$

Remark 1. In many cases the power-conserving interconnection is actually *modulated* by the energy variables, in which case the above definition of a constant Dirac structure on a vector space has to be generalized to a general Dirac structure on a manifold, see (Courant, 1990), (Dorfman, 1993).

As an immediate consequence of the definition of a Dirac structure, let $(f, e) \in \mathcal{D} = \mathcal{D}^\perp$. Then by (2)

$$0 = \langle\langle (f, e), (f, e) \rangle\rangle = 2\langle e \mid f \rangle. \qquad (6)$$

Thus for all $(f, e) \in \mathcal{D}$ we obtain

$$\langle e \mid f \rangle = 0. \qquad (7)$$

Hence a Dirac structure $\mathcal{D}$ on $\mathcal{V}$ defines a power-conserving relation between the power variables $(f, e) \in \mathcal{V} \times \mathcal{V}^*$.

Remark 2. Furthermore, it immediately follows that the dimension of any Dirac structure $\mathcal{D}$ on a *finite-dimensional* linear space $\mathcal{V}$ is equal to $dim\mathcal{V}$. This is intimately related to the usually expressed statement that a physical interconnection can *not* determine at the same time both the flow and effort (e.g. current *and* voltage, or velocity *and* force).

It is well-known that a Dirac structure is a generalization both of a *symplectic structure* and a *Poisson structure*(Courant, 1990), (Dorfman, 1993), which are the usual geometric building blocks in the definition of a Hamiltonian system. In our previous papers we used Dirac structures in order to also formulate *implicit* systems as Hamiltonian systems (van der Schaft and Maschke, 1995), (Maschke and van der Schaft, 1996), (Maschke and van der Schaft, 1997), (Maschke and van der Schaft, 1998). Implicit system descriptions naturally arise from network modelling, and the corresponding Dirac structure captures the power conserving interconnection structure of the network. A strongly related argument in favor of the systematic use of Dirac structures in the description of Hamiltonian systems is the fact that the power-conserving interconnection of Dirac structures again defines a Dirac structure (van der Schaft, 1999). In the next section we show how the notion of a Dirac structure (on infinite-dimensional spaces) is also instrumental in the definition of standard distributed parameter systems, which are only implicit in the way the boundary conditions enter the system description.

In order to facilitate the understanding of the definition of Hamiltonian dynamics of distributed parameter systems in the next section we briefly recall the definition of a finite-dimensional implicit PCHD system (van der Schaft and Maschke, 1995), (Maschke and van der Schaft, 1996), (Maschke and van der Schaft, 1997), (Maschke and van der Schaft, 1998), (van der Schaft, 2000). We consider a constant Dirac structure $\mathcal{D}$ on the finite-dimensional linear space $\mathcal{V} := \mathcal{F}_S \times \mathcal{F}_R \times \mathcal{F}_P$, with $\mathcal{F}_S$ denoting the space of flows f_S connected to the energy-storing elements, $\mathcal{F}_R$ denoting the space of flows f_R connected to the dissipative (resistive) elements, and $\mathcal{F}_P$ the space of external flows f_P which can be connected to the environment. Dually, we write $\mathcal{V}^* = \mathcal{F}_S^* \times \mathcal{F}_R^* \times \mathcal{F}_P^*$, with $e_S \in \mathcal{F}_S^*$ the efforts connected to the energy-storing elements, $e_R \in \mathcal{F}_R^*$ the efforts connected to the resistive elements, and $e_P \in \mathcal{F}_P^*$ the efforts to be connected to the environment of the system. The flow variables of the energy-storing elements are given as $\dot{x}(t) = \frac{dx}{dt}(t), t \in \mathbb{R}$, and the effort variables of the energy-storing elements as $\frac{\partial H}{\partial x}(x(t))$ (implying that $< \frac{\partial H}{\partial x}(x(t)) | \dot{x}(t) >= \frac{dH}{dt}(x(t))$ is the increase in energy). In order to have a consistent sign convention for energy flow we put

$$f_S = -\dot{x}$$
$$e_S = \frac{\partial H}{\partial x}(x) \tag{8}$$

Then the equations for the implicit PCHD system take the form

$$(-\dot{x}(t), f_R(t), f_P(t), \frac{\partial H}{\partial x}(x(t)), e_R(t), e_P(t)) \in \mathcal{D}, \tag{9}$$

where the flow and effort variables connected to the resistive elements are related by certain equations

$$f_R = -R(e_R), \tag{10}$$

where $e_R^T R(e_R) \geq 0$.

3. DIRAC STRUCTURE ASSOCIATED WITH THE EXTERIOR DERIVATIVE

In this section we provide the basic framework for describing distributed parameter systems as port-controlled Hamiltonian systems. Basic ingredients are the identification of a suitable space of energy variables, closely connected to the *geometry* of the spatial variables of the distributed parameter system, and the definition of a suitable *Dirac structure* on the space of energy variables.

In order to define the space of energy variables we first need some preliminaries. Let N be an n-dimensional manifold with boundary ∂N (of dimension $n-1$), representing the space of *spatial variables*. We denote by $\Omega^k(N)$, $k = 0, 1, \ldots, n$, the space of k-forms on N, and by $\Omega^k(\partial N)$, $k = 0, 1, \ldots, n-1$, the space of k-forms on ∂N. (The space $\Omega^0(N)$, respectively $\Omega^0(\partial N)$, is the space of smooth functions on N, respectively ∂N.) We note that $\Omega^k(N)$ and $\Omega^k(\partial N))$ are (infinite-dimensional) *linear* spaces with respect to addition and multiplication by elements in $\mathbb{R}$. The dual linear space $(\Omega^k(N))^*$ can be naturally identified with $\Omega^{n-k}(N)$, and similarly the dual space $(\Omega^k(\partial N))^*$ with $\Omega^{n-1-k}(\partial N)$, as stated in the following proposition.

Proposition 1. $(\Omega^k(N))^*$ can be identified with $\Omega^{n-k}(N)$, replacing the duality product between $\Omega^k(N)$ and $(\Omega^k(N))^*$ by

$$\langle \beta \mid \alpha \rangle := \int_N \alpha \wedge \beta, \quad \alpha \in \Omega^k(N),\ \beta \in \Omega^{n-k}(N) \tag{11}$$

This seems to be a basic fact in the theory of differential forms, although we could not find a reference to this result in the literature. A possible proof for Proposition 1 runs as follows. Equip N with a Riemannian metric (this is always possible), and denote by $\star$ the Hodge star operator. It is well-known (see e.g. Definition 2.7.13 in (Abraham and Marsden, 1978)) that

$$(\alpha_1, \alpha_2) := \int_N \alpha_1 \wedge \star\alpha_2, \quad \alpha_1, \alpha_2 \in \Omega^k(N); \tag{12}$$

defines an inner product on $\Omega^k(N)$. Then by the Riesz representation theorem there exists for any $\alpha^* \in (\Omega^k(N))^*$ a $\gamma \in \Omega^k(N)$ such that

$$\langle \alpha^* \mid \alpha \rangle = (\alpha, \gamma), \quad \alpha \in \Omega^k(N) \tag{13}$$

Denoting $\star\gamma \in \Omega^{n-k}(N)$ by β the claimed result follows.

Consider now as space of energy variables the linear space $\mathcal{V}$ defined as follows:

$$\mathcal{V} := \Omega^p(N) \times \Omega^q(N) \times \Omega^{n-q}(\partial N) \tag{14}$$

for p and q positive integers satisfying

$$p + q = n + 1. \tag{15}$$

By linearity $\mathcal{V}$ is also the space of *flows* (the rate energy variables). The space $\mathcal{V}$ will be the carrier space for the constant Dirac structure representing the interconnection structure of distributed parameter systems.

Its dual space $\mathcal{V}^*$ can be identified as in Proposition 1 with the linear space

$$\mathcal{V}^* \simeq \Omega^{n-p}(N) \times \Omega^{n-q}(N) \times \Omega^{n-p}(\partial N) \tag{16}$$

(note that $(n-1)-(n-q) = n-p$). The dual space $\mathcal{V}^*$ will represent the space of efforts, or co-energy variables, of the system.

In the present paper we treat for simplicity only the symmetric case $p = q = k$, since this will be sufficient for all the examples treated in Section 4. However, for the treatment of examples from fluid dynamics and elastodynamics it will be necessary to consider the general case, as will be done in a future paper.

Note that in the symmetric case $2k = n + 1$, whence it follows that n is necessarily *odd*. In fact, the two cases of primary interest for us will be $n = 3$, $k = 2$, and $n = 1$, $k = 1$.

Using the identification of Proposition 1 the bilinear form $\langle\langle \cdot, \cdot \rangle\rangle$ on $\mathcal{V} \times \mathcal{V}^*$ takes the form

$$\begin{aligned} &\langle\langle (f_E^1, f_M^1, f_b^1, e_E^1, e_M^1, e_b^1), (f_E^2, f_M^2, f_b^2, e_E^2, e_M^2, e_b^2) \rangle\rangle \\ &:= \int_N (f_E^1 \wedge e_E^2 + f_M^1 \wedge e_M^2 + f_E^2 \wedge e_E^1 + f_M^2 \wedge e_M^1) \\ &+ \int_{\partial N} (f_b^1 \wedge e_b^2 + f_b^2 \wedge e_b^1) \quad (\in \mathbb{R}) \end{aligned} \tag{17}$$

with for $i = 1, 2$

$$
\begin{array}{lll}
(f_E^i, f_M^i) & \in \Omega^k(N) \times \Omega^k(N) & \\
f_b^i & \in \Omega^{n-k}(\partial N) & \\
(e_E^i, e_M^i) & \in \Omega^{n-k}(N) \times \Omega^{n-k}(N) & (18) \\
 & \simeq (\Omega^k(N))^* \times (\Omega^k(N))^* & \\
e_b^i & \in \Omega^{n-2-k}(\partial N) \simeq (\Omega^{k-1}(\partial N))^* &
\end{array}
$$

The letters "E" and "M" here stand for "electric" and "magnetic", corresponding to the two different energy domains in the examples of Maxwell's equations and the telegrapher's equations. However, more generally they will be used as mnemonic notations to denote any two interacting pair of energy domains (as in the vibrating string example, where the energy domains are kinetic and potential energy). Of course, the letter "b" stands for "boundary".

On $\mathcal{V}$ we now define the following (constant) Dirac structure:

Theorem 2. Define the following linear subspace of $\mathcal{V} \times \mathcal{V}^*$

$$
\begin{aligned}
\mathcal{D} = \{ & (f_E, f_M, f_b, e_E, e_M, e_b) \in \mathcal{V} \times \mathcal{V}^* \mid \\
& \begin{bmatrix} f_E \\ f_M \end{bmatrix} = \begin{bmatrix} 0 & (-1)^{n-k} d \\ d & 0 \end{bmatrix} \begin{bmatrix} e_E \\ e_M \end{bmatrix}, \\
& \begin{bmatrix} f_b \\ e_b \end{bmatrix} = \begin{bmatrix} 0 & (-1)^k \\ 1 & 0 \end{bmatrix} \begin{bmatrix} e_E|_{\partial N} \\ e_M|_{\partial N} \end{bmatrix} \}
\end{aligned} \tag{19}
$$

with $d : \Omega^p(N) \to \Omega^{p+1}(N)$ the usual exterior derivative. Then $\mathcal{D} \subset \mathcal{V} \times \mathcal{V}^*$ is a constant Dirac structure.

Proof. (i) $\mathcal{D} \subset \mathcal{D}^\perp$: Let $(f_E^1, f_M^1, f_b^1, e_E^1, e_M^1, e_b^1) \in \mathcal{D}$ and consider any $(f_E^2, f_M^2, f_b^2, e_E^2, e_M^2, e_b^2) \in \mathcal{D}$. By substitution of (19) into (17) the right-hand side of (17) becomes

$$
\begin{aligned}
& \int_N [\ (-1)^{n-k} de_M^1 \wedge e_E^2 + de_E^1 \wedge e_M^2 \\
& \qquad +(-1)^{n-k} de_M^2 \wedge e_E^1 + de_E^2 \wedge e_M^1] \\
& + \int_{\partial N} (-1)^k [e_M^1|_{\partial N} \wedge e_E^2|_{\partial N} + e_M^2|_{\partial N} \wedge e_E^1|_{\partial N}]
\end{aligned} \tag{20}
$$

From the properties of the exterior derivative it follows that

$$
\begin{aligned}
d(e_E^1 \wedge e_M^2) &= de_E^1 \wedge e_M^2 + (-1)^{n-k} e_E^1 \wedge de_M^2 \\
d(e_E^2 \wedge e_M^1) &= de_E^2 \wedge e_M^1 + (-1)^{n-k} e_E^2 \wedge de_M^1
\end{aligned} \tag{21}
$$

Note that

$$
\begin{aligned}
(-1)^{n-k} e_E^1 \wedge de_M^2 &= (-1)^{n-k}(-1)^{(n-k)(n-k+1)} de_M^2 \wedge e_E^1 \\
&= (-1)^{(n-k)(n-k+2)} de_M^2 \wedge e_E^1 \\
&= (-1)^{n-k} de_M^2 \wedge e_E^1
\end{aligned}
$$

since $(n-k)(n-k+2)$ is odd (respectively even) iff $n-k$ is odd (respectively even). Similarly,

$$
(-1)^{n-k} e_E^2 \wedge de_M^1 = (-1)^{n-k} de_M^1 \wedge e_E^2 \tag{22}
$$

$$
\begin{aligned}
d(e_E^1 \wedge e_M^2) &= (-1)^{n-k} d(e_M^2 \wedge e_E^1) \\
d(e_E^2 \wedge e_M^1) &= (-1)^{n-k} d(e_M^1 \wedge e_E^2)
\end{aligned} \tag{23}
$$

Substitution of (3),(22) into (21) yields

$$
\begin{aligned}
& (-1)^{n-k} d(e_M^2 \wedge e_E^1) = \\
& de_E^1 \wedge e_M^2 + (-1)^{n-k} de_M^2 \wedge e_E^1 \\
& (-1)^{n-k} d(e_M^1 \wedge e_E^2) = \\
& de_E^2 \wedge e_M^1 + (-1)^{n-k} de_M^1 \wedge e_E^2
\end{aligned} \tag{24}
$$

Finally, by Stokes' theorem

$$
\begin{aligned}
\int_N d(e_M^2 \wedge e_E^1) &= \int_{\partial N} e_M^2|_{\partial N} \wedge e_E^1|_{\partial N} \\
\int_N d(e_M^1 \wedge e_E^2) &= \int_{\partial N} e_M^1|_{\partial N} \wedge e_E^2|_{\partial N}
\end{aligned} \tag{25}
$$

Substitution of (24),(25) into (20) yields

$$
\begin{aligned}
& \int_{\partial N} [(-1)^{n-k} + (-1)^k] \\
& (e_M^1|_{\partial N} \wedge e_E^2|_{\partial N} + e_M^2|_{\partial N} \wedge e_E^1|_{\partial N}
\end{aligned} \tag{26}
$$

which is zero since for n odd, $(-1)^{n-k} = -(-1)^k$. Hence $(f_E^1, f_M^1, f_b^1, e_E^1, e_M^1, e_b^1) \in \mathcal{D}^\perp$.

(ii) $\mathcal{D}^\perp \subset \mathcal{D}$: Let $(f_E^1, f_M^1, f_b^1, e_E^1, e_M^1, e_b^1) \in \mathcal{D}^\perp$, implying that for all $(f_E^2, f_M^2, f_b^2, e_E^2, e_M^2, e_b^2) \in \mathcal{D}$, (17) equals zero, and hence, by (19) for all

$$
((-1)^{n-k} de_M^2, de_E^2, (-1)^k e_M^2|_{\partial N}, e_E^2, e_M^2, e_E^2|_{\partial N}) \tag{27}
$$

Therefore for all e_E^2, e_M^2:

$$
\begin{aligned}
& \int_N (f_E^1 \wedge e_E^2 + f_M^1 \wedge e_M^2 + \\
& (-1)^{n-k} de_M^2 \wedge e_E^1 + de_E^2 \wedge e_M^1) + \\
& (-1)^k \int_{\partial N} (f_b^1 \wedge e_E^2|_{\partial N} + e_M^2|_{\partial N} \wedge e_b^1) = 0
\end{aligned} \tag{28}
$$

Now, first consider $(n-k)$-forms e_E^1, e_M^2 with $e_E^2|_{\partial N} = e_M^2|_{\partial N} = 0$. Then the second term of (28) automatically vanishes. Substitution of (24) into the first term of (28) now yields by Stokes' theorem (since $e_E^2|_{\partial N} = e_M^2|_{\partial N} = 0$)

$$
\begin{aligned}
\int_N & (f_E^1 \wedge e_E^2 + f_M^1 \wedge e_M^2 - de_E^1 \wedge e_M^2 - \\
& (-1)^{n-k} de_M^1 \wedge e_E^2) = 0
\end{aligned} \tag{29}
$$

for all e_E^2, e_M^2 with $e_E^2|_{\partial N} = e_M^2|_{\partial N} = 0$. Hence

$$
\begin{aligned}
\int_N (f_E^1 - (-1)^{n-k} de_M^1) \wedge e_E^2 &= 0 \\
\int_N (f_M^1 - de_E^1) \wedge e_M^2 &= 0
\end{aligned} \tag{30}
$$

for all e_E^2, e_M^2 with $e_E^2|_{\partial N} = e_M^2|_{\partial N} = 0$. This implies

$$f_E^1 = (-1)^{n-k} de_M^1, \quad f_M^1 = de_E^1 \tag{31}$$

Substitution of (31) into (28) then yields, after application of (24),(25),

$$(-1)^{n-k} \int_{\partial N} (e_M^1|_{\partial N} \wedge e_E^2|_{\partial N} + e_M^2|_{\partial N} \wedge e_E^1|_{\partial N}) + \int_N (f_b^1 \wedge e_E^2|_{\partial N} + (-1)^k e_M^2|_{\partial N} \wedge e_b^1) = 0 \tag{32}$$

for all $e_E^2|_{\partial N}, e_M^2|_{\partial N}$, showing that

$$\begin{aligned} f_b^1 &= -(-1)^{n-k} e_M^1|_{\partial N} \\ (-1)^k e_b^1 &= -(-1)^{n-k} e_E^1|_{\partial N} \end{aligned} \tag{33}$$

or equivalently (since n is odd)

$$f_b^1 = (-1)^k e_M^1|_{\partial N}, \quad e_b^1 = e_E^1|_{\partial N} \tag{34}$$

Thus, (31) together with (34) yields $(f_E^1, f_M^1, f_b^1, e_E^1, e_M^1, e_b^1) \in \mathcal{D}$, showing that $\mathcal{D}^\perp \subset \mathcal{D}$.

Remark 3. Note that for the two cases of primary interest ($n = 3$, $k = 2$, and $n = 1$, $k = 1$) the Dirac structure $\mathcal{D}$ amounts for $n = 3, k = 2$ to

$$\begin{bmatrix} f_E \\ f_M \end{bmatrix} = \begin{bmatrix} 0 & -d \\ d & 0 \end{bmatrix} \begin{bmatrix} e_E \\ e_M \end{bmatrix}, \qquad \begin{bmatrix} f_b \\ e_b \end{bmatrix} = \begin{bmatrix} 0 & 1 \\ 1 & 0 \end{bmatrix} \begin{bmatrix} e_E|_{\partial N} \\ e_M|_{\partial N} \end{bmatrix} \tag{35}$$

and for $n = 1, k = 1$ to:

$$\begin{bmatrix} f_E \\ f_M \end{bmatrix} = \begin{bmatrix} 0 & d \\ d & 0 \end{bmatrix} \begin{bmatrix} e_E \\ e_M \end{bmatrix}, \qquad \begin{bmatrix} f_b \\ e_b \end{bmatrix} = \begin{bmatrix} 0 & -1 \\ 1 & 0 \end{bmatrix} \begin{bmatrix} e_E|_{\partial N} \\ e_M|_{\partial N} \end{bmatrix} \tag{36}$$

Remark 4. The vanishing of the bilinear form (17) restricted to $\mathcal{D}$ should be interpreted as a generalized form of power conservation. Indeed, for

$$(f_E^i, f_M^i, f_b^i, e_E^i, e_M^i, e_b^i) = (f_E, f_M, f_b, e_E, e_M, e_b) \in \mathcal{D},$$

$i = 1, 2$, we obtain

$$\int_N (f_E \wedge e_E + f_M \wedge e_M) + \int_{\partial N} (f_b \wedge e_b) = 0 \tag{37}$$

The first term in (37) represents incoming power via the energy-storing elements in the domain N, while the second term represents the incoming power (originating from the environment) via the boundary ∂N.

Remark 5. The *compositionality* properties of the Dirac structure defined in Theorem 2 follow immediately. Indeed, consider two manifolds with boundary N_1 and N_2, such that

$$\begin{aligned} \partial N_1 &= \Gamma \cup \Gamma_1, \quad \Gamma \cap \Gamma_1 = \emptyset \\ \partial N_2 &= \Gamma \cup \Gamma_2, \quad \Gamma \cap \Gamma_2 = \emptyset \end{aligned} \tag{38}$$

(that is, N_1 and N_2 have boundary Γ in common). Then the Dirac structures $\mathcal{D}_1$, $\mathcal{D}_2$ on N_1, respectively N_2, compose to the Dirac structure $\mathcal{D}$ on $N = N_1 \cup N_2$ with boundary $\Gamma_1 \cup \Gamma_2$, if we equate f_b^1 on Γ with $-f_b^2$ on Γ, and e_b^1 on Γ with e_b^2 on Γ. (That is, the power flowing into N_1 via Γ should be equal to the power flowing *out* of N_2 via Γ.)

The definition of a distributed parameter port-controlled Hamiltonian systems now follows immediately. Consider a *Hamiltonian density* (energy per volume element)

$$H : \Omega^k(N) \times \Omega^k(N) \times N \to \Omega^n(N) \tag{39}$$

resulting in the total energy

$$\mathcal{H} := \int_N H \quad \in \mathbb{R} \tag{40}$$

We throughout assume H to be differentiable, with gradient vector denoted as

$$\text{grad } H = (\delta_E H, \delta_M H) \in (\Omega^k(N) \times \Omega^k(N))^* \tag{41}$$

Using the identification of Proposition 1 we thus obtain the co-energy variables

$$\begin{aligned} \delta_E H &\in \Omega^{n-k}(N) \\ \delta_M H &\in \Omega^{n-k}(N) \end{aligned} \tag{42}$$

Now, consider time-functions

$$(\alpha_E(t), \alpha_M(t)) \in \Omega^k(N) \times \Omega^k(N), \quad t \in \mathbb{R} \tag{43}$$

and the Hamiltonian density $H(\alpha_E(t), \alpha_M(t))$, $t \in \mathbb{R}$, evaluated along this trajectory. It follows that

$$\begin{aligned} \frac{dH}{dt} &= \langle \text{grad } H \mid (\frac{\partial \alpha_E}{\partial t}, \frac{\partial \alpha_M}{\partial t}) \rangle \\ &= \frac{\partial \alpha_E}{\partial t} \wedge \delta_E H + \frac{\partial \alpha_M}{\partial t} \wedge \delta_M H \end{aligned} \tag{44}$$

and hence for the total energy $\mathcal{H}$,

$$\frac{d\mathcal{H}}{dt} = \int_N (\frac{\partial \alpha_E}{\partial t} \wedge \delta_E H + \frac{\partial \alpha_M}{\partial t} \wedge \delta_M H) \tag{45}$$

The k-forms $\frac{\partial \alpha_E}{\partial t}$, $\frac{\partial \alpha_M}{\partial t}$ represent the (infinite-dimensional) generalized velocities corresponding to the energy storage in N. They are connected to the Dirac structure D by setting (analogous to (8))

$$f_E = -\frac{\partial \alpha_E}{\partial t}$$
$$f_M = -\frac{\partial \alpha_M}{\partial t} \qquad (46)$$

(The minus sign is included in order to have a consistent energy flow description.) Furthermore, we set (analogously to (8))

$$e_E = \delta_E H$$
$$e_M = \delta_M H \qquad (47)$$

Definition 2. The distributed parameter port-controlled Hamiltonian system with manifold of spatial variables N, state space $\Omega^k(N) \times \Omega^k(N)$, Dirac structure D on $\Omega^k(N) \times \Omega^k(N) \times \Omega^{k-1}(\partial N)$ given by (19), and Hamiltonian density H, is given as

$$\begin{bmatrix} -\frac{\partial \alpha_E}{\partial t} \\ -\frac{\partial \alpha_M}{\partial t} \end{bmatrix} = \begin{bmatrix} 0 & (-1)^{n-k}d \\ d & 0 \end{bmatrix} \begin{bmatrix} \delta_E H \\ \delta_M H \end{bmatrix}$$
$$\begin{bmatrix} f_b \\ e_b \end{bmatrix} = \begin{bmatrix} 0 & (-1)^k \\ 1 & 0 \end{bmatrix} \begin{bmatrix} \delta_E H \\ \delta_M H \end{bmatrix} \qquad (48)$$

with $f_b, e_b \in \Omega^{k-1}(\partial N)$ denoting the boundary variables.

Note that (48) defines a (nonlinear) *boundary control* system in the sense of e.g. (Fattorini, 1968), with *inputs*, say, f_b, and *outputs* e_b. It immediately follows from (45) and the power-conservation property of the Dirac structure (see Remark 4) that any distributed parameter port-controlled Hamiltonian system satisfies along its trajectories the energy-balance

$$\frac{d\mathcal{H}}{dt} = \int_{\partial N} f_b \wedge e_b, \qquad (49)$$

expressing that the increase in internally stored energy in N equals the incoming power via the boundary ∂N.

Energy exchange through the boundary is not the only way a distributed parameter system may interact with its environment. An example of a different situation is formed by Maxwell's equations (see Section 4.2), where interaction may also take place via the current density J which directly affects the electric charge distribution in the domain N. In order to cope with this other possibility we extend the space $\mathcal{V}$ given by

$$\Omega^k(N) \times \Omega^k(N) \times \Omega^{k-1}(\delta N) \qquad (50)$$

to

$$\Omega^k(N) \times \Omega^k(N) \times \Omega^{k-1}(\delta N) \times \Omega^k(N) \qquad (51)$$

with the last component denoting the external distributed flow $f_d \in \Omega^k(N)$, with dual variable the distributed effort $e_d \in \Omega^{n-k}(N)$. In this case, the first line of the definition of the Dirac structure given in Theorem 2 is extended to

$$\begin{bmatrix} f_E \\ f_M \end{bmatrix} = \begin{bmatrix} 0 & (-1)^{n-k}d \\ d & 0 \end{bmatrix} \begin{bmatrix} e_E \\ e_M \end{bmatrix} + G f_d$$
$$e_d = G^* \begin{bmatrix} e_E \\ e_M \end{bmatrix} \qquad (52)$$

for some linear map $G : \Omega^k(N) \to \Omega^k(N) \times \Omega^k(N)$, with dual map $G^* : \Omega^{n-k}(N) \times \Omega^{n-k}(N) \to \Omega^{n-k}(N)$. If the map G is modulated by the state variables α_E, α_M, then the resulting Dirac structure is not constant anymore. The port variables are now (f_b, f_d, e_b, e_d), with f_b, e_b the boundary port variables, and f_d, e_d the distributed port variables. Finally, energy dissipation can be incorporated in the framework by terminating some of the ports (boundary or distributed) with a resistive relation as in (10).

4. ELECTRODYNAMICS AND THE VIBRATING STRING

4.1 *The telegrapher's equations*

In this example we shall consider the telegrapher's equations, i.e. the transverse electromagnetic wave propagating in the dielectric of an ideal lossless transmission line (Dworsky, 1979). Let us first briefly recall the classical formulation of the telegrapher's equations using real functions. The spatial variable z belongs to some segment, for instance $N = [0, L]$ of the real line $\mathbb{R}$. The energy variables are the charge per unit length $q(z,t)$ and the magnetic flux per unit length $\phi(z,t)$. The lossless transmission line is characterized by the distributed capacitance $C(z)$ and the distributed inductance $L(z)$ which defines the electromagnetic energy density:

$$E(q, \phi, z) = \frac{1}{2}\frac{q^2}{C} + \frac{1}{2}\frac{\phi^2}{L} \qquad (53)$$

The coenergy variables are hence the voltage $V(z)$:

$$V(z) = \frac{\partial H}{\partial q} = \frac{q}{C} \qquad (54)$$

and the current $I(z)$:

$$I(z) = \frac{\partial H}{\partial \phi} = \frac{\phi}{L} \qquad (55)$$

Then Maxwell's equations in the case of a planar transverse electromagnetic wave become the telegrapher's equations:

$$\begin{aligned} \frac{\partial q}{\partial t} &= -\frac{\partial}{\partial z} I(z) \\ \frac{\partial \phi}{\partial t} &= -\frac{\partial}{\partial z} V(z) \end{aligned} \qquad (56)$$

In the formalism of port controlled Hamiltonian systems with power flow through the boundaries proposed in Section 3, the telegrapher's equations are formulated as follows. The energy variables (electric charge and magnetic flux) are the 1-forms:

$$\begin{aligned} \alpha_E(t) &= q(z,t)dz \in \Omega^1([0,L]) \\ \alpha_M(t) &= \phi(z,t)dz \in \Omega^1([0,L]) \end{aligned} \qquad (57)$$

The co-energy variables (voltage and current) are the following 0-forms, (functions) $V(z) \in \Omega^0([0,L])$ and $I(z) \in \Omega^0([0,L])$ which are related to the energy variables using the Hodge star product (associated with the canonical inner product on $\mathbb{R}$) and the characteristic capacitance and inductance of the line, as follows:

$$\begin{aligned} V &= \frac{1}{C} \star \alpha_E \\ I &= \frac{1}{L} \star \alpha_M \end{aligned} \qquad (58)$$

Then the energy density E in (53) becomes the following one-form:

$$H_{tl}(\alpha_E, \alpha_M) = \frac{1}{2}(\alpha_E \wedge V + \alpha_M \wedge I) \qquad (59)$$

which by definition of the Hodge star operator $\star$ may be expressed as a quadratic form on α_E and α_M. Moreover, by construction, the energy density H_{tl} in (59) satisfies Maxwell's reciprocity conditions. The total electromagnetic energy of the transmission line is then:

$$\mathcal{H}_{tl} = \int_N H_{tl} \qquad (60)$$

Then the telegrapher's equations may be expressed as a distributed port controlled Hamiltonian system according to definition 2 with power flow through the boundary $\delta N = \{0, L\}$ of N and boundary port variables being the current at the terminal points of the line: $f_b = -I|_{\delta N}$ and the voltage: $e_b = V|_{\delta N}$. This port controlled Hamiltonian system is defined with respect to the Dirac structure defined in Theorem 2 with $n = k = 1$, and the Hamiltonian functional being the total electromagnetic energy defined in (59) and (60).

The power balance (49) becomes:

$$\frac{d\mathcal{H}_{tl}}{dt} = \int_{\delta N} f_b \wedge e_b = V(0)I(0) - V(L)I(L) \qquad (61)$$

and says that the time derivative of the total magnetic energy is equal to the electromagnetic power ingoing the line at the point 0 minus the electromagnetic power outgoing the line at L.

4.2 *Maxwell's equations*

In this paragraph we shall briefly present the port controlled Hamiltonian description of Maxwell's equations following closely the formulation in terms of differential forms presented in (Ingarden and Jamiolkowski, 1985).

We consider some connected closed domain N with non void interior, of the three-dimensional oriented Euclidean space E^3 which defines the spatial variable $x \in N$ and consider the electromagnetic field in the medium in N.

The *energy variables* are the *electric field induction 2-forms* $\alpha_E = \mathcal{D} \in \Omega^2(N)$:

$$\mathcal{D} = \frac{1}{2} D_{ij}(x,t) dx^i \wedge dx^j \qquad (62)$$

and the *magnetic field induction 2-form* $\alpha_M = \mathcal{B} \in \Omega^2(N)$:

$$\mathcal{B} = \frac{1}{2} B_{ij}(x,t) dx^i \wedge dx^j \qquad (63)$$

Note that using the Hodge star operator the two 2-forms may be transformed to 1-forms usually called *electric field induction vector* $\star\mathcal{D} = D_i(x,t)dx^i$ and the *magnetic field induction vector* $\star\mathcal{B} = B_i(x,t)dx^i$.

The *coenergy variables* are the *electric field intensity 1-forms* $\mathcal{E} \in \Omega^1(N)$:

$$\mathcal{E} = E_i(x,t)dx^i \qquad (64)$$

and the *magnetic field intensity 1-forms* $\mathcal{H} \in \Omega^1(N)$:

$$\mathcal{H} = H_i(x,t)dx^i \qquad (65)$$

These two 1-forms are usually called vectors of electric and magnetic field intensity.

The *constitutive relations* of the medium define the relations between the coenergy and the energy variables:

$$\begin{aligned} \mathcal{E} &= \epsilon^{-1} \star \mathcal{D} \\ \mathcal{H} &= \mu^{-1} \star \mathcal{B} \end{aligned} \qquad (66)$$

where $\epsilon(x)$ denotes the (symmetric positive) *electric permittivity tensor* and $\mu(x)$ denotes the (symmetric positive) *magnetic permittivity tensor*.

Hence the *electromagnetic energy density 3-form* $\mathcal{H}_em \in \Omega^3(N)$ becomes:

$$H_{em}(\mathcal{D}, \mathcal{B}) = \frac{1}{2}(\mathcal{E} \wedge \mathcal{D} + \mathcal{H} \wedge \mathcal{B}) \qquad (67)$$

One may note again that, by definition of the Hodge star product, the electromagnetic energy density may be expressed as a quadratic form on $\mathcal{D}$ and $\mathcal{B}$. Moreover, by symmetry of the electric and magnetic permittivity tensors, the energy density H_{em} in (67) satisfies Maxwell's reciprocity conditions. The total electromagnetic energy in the medium in N is then:

$$\mathcal{H}_{em} = \int_N H_{em} \tag{68}$$

Assuming that there is no current in the medium, Maxwell's equations may then be written as follows:

$$\begin{aligned} \frac{\partial}{\partial t}\mathcal{D} &= d\mathcal{H} \\ \frac{\partial}{\partial t}\mathcal{B} &= -d\mathcal{E} \end{aligned} \tag{69}$$

Note that the d denotes the exterior derivative and is applied to 1-forms in $\Omega^1(N)$. Using the Hodge star product and the resulting identification of 1- and 2-forms with vectors, the exterior derivative is then simply the curl of a vector.

Using the definition of the energy and coenergy variables as differential forms, Maxwell's equations may hence be expressed as a distributed parameter port controlled Hamiltonian system according to Definition 2, with power flow through the boundary δN of N and boundary port variables being the electric field intensity at the boundary: δN $f_b = \mathcal{E}|_{\delta N}$ and the magnetic field intensity at the boundary δN: $e_b = \mathcal{H}|_{\delta N}$. This port controlled Hamiltonian system is defined with respect to the Dirac structure defined in Theorem 2 with $n = 3$ and $k = 2$, and is generated by the Hamiltonian functional being the total electromagnetic energy defined in (67) and (68).

The power balance (49) becomes:

$$\frac{d\mathcal{H}_{tl}}{dt} = \int_{\delta N} f_b \wedge e_b = \int_{\delta N} \mathcal{E} \wedge \mathcal{H} \tag{70}$$

and says that the time derivative of the total electromagnetic energy in N is equal to the flow of electromagnetic power radiating through the boundary δN. It may be noted that the power through the boundary is defined by the 2-form $\mathcal{S} \in \Omega^2(N)$, the so-called *Poynting vector*:

$$\mathcal{S} = \mathcal{E} \wedge \mathcal{H} \tag{71}$$

Assuming that there exists a current density in the domain N, for instance if the medium is conducting due to Ohm's law or due to the diffusion of the charges or to some temperature gradient, then there exist an energy exchange with some other physical domain than the electromagnetic field, for instance with the thermic domain. This energy exchange does not go through the boundary of the domain N but is distributed in N. The *current density* may be defined as a 2-form $\mathcal{J} \in \Omega^2(N)$ which is related to the classically defined current density vector by the Hodge star operator: $\star\mathcal{J} = J_i(x,t)dx^i$. The dynamics of the electromagnetic field is now described by the following port controlled Hamiltonian system with power exchange through the boundary and in the domain:

$$\begin{aligned} \begin{bmatrix} \frac{\partial \mathcal{D}}{\partial t} \\ \frac{\partial \mathcal{B}}{\partial t} \end{bmatrix} &= \begin{bmatrix} 0 & d \\ -d & 0 \end{bmatrix} \begin{bmatrix} \delta_{\mathcal{D}} H_{em} \\ \delta_{\mathcal{B}} H_{em} \end{bmatrix} + \begin{bmatrix} 1 \\ 0 \end{bmatrix} \mathcal{J} \\ \begin{bmatrix} f_b \\ e_b \end{bmatrix} &= \begin{bmatrix} 0 & 1 \\ 1 & 0 \end{bmatrix} \begin{bmatrix} \delta_{\mathcal{D}} H_{em}|_{\delta N} \\ \delta_{\mathcal{B}} H_{em}|_{\delta N} \end{bmatrix} \\ e_d &= [1\ 0] \begin{bmatrix} \delta_{\mathcal{D}} H_{em} \\ \delta_{\mathcal{B}} H_{em} \end{bmatrix} = \mathcal{E} \end{aligned} \tag{72}$$

One may note that the conjugated port variable to the current density is simply the electric intensity 2-form: $e_d = \mathcal{E}$.

The power balance (70) becomes:

$$\frac{d\mathcal{H}_{tl}}{dt} = \int_{\delta N} \mathcal{E} \wedge \mathcal{H} + \int_N \mathcal{E} \wedge \mathcal{J} \tag{73}$$

and says that the variation of the total electromagnetic energy in N is equal to the flow of electromagnetic power radiating through the boundary δN plus the power exchanged in the volume (for instance dissipated by Ohm's law in the case of a conducting medium).

4.3 *Vibrating string*

Consider now an elastic string subject to traction forces at its ends. The spatial variable z belong to some segment, for instance $N = [0, L]$ of the real line $\mathbb{R}$. The dynamics of the string arise from the interaction of the elastic-potential energy and the kinetic energy of the string. Let us denote by $u(z,t)$ the displacement of the string. The elastic potential energy is a function of the *strain variable*, the 1-form:

$$\alpha_E(t) = \epsilon(z,t)dz \in \Omega^1([0,L]) \tag{74}$$

where $\epsilon(z,t) = \frac{\partial}{\partial z}u(z,t)$.

The associated coenergy variable is the *stress variable* which is the 0-form (function) $\sigma(z) \in \Omega^0([0,L])$, which is related to the strain variable α_E using the Hodge star operator (associated with the canonical inner product on $\mathbb{R}$) and the characteristic elasticity modulus T:

$$\sigma = T \star \alpha_E \tag{75}$$

The kinetic energy is defined through the energy variable given as the kinetic momentum, which is a 1-form:

$$\alpha_M(t) = p(z,t)dz \in \Omega^1([0,L]) \tag{76}$$

with the co-energy variable being the velocity $v(z,t) = \frac{\partial}{\partial t}u(z,t)$ at z, interpreted as a 0-form

$v \in \Omega^0([0, L])$, related to the kinetic momentum by:

$$v = \frac{1}{\mu} \star \alpha_E \tag{77}$$

where μ is the mass density of the string.

Then the total energy density (sum of the elastic potential and kinetic energy density) is then the following one-form:

$$H_{string}(\alpha_E, \alpha_M) = \frac{1}{2}(\alpha_E \wedge \sigma + \alpha_M \wedge v) \tag{78}$$

which by definition of the Hodge star product may be expressed as a quadratic form on α_E and α_M. The total energy of the string is then:

$$\mathcal{H}_{string} = \int_N H_{string} \tag{79}$$

The elastodynamic equations of the string may be expressed as a distributed port controlled Hamiltonian system according to Definition 2 with power flow through the boundary $\delta N = \{0, L\}$ of N and boundary port variables being the stress at the terminal points of the line: $f_b = -\sigma|_{\delta N}$ and the velocity: $e_b = v|_{\delta N}$. This port controlled Hamiltonian system is defined with respect to the Dirac structure defined in Theorem 2 with $n = k = 1$ and the Hamiltonian functional being the total energy of the string defined in (78) and (79).

The power balance (49) becomes:

$$\frac{d\mathcal{H}_{tl}}{dt} = \int_{\delta N} f_b \wedge e_b = v(0)\sigma(0) - v(L)\sigma(L) \tag{80}$$

and says that the time derivative of the total magnetic energy is equal to the balance of the mechanical work at the points 0 and L.

5. CONCLUSION

We have presented a port controlled Hamiltonian formulation of the dynamics of distributed parameter systems which allows to represent the energy flows through the boundary of the domain of the system.

The state space of the distributed port controlled Hamiltonian system is a space of differential forms on the domain of the spatial variables and corresponds to the energy variables of the system. The boundary variables are similarly defined as differential forms on the boundary of the domain. From this geometric definition of the state variables and the boundary variables, we have derived a Dirac structure associated with the exterior derivative and based on Stokes' theorem.

Then the dynamics of the distributed parameter system was formulated as an implicit port controlled Hamiltonian system defined with respect to this Dirac structure. This was illustrated on the examples of the telegrapher's equations, Maxwell's equations and the vibrating string.

In this paper we have restricted the presentation to the symmetric case where the state space consists of the product space of differential forms of the same degree. However the construction may be generalized in a straightforward way to the asymmetric case which arises for instance in the dynamics of compressible fluids, as will be presented in a forthcoming paper. The same examples also necessitate to consider Dirac structures not only derived from the exterior derivative but augmented with a Lie-Poisson bracket. Furthermore, Dirac structures offer an ideal frame to treat geometrically constrained distributed parameter systems as they appear in models of flexible beams (Saintellier, 1993).

Finally, an important goal of the Hamiltonian formulation, apart from the system theoretic, modelling and simulation purposes, is the design of passivity-based and physically interpretable stabilizing control stategies for distributed parameter systems using an extension of control schemes proposed for finite-dimensional systems (Ortega *et al.*, 1999*a*; Ortega *et al.*, 1999*b*).

6. REFERENCES

Abraham, R. and J.E. Marsden (1978). *Foundations of Mechanics.* second ed.. Benjamin/Cummings Publishing Company.

Breedveld, P.C. (1984). Physical Systems Theory in Terms of Bond Graphs. PhD thesis. Technische Hogeschool Twente. Enschede, The Netherlands.

Courant, T. (1990). Dirac manifolds. *Trans. American Math. Soc.* **319**, 631–661.

Dalsmo, M. and A.J. van der Schaft (1999). On representations and integrability of mathematical structures in energy-conserving physical systems. *SIAM J. Cont. Opt.* **37**(1), 54–91.

Dorfman, I. (1993). *Dirac Structures and Integrability of Nonlinear Evolution Equations.* Chichester: John Wiley.

Dworsky, L.N. (1979). *Modern Transmission Line Theory and Applications.* Wiley.

Fattorini, H.O. (1968). Boundary control systems. *SIAM J. Control* **6**, 349–385.

Holm, D.D., J.E. Marsden, T.E. Ratiu and A. Weinstein (1985). Nonlinear stability of fluid and plasma equilibria. *Phys.Rep.* **123**, 1–116.

Ingarden, R.S. and A. Jamiolkowski (1985). *Classical Electrodynamics.* PWN-Polish Sc. Publ., Warszawa, Elsevier.

Lewis, D., J.E. Marsden, R. Montgomery and R. Ratiu (1986). The Hamiltonian structure for dynamic free boundary problems. *Physica D* **18**, 391–404.

Maschke, B.M., A.J. van der Schaft and P.C. Breedveld (1992). An intrinsic Hamiltonian formulation of network dynamics: Non-standard poisson structures and gyrators. *Journal of the Franklin Institute* **329**(5), 923–966.

Maschke, B.M. and A.J. van der Schaft (1996). Interconnection of systems: the network paradigm. In: *Proc. of the IEEE Int. Conf. on Decision and Control, CDC'96*. Kobe, Japan. pp. 207–212.

Maschke, B.M. and A.J. van der Schaft (1997). Interconnected mechanical systems, part II: the dynamics of spatial mechanical networks. In: *Modelling and Control of Mechanical Systems* (A. Astolfi, D.J.N Limebeer, C. Melchiorri, A. Tornambè and R.B. Vinter, Eds.). Imperial College Press. pp. 17–30.

Maschke, B.M. and A.J. van der Schaft (1998). Note on the dynamics of LC circuits with elements in excess. Memorandum 1426. University of Twente. Faculty of Applied Mathematics.

Olver, P.J. (1993). *Applications of Lie Groups to Differential Equations*. second ed.. Springer-Verlag.

Ortega, R., A.J. van der Schaft and B.M Maschke (1999*a*). Stabilization of port controlled Hamiltonian systems. In: *Stability and Stabilization of Nonlinear Systems* (D.Aeyels, F.Lamnabhi-Lagarrigue and A.J. van der Schaft, Eds.). Vol. 246 of *LNCIS*. Springer. pp. 239–260.

Ortega, R., A.J. van der Schaft, B.M. Maschke and G. Escobar (1999*b*). Stabilization of port-controlled hamiltonian systems: Passivation and energy-balancing. submitted to Automatica. University of Twente.

Paynter, H. M. (1961). *Analysis and Design of Engineering Systems*. M.I.T. Press. Cambridge, Massachusetts.

Saintellier, F. (1993). Modelling and Simulation of Robots with Flexible Links using Bond Graphs for their Control (in french). Master's thesis. Conservatoire national des art et métiers. Paris, France.

van der Schaft, A.J. (1999). Interconnection and Geometry. In: *From Intelligent Control to Behavioral Systems* (J.W. Polderman and H.L. Trentelman, Eds.). Groningen. pp. 203–218.

van der Schaft, A.J. (2000). *L_2-Gain and Passivity Techniques in Nonlinear Control*. 2nd revised edition, Springer.

van der Schaft, A.J. and B.M Maschke (1995). The Hamiltonian formulation of energy conserving physical systems with external ports. *Archiv für Elektronik und Übertragungstechnik* **49**, 362–371.

HAMILTONIAN REALIZATIONS OF NONLINEAR ADJOINT OPERATORS

Kenji Fujimoto[*,] Jacquelien M.A. Scherpen[*]**
W. Steven Gray[***]

[*] *Faculty of Information Technology and Systems*
Department of Electrical Engineering
Delft University of Technology
P.O.Box 5031, 2600 GA Delft, The Netherlands
[**] *Department of Systems Science*
Graduate School of Informatics, Kyoto University
Uji, Kyoto 611-0011 Japan
[***] *Department of Electrical and Computer Engineering*
Old Dominion University
Norfolk, Virginia 23529-0246, U.S.A.
fujimoto@i.kyoto-u.ac.jp
J.M.A.Scherpen@its.tudelft.nl, gray@ece.odu.edu

Abstract: This paper addresses state-space realizations for nonlinear adjoint operators. In particular the relationship among nonlinear Hilbert adjoint operators, Hamiltonian extensions and port-controlled Hamiltonian systems are clarified. The characterization of controllability, observability and Hankel operators, and controllability and observability functions will be derived based on it. Furthermore a duality between the controllability and observability functions will be proven. The state-space realizations of such operators provide new insights to nonlinear control systems theory.

Keywords: Nonlinear control systems, adjoint operators, Hamiltonian control systems, Hamiltonian extensions, Legendre transformations.

1. INTRODUCTION

Adjoint operators play an important role in the linear control systems theory. They provide duality between inputs and outputs of linear systems. The properties with respect to input, e.g. controllability and stabilizability issues, of linear systems directly reduce to the dual results with respect to output, observability and detectability issues. Let us consider a linear operator (transfer function) $\Sigma(s) : E \to F$ with Hilbert spaces E and F. Then its adjoint operator $\Sigma'(s) : F' \to E'$ is isomorphic to $\Sigma^T(-s) : F \to E$. Thus the adjoint can be easily described by a state-space realization if the operator $\Sigma(s)$ has a finite dimensional state-space realization. The objective of this paper is to provide the nonlinear extension of such adjoint operators.

Nonlinear adjoint operators can be found in the mathematics literature, e.g. (Batt, 1970), and they are expected to play a similar role in the nonlinear control systems theory. So called nonlinear Hilbert adjoint operators are introduced in (Gray and Scherpen, 1998; Scherpen and Gray, 1999) as a special class of nonlinear adjoint operators. The definition of them is as follows: The *nonlinear Hilbert adjoint* of $\Sigma : E \to F$ with Hilbert spaces E and F is an operator $\Sigma^* : F \times E \to E$ such that

$$\langle \Sigma(u), y \rangle_F = \langle u, \Sigma^*(y, u) \rangle_E \qquad (1)$$

holds for $\forall u \in E$, $\forall y \in F$. The existence of such operators in input-output sense was shown in (Gray and Scherpen, 1999) but their state-space realizations are not available so far.

On the other hand, Hamiltonian extensions (Crouch and van der Schaft, 1987) are used to characterize state-space adjoints of nonlinear control systems. In (Scherpen and van der Schaft, 1994; Ball and van der Schaft, 1996) Hamiltonian extensions are used extensively as the realizations of adjoint operators to characterize norm preserving properties. Indeed in the linear case the Hamiltonian extension of a given system is the Hilbert adjoint system. However, in the nonlinear case, this is not such a straightforward issue.

In this paper we give the state-space realization of nonlinear Hilbert adjoint operators. Firstly we show some relationship between the nonlinear Hilbert adjoint operators and Hamiltonian extensions but this does not give a complete characterization of adjoint operators. Secondly we derive a more complete characterization of the state-space realizations of nonlinear Hilbert adjoint operators which have the form of port-controlled Hamiltonian systems (Maschke and van der Schaft, 1992). The state-space realizations of such operators provide a characterization of the observability functions, controllability functions and Hankel operators which supplies a set of similarity invariants related to input-state and state-output behaviours of nonlinear systems (Scherpen, 1993; Gray and Scherpen, 1998; Scherpen and Gray, 1999). Hence it is expected that the adjoint operators in this paper will extend the existing results on the realization theory for nonlinear control systems a bit further. Furthermore we show a duality between the observability and controllability functions.

2. LINEAR SYSTEMS AS A PARADIGM

This section gives some examples of linear adjoint operators which play an important role in the linear systems theory, see e.g. (Zhou *et al.*, 1996). We present them here in a way that clarifies the line of thinking in the nonlinear case. Consider a causal linear input-output system $\Sigma : L_2^m[0,\infty) \to L_2^r[0,\infty)$ with a state-space realization

$$u \mapsto y = \Sigma(u) : \begin{cases} \dot{x} = Ax + Bu \\ y = Cx \end{cases} \tag{2}$$

where $x(0) = 0$. The Laplace transformation gives its transfer function matrix

$$\Sigma(s) := C(sI - A)^{-1}B. \tag{3}$$

Its adjoint operator is isomorphic to $\Sigma^* : L_2^r[0,\infty) \to L_2^m[0,\infty)$ given by

$$\Sigma^*(s) := \Sigma^T(-s) = B^T(-sI - A^T)^{-1}C^T \tag{4}$$

with a state-space realization

$$u_a \mapsto y_a = \Sigma^*(u_a) : \begin{cases} \dot{x} = -A^T x - C^T u_a \\ y_a = B^T x \end{cases} \tag{5}$$

where $x(\infty) = 0$. Here u_a and y_a have the same dimensions as y and u respectively. It satisfies the definition for Hilbert adjoint operators, namely,

$$\langle \Sigma(u), u_a \rangle_{L_2^r} = \langle u, \Sigma^*(u_a) \rangle_{L_2^m}. \tag{6}$$

Since u_a has the same dimension as y we can calculate the magnitude of operators as

$$\|\Sigma(u)\|_{L_2^r}^2 = \langle \Sigma(u), \Sigma(u) \rangle_{L_2^r} = \langle u, \Sigma^* \circ \Sigma(u) \rangle_{L_2^m}$$

by substituting $u_a = \Sigma(u)$. This relation can be utilized to derive the singular values of the corresponding input-output map.

The above representation is for general linear systems on L_2. We can relate it to the observability and controllability operators. They are given by $\mathcal{O} : \mathbb{R}^n \to L_2^r[0,\infty)$ and $\mathcal{C} : L_2^m[0,\infty) \to \mathbb{R}^n$

$$x^0 \mapsto y = \mathcal{O}(x^0) := Ce^{At}x^0 \tag{7}$$
$$u \mapsto x^0 = \mathcal{C}(u) := \int_0^\infty e^{A\tau}Bu(\tau)d\tau. \tag{8}$$

Note that these operators $\mathcal{O}$ and $\mathcal{C}$ are also operators on Hilbert spaces, hence their adjoint operators are given by $\mathcal{O}^* : L_2^m[0,\infty) \to \mathbb{R}^n$ and $\mathcal{C}^* : \mathbb{R}^n \to L_2^r[0,\infty)$

$$u_a \mapsto x^0 = \mathcal{O}^*(u_a) := \int_0^\infty e^{A^T\tau}C^T u_a(\tau)d\tau \tag{9}$$
$$x^0 \mapsto y_a = \mathcal{C}^*(x^0) := B^T e^{A^T t}x^0. \tag{10}$$

It can be easily checked that they satisfy

$$\langle \mathcal{O}(x^0), u_a \rangle_{L_2^r} = \langle x^0, \mathcal{O}^*(u_a) \rangle_{\mathbb{R}^n} \tag{11}$$
$$\langle \mathcal{C}(u), x^0 \rangle_{\mathbb{R}^n} = \langle u, \mathcal{C}^*(x^0) \rangle_{L_2^m}. \tag{12}$$

These adjoint operators can be used to calculate the observability and controllability Gramians:

$$\begin{aligned} \|\mathcal{O}(x^0)\|_{L_2^r}^2 &= \langle x^0, \mathcal{O}^* \circ \mathcal{O}(x^0) \rangle_{\mathbb{R}^n} \\ &= \langle x^0, \textstyle\int_0^\infty CA^\tau A^{T\tau} C^T d\tau\ x^0 \rangle_{\mathbb{R}^n} \\ &= \langle x^0, Q\ x^0 \rangle_{\mathbb{R}^n} \end{aligned} \tag{13}$$
$$\begin{aligned} \|\mathcal{C}^*(x^0)\|_{L_2^m}^2 &= \langle x^0, \mathcal{C}^{**} \circ \mathcal{C}^*(x^0) \rangle_{\mathbb{R}^n} \\ &= \langle x^0, \textstyle\int_0^\infty B^T A^{T\tau} A^\tau B d\tau\ x^0 \rangle_{\mathbb{R}^n} \\ &= \langle x^0, P\ x^0 \rangle_{\mathbb{R}^n} \end{aligned} \tag{14}$$

These imply $Q = \mathcal{O}^* \circ \mathcal{O}$ and $P = \mathcal{C}^{**} \circ \mathcal{C}^*$.

3. STATE-SPACE REALIZATION OF NONLINEAR ADJOINT OPERATORS

This section is devoted to the state-space characterization of nonlinear Hilbert adjoint operators. We will show some relationship between nonlinear Hilbert adjoint operators and Hamiltonian extensions in section 3.1 and give the state-space realization of adjoint operators based on port-controlled Hamiltonian systems in section 3.2.

3.1 *Adjoint operators and Hamiltonian extensions*

This subsection shows the relationship between nonlinear Hilbert adjoint operators and Hamiltonian extensions. Let us consider an input-output system $\Sigma : L_2^m(\Omega) \to L_2^r(\Omega)$ defined on a (possibly infinite) time interval $\Omega = [t^0, t^1] \subseteq \mathbb{R}$ which has a state-space realization

$$u \mapsto y = \Sigma(u) : \begin{cases} \dot{x} = f(x,u) \; x(t^0) = 0 \\ y = h(x,u) \end{cases} \quad (15)$$

with $x(t) \in \mathbb{R}^n$, $u(t) \in \mathbb{R}^m$ and $y(t) \in \mathbb{R}^r$. Here we assume the origin is an equilibrium, i.e. $f(0,0) = 0$ and $h(0,0) = 0$ hold and that all signals and functions are sufficiently smooth.

The Hamiltonian extension of Σ is given by a Hamiltonian control system (Crouch and van der Schaft, 1987)

$$\begin{cases} \dot{x} = \frac{\partial H}{\partial p}^T = f(x,u) \\ \dot{p} = -\frac{\partial H}{\partial x}^T = -\left(\frac{\partial f}{\partial x}^T p + \frac{\partial h}{\partial x}^T u_a\right) \\ y_a = \frac{\partial H}{\partial u}^T = \frac{\partial f}{\partial u}^T p + \frac{\partial h}{\partial u}^T u_a \\ y = \frac{\partial H}{\partial u_a}^T = h(x,u) \end{cases} \quad (16)$$

with initial conditions $x^1 := x(t^1) = 0$, $p^0 := p(t^0) = 0$ and the Hamiltonian

$$H(x,p,u,u_a) := p^T f(x,u) + u_a^T h(x,u). \quad (17)$$

We now prove some properties of this system which are related to nonlinear Hilbert adjoint operators.

Proposition 1 *Consider the Hamiltonian extension (16) of Σ as in (15). Define scalar functions H_1, H_2 and H_3 as*

$$H_1(x,p,u) := H - \frac{\partial H}{\partial u_a} u_a \quad (18)$$
$$H_2(x,p,u,u_a) := H - \frac{\partial H}{\partial u} u \quad (19)$$
$$H_3(x,p,u,u_a) := H - \frac{\partial H}{\partial u} u - \frac{\partial H}{\partial u_a} u_a. \quad (20)$$

Then the following relations hold.

$$\frac{dH}{dt} = y_a^T \dot{u} + y^T \dot{u}_a \quad (21)$$
$$\frac{dH_1}{dt} = y_a^T \dot{u} - \dot{y}^T u_a \quad (22)$$
$$\frac{dH_2}{dt} = -\dot{y}_a^T u + y^T \dot{u}_a \quad (23)$$
$$\frac{dH_3}{dt} = -\dot{y}_a^T u - \dot{y}^T u_a \quad (24)$$

Proof. Equation (21) follows from

$$\frac{dH}{dt} = \frac{\partial H}{\partial x}\dot{x} + \frac{\partial H}{\partial p}\dot{p} + \frac{\partial H}{\partial u}\dot{u} + \frac{\partial H}{\partial u_a}\dot{u}_a \quad (25)$$
$$= \frac{\partial H}{\partial x}\frac{\partial H}{\partial p}^T - \frac{\partial H}{\partial p}\frac{\partial H}{\partial x}^T + y_a^T \dot{u} + y^T \dot{u}_a \quad (26)$$
$$= y_a^T \dot{u} + y^T \dot{u}_a. \quad (27)$$

Hence the time derivative of the other functions are obtained by

$$\frac{dH_1}{dt} = \frac{d}{dt}\left(H - y^T u_a\right) = y_a^T \dot{u} - \dot{y}^T u_a \quad (28)$$
$$\frac{dH_2}{dt} = \frac{d}{dt}\left(H - y_a^T u\right) = -\dot{y}_a^T u + y^T \dot{u}_a \quad (29)$$
$$\frac{dH_3}{dt} = \frac{d}{dt}\left(H - y_a^T u - y^T u_a\right) = -\dot{y}_a^T u - \dot{y}^T u_a. \quad (30)$$

This proves the proposition. □

This proposition shows that the Hamiltonian extension has a close relationship to nonlinear Hilbert adjoint operators. Roughly speaking, e.g. (21) shows that the mapping $(\dot{u}_a, \dot{u}) \mapsto (-y_a)$ is a nonlinear Hilbert adjoint of the map $\check{\Sigma} : \dot{u} \mapsto y$ on $L_2[t^0, t^1]$ provided $H|_{t=t^0} = H|_{t=t^1}$ holds, because

$$\langle \check{\Sigma}(\dot{u}), \dot{u}_a \rangle_{L_2^r} := \langle y, \dot{u}_a \rangle_{L_2^r} = \langle \dot{u}, -y_a \rangle_{L_2^m} =: \langle \dot{u}, \check{\Sigma}^*(\dot{u}_a, \dot{u}) \rangle_{L_2^m}$$

holds. However we cannot obtain the nonlinear Hilbert adjoint Σ^* itself directly from Hamiltonian extensions.

The equation (22) describes an intrinsic property of Hamiltonian extensions. The mapping $u_a \mapsto y_a$ is the nonlinear Hilbert adjoint of the variational mapping $\dot{u} \mapsto \dot{y}$ of the original mapping $u \mapsto y$ and this corresponds to the original definition of Hamiltonian extensions (Crouch and van der Schaft, 1987).

Furthermore the property (23) shows that the mapping of the original system $u \mapsto y$ corresponds to the adjoint of the variational map $\dot{u}_a \mapsto \dot{y}_a$.

Property (24) corresponds to the so called *energy balancing* property of *physical* Hamiltonian control systems (Crouch and van der Schaft, 1987).

Also the relation (23) can be utilized to obtain a state-space realization of the nonlinear Hilbert adjoint of input-affine nonlinear systems. Details about this formulation are omitted for the reason of space. This state-space realization has a $(2n + m)$-dimensional state and it corresponds to

$$\left(s\,\Sigma(s)\,\tfrac{1}{s}\right)^* = s\,\Sigma^T(-s)\,\tfrac{1}{s} \quad (31)$$

in the linear case.

3.2 *Adjoint operators and port-controlled Hamiltonian systems*

Hamiltonian extensions have some relationships with nonlinear adjoint operators but their complete state-space characterization is not obtained. This section will give a more general state-space formulation based on port-controlled Hamiltonian systems (Maschke and van der Schaft, 1992).

Let us consider a possibly time-varying system $\Sigma : L_2^m(\Omega) \to L_2^r(\Omega)$ defined on a time interval $\Omega = [t^0, t^1] \subseteq \mathbb{R}$ which has a state-space realization

$$\Sigma : \begin{cases} \dot{x} = f(x,u,t) \; x(t^0) = x^0 \\ y = h(x,u,t) \end{cases} . \quad (32)$$

Here we assume the origin is an equilibrium, i.e. $f(0,0,t)=0$ and $h(0,0,t)=0$ hold for $\forall t \in \mathbb{R}$.

This system can be regarded as an operator $\hat{\Sigma}$: $\mathbb{R}^n \times L_2^m(\Omega) \to \mathbb{R}^n \times L_2^r(\Omega)$ with

$$(x^0,u) \mapsto (x^1,y) = \hat{\Sigma}(x^0,u) : \begin{cases} \dot{x} = f(x,u,t) \; x(t^0) = x^0 \\ y = h(x,u,t) \\ x^1 = x(t^1) \end{cases} \quad (33)$$

Note that $\mathbb{R}^n \times L_2^m(\Omega)$ is a Hilbert space with the inner product $\langle (x^0,u),(x^1,\tilde{u})\rangle_{\mathbb{R}^n \times L_2^m(\Omega)} := \langle x^0,x^1\rangle_{\mathbb{R}^n} + \langle u,\tilde{u}\rangle_{L_2^m(\Omega)}$. The following proposition gives a state-space realization of the nonlinear Hilbert adjoint of $\hat{\Sigma}$.

Proposition 2 *Consider an operator $\hat{\Sigma}$: $\mathbb{R}^n \times L_2^m(\Omega) \to \mathbb{R}^n \times L_2^r(\Omega)$ as in (33) with $\Omega = [t^0,t^1] \subseteq \mathbb{R}$ and the corresponding port-controlled Hamiltonian system $H_{\hat{\Sigma}} : \mathbb{R}^{2n} \times L_2^{m+r}(\Omega) \to \mathbb{R}^{2n} \times L_2^{m+r}(\Omega)$ given by*

$$(x^0,p^1,\hat{u}) \mapsto (x^1,p^0,\hat{y}) = H_{\hat{\Sigma}}(x^0,p^1,\hat{u}) : \begin{cases} \dot{\hat{x}} = J(\hat{x},\hat{u},t)\frac{\partial H}{\partial \hat{x}}^T + g(\hat{x},\hat{u},t)\,\hat{u} \\ \hat{y} = g^T(\hat{x},\hat{u},t)\frac{\partial H}{\partial \hat{x}}^T + \hat{D}(\hat{x},\hat{u},t)\,\hat{u} \\ \hat{x}(t^0) = (x^0,p^0) \\ \hat{x}(t^1) = (x^1,p^1) \end{cases} \quad (34)$$

with $\hat{x} := (x,p) \in \mathbb{R}^n \times \mathbb{R}^n$, $\hat{u} := (u,u_a) \in \mathbb{R}^m \times \mathbb{R}^r$, $\hat{y} := (y_a,-y) \in \mathbb{R}^m \times \mathbb{R}^r$ and

$$H(\hat{x}) := p^T x \quad (35)$$

$$J(\hat{x},\hat{u},t) := \begin{pmatrix} 0 & A(x,u,t) \\ -A^T(x,u,t) & 0 \end{pmatrix} \quad (36)$$

$$g(\hat{x},\hat{u},t) := \begin{pmatrix} B(x,u,t) & 0 \\ 0 & -C^T(x,u,t) \end{pmatrix} \quad (37)$$

$$\hat{D}(\hat{x},\hat{u},t) := \begin{pmatrix} 0 & D^T(x,u,t) \\ -D(x,u,t) & 0 \end{pmatrix}. \quad (38)$$

Here $A(x,u,t) \in \mathbb{R}^{n\times n}$, $B(x,u,t) \in \mathbb{R}^{n\times m}$, $C(x,u,t) \in \mathbb{R}^{r\times n}$ and $D(x,u,t) \in \mathbb{R}^{r\times m}$ are appropriate matrices such that

$$f(x,u,t) = A(x,u,t)\,x + B(x,u,t)\,u \quad (39)$$

$$h(x,u,t) = C(x,u,t)\,x + D(x,u,t)\,u \quad (40)$$

hold. Suppose that

$$|(x^0,p^1)| < \infty,\; u,u_a \in L_2(\Omega) \Rightarrow |(x^1,p^0)| < \infty. \quad (41)$$

Then the mapping $(x^0,p^1,u,u_a) \mapsto (p^0,y_a)$ corresponding to the state-space realization (34) is a state-space realization of the nonlinear Hilbert adjoint operator $\hat{\Sigma}^*$: $\mathbb{R}^{2n} \times L_2^{m+r}(\Omega) \to \mathbb{R}^n \times L_2^r(\Omega)$ of $\hat{\Sigma}$.

By direct calculation the port-controlled Hamiltonian system (34) reduces down to

$$H_{\hat{\Sigma}} : \begin{cases} \dot{x} = f(x,u,t) \\ \dot{p} = -A^T(x,u,t)\,p - C^T(x,u,t)u_a \\ y_a = B^T(x,u,t)\,p + D^T(x,u,t)u_a \\ y = h(x,u,t) \\ x^1 = x(t^1) \\ p^0 = p(t^0) \end{cases} \quad (42)$$

with $x(t_0) = x^0$ and $p(t_1) = p^1$ as "initial" conditions. The mapping $(x^0,p^1,u,u_a) \mapsto (p^0,y_a)$ is a state-space realization of the nonlinear Hilbert adjoint operator $\hat{\Sigma}^*$. Since the assumption (41) can be regarded as a stability requirement for the x-subsystem and an anti-stability requirement for the p-subsystem, then, as in the linear case, (41) it is always satisfied locally when Ω is finite or when the given system Σ in (32) with $u=0$ is locally exponentially stable at least.

Proof of Proposition 2. The property of (time-varying) port-controlled Hamiltonian systems (see e.g. (Fujimoto and Sugie, 1998)) proves

$$\begin{aligned} \tfrac{dH}{dt} &= \tfrac{\partial H}{\partial \hat{x}}\left(J\tfrac{\partial H}{\partial \hat{x}}^T + g\hat{u}\right) = \tfrac{\partial H}{\partial \hat{x}} g\hat{u} \\ &= \left(\tfrac{\partial H}{\partial \hat{x}} g + \hat{u}^T\hat{D}^T\right)\hat{u} = \hat{y}^T\hat{u} = y_a^T u - y^T u_a. \end{aligned}$$

This reduces to

$$\begin{aligned} &\langle y_a,u\rangle_{L_2^m(\Omega)} - \langle y,u_a\rangle_{L_2^r(\Omega)} \\ &= \textstyle\int_{t^0}^{t^1}(y_a^T u - y^T u_a)dt = \int_{t^0}^{t^1}\frac{dH}{dt}dt \\ &= H(\hat{x}(t^1)) - H(\hat{x}(t^0)) \\ &= \langle x^1,p^1\rangle_{\mathbb{R}^n} - \langle x^0,p^0\rangle_{\mathbb{R}^n}. \end{aligned}$$

Hence

$$\begin{aligned} &\langle (x^1,y),(p^1,u_a)\rangle_{\mathbb{R}^n \times L_2^r(\Omega)} \\ &= \langle (x^0,u),(p^0,y_a)\rangle_{\mathbb{R}^n \times L_2^m(\Omega)} \end{aligned} \quad (43)$$

holds. Substituting $(x^1,y) = \hat{\Sigma}(x^0,u)$ and $(p^0,y_a) = \hat{\Sigma}^*((p^1,u_a),(x^0,u))$ yields the definition of nonlinear Hilbert adjoint operators

$$\begin{aligned} &\langle \hat{\Sigma}(x^0,u),(p^1,u_a)\rangle_{\mathbb{R}^n \times L_2^r(\Omega)} \\ &= \langle (x^0,u),\hat{\Sigma}^*((p^1,u_a),(x^0,u))\rangle_{\mathbb{R}^n \times L_2^m(\Omega)}. \end{aligned} \quad (44)$$

This proves the proposition. □

This result provides a useful tool to analyze the properties of nonlinear input-output systems with state-space realizations. It should be noted that the characterization given here uses a coordinate dependent expression because we employ the inner product on $\mathbb{R}^n$ which is intrinsically coordinate dependent. However, it may be argued that it provides very natural state-space realizations of adjoint operators because it only requires rather mild assumptions indeed.

Proposition 2 directly derives the state-space characterization of the nonlinear Hilbert adjoint operators of $L_2^m(\Omega) \to L_2^r(\Omega)$ in the usual setting.

Corollary 1 *Consider the system Σ as in (32) with the initial condition $x^0 = 0$ and let $\Sigma : L_2^m(\Omega) \to L_2^r(\Omega)$ denote the mapping $u \mapsto y$. Suppose the assumption (41) holds. Then a state-space realization of the nonlinear Hilbert adjoint $\Sigma^* : L_2^{m+r}(\Omega) \to L_2^m(\Omega)$ of Σ is given by*

$$(u_a, u) \mapsto y_a = \Sigma^*(u_a, u) :$$
$$\begin{cases} \dot{x} = f(x,u,t) & x(t^0) = 0 \\ \dot{p} = -A^T(x,u,t)\, p - C^T(x,u,t)\, u_a & p(t^1) = 0\,. \\ y_a = B^T(x,u,t)\, p + D^T(x,u,t)\, u_a \end{cases}$$

The results presented in this section are useful for obtaining state-space characterizations of adjoints of some characteristic operators appearing in nonlinear realization theory. This matter is the topic of the next section.

4. ENERGY FUNCTIONS AND OPERATORS

4.1 *Observability, controllability and Hankel operators*

This section gives the state-space realizations for nonlinear Hilbert adjoint of some energy functions and operators. We only consider a causal L_2-stable time invariant input-affine nonlinear system without direct feed-through in the form of

$$\Sigma : \begin{cases} \dot{x} = f(x) + g(x)u \\ y = h(x) \end{cases} \tag{45}$$

defined on the time interval $\Omega := (-\infty, \infty)$.

Let us consider the observability and controllability operators $\mathcal{O} : \mathbb{R}^n \to L_2^r(\Omega_+)$ and $\mathcal{C} : L_2^m(\Omega_+) \to \mathbb{R}^n$ with $\Omega_+ := [0, \infty)$ of Σ defined by

$$x^0 \mapsto y = \mathcal{O}(x^0) : \begin{cases} \dot{x} = f(x) & x(0) = x^0 \\ y = h(x) \end{cases} \tag{46}$$

$$u \mapsto x^1 = \mathcal{C}(u) : \begin{cases} \dot{x} = f(x) + g(x)\mathcal{F}_-(u) \\ x^1 = x(0) \\ 0 = x(-\infty) \end{cases} \tag{47}$$

These operators are originally defined (Gray and Scherpen, 1998) by using Chen-Fliess expansions for analytic nonlinear systems, see e.g.(Fliess, 1974; Isidori, 1995). These are natural generalizations of the linear operators (7) and (8). Furthermore the Hankel operator (Gray and Scherpen, 1998; Scherpen and Gray, 1999) $\mathcal{H} : L_2^m(\Omega_+) \to L_2^r(\Omega_+)$ of Σ is given by

$$\mathcal{H} := \Sigma \circ \mathcal{F}_-. \tag{48}$$

Here $\mathcal{F}_- : L_2^m(\Omega_+) \to L_2^m(\Omega_-)$ with $\Omega_- := (-\infty, 0]$ denotes the so called *flipping operator* defined by

$$\mathcal{F}_-(u)(t) = \begin{cases} u(-t) & t \in \Omega_- \\ 0 & t \in \Omega_+ \end{cases}. \tag{49}$$

The state-space realizations of the nonlinear Hilbert adjoint operators of $\mathcal{O}$, $\mathcal{C}$ and $\mathcal{H}$ are given by the following proposition.

Proposition 3 *Consider the operator Σ as in (45). Suppose that the assumption (41) in Proposition 2 holds for the relevant port-controlled Hamiltonian system (34). Then state-space realizations of $\mathcal{O}^* : L_2^r(\Omega_+) \times \mathbb{R}^n \to \mathbb{R}^n$, $\mathcal{C}^* : \mathbb{R}^n \times L_2^m(\Omega_+) \to L_2^m(\Omega_+)$ and $\mathcal{H}^* : L_2^r(\Omega_+) \times L_2^m(\Omega_+) \to L_2^m(\Omega_+)$ are given by*

$$(x^0, u_a) \mapsto p^0 = \mathcal{O}^*(x^0, u_a) :$$
$$\begin{cases} \dot{x} = f(x) & x(0) = x^0 \\ \dot{p} = -A^T(x)\, p - C^T(x)\, u_a & p(\infty) = 0 \\ p^0 = p(0) \end{cases} \tag{50}$$

$$(p^1, u) \mapsto y_a = \mathcal{C}^*(p^1, u) :$$
$$\begin{cases} \dot{x} = f(x) + g(x)\mathcal{F}_-(u) & x(-\infty) = 0 \\ \dot{p} = -A^T(x)\, p & p(0) = p^1 \\ y_a = \mathcal{F}_+(g^T(x)\, p) \end{cases} \tag{51}$$

$$(u_a, u) \mapsto y_a = \mathcal{H}^*(u_a, u) :$$
$$\begin{cases} \dot{x} = f(x) + g(x)\, \mathcal{F}_-(u) & x(-\infty) = 0 \\ \dot{p} = -A^T(x)\, p - C^T(x)\, u_a & p(\infty) = 0 \\ y_a = \mathcal{F}_+(g^T(x)\, p) \end{cases} \tag{52}$$

respectively with matrices $A(x) \in \mathbb{R}^{n \times n}$ and $C(x) \in \mathbb{R}^{r \times n}$ such that $f(x) \equiv A(x)x$ and $h(x) \equiv C(x)x$ hold. Here $\mathcal{F}_+ : L_2^m(\Omega_-) \to L_2^m(\Omega_+)$ denotes another flipping operator defined by

$$\mathcal{F}_+(u)(t) = \begin{cases} 0 & t \in \Omega_- \\ u(-t) & t \in \Omega_+ \end{cases}. \tag{53}$$

The proofs of Proposition 3 and 4 are omitted for the reason of space. It is expected that the state-space characterization of $\mathcal{O}^*$, $\mathcal{C}^*$ and $\mathcal{H}^*$ will provide further developments in the realization theory of nonlinear control systems.

4.2 *Observability and controllability functions*

In this subsection some relationship between the observability and controllability functions, operators and Gramians are developed.

Definition The observability function $L_o(x)$ and the controllability function $L_c(x)$ of Σ as in (45) are defined by

$$L_o(x^0) := \tfrac{1}{2}\int_0^\infty \|y(t)\|^2 dt, \quad x(0) = x^0, \quad u(t) \equiv 0$$
$$L_c(x^1) := \min_{\substack{u \in L_2^m(\Omega_-) \\ x(-\infty) = 0,\, x(0) = x^1}} \tfrac{1}{2}\int_{-\infty}^0 \|u(t)\|^2 dt$$

respectively.

These functions are closely related to observability and controllability operators and Gramians in the linear case (13) and (14). At first we present the relation between the observability function, operator and Gramian.

$$L_o(x^0) = \tfrac{1}{2}\|\mathcal{O}(x^0)\|_{L_2^r}^2 = \tfrac{1}{2}\langle x^0, \mathcal{O}^*(\mathcal{O}(x^0), x^0)\rangle_{\mathbb{R}^n}$$
$$= \tfrac{1}{2}\langle x^0, p^0\rangle_{\mathbb{R}^n} =: \tfrac{1}{2}\langle x^0, \phi(x^0)\rangle_{\mathbb{R}^n} \qquad (54)$$

Here $p^0 = p(0)$ is the initial state of the state-space realization of $\mathcal{O}^*$ in (50) with the input $(u_a, x^0) = (\mathcal{O}(x^0), x^0)$. The function $\phi(x^0)$ can always be rewritten by $\phi(x^0) = Q(x^0)\, x^0$ using a square symmetric matrix $Q(x^0)$. This matrix coincides with the observability Gramian in the linear case. Further it should be noticed that

$$L_o(x(t)) = \tfrac{1}{2}\langle x(t), p(t)\rangle_{\mathbb{R}^n} \qquad (55)$$

holds along the trajectory of the state-space realization (50) of $\mathcal{O}^*$.

In the controllability case, there does not hold such a relation as in the observability case. As for the equation (54) it does follow that

$$L_c(x^1) = \tfrac{1}{2}\|\mathcal{C}^\dagger(x^1)\|_{L_2^m}^2 = \tfrac{1}{2}\langle x^1, \mathcal{C}^{\dagger *}(\mathcal{C}^\dagger(x^1), x^1)\rangle_{\mathbb{R}^n}$$
$$=: \tfrac{1}{2}\langle x^1, \varphi(x^1)\rangle_{\mathbb{R}^n} \qquad (56)$$

with $\mathcal{C}^\dagger : \mathbb{R}^n \to L_2^m(\Omega_+)$ the pseudo-inverse of $\mathcal{C}$ such that

$$\mathcal{C}^\dagger(x^1) := \arg \min_{\mathcal{C}(u)=x^1} \|u\|_{L_2^m} \qquad (57)$$

holds. The state-space realization of $\mathcal{C}^\dagger$ can be easily obtained and $L_c(x)$ can be calculated in a similar way as $L_o(x)$. This relation is slightly different from the linear case (14), since here we deal with the "inverse" of the controllability Gramian. The correspondence with the linear case can however be explored in terms of duality as follows.

Proposition 4 *Consider the system Σ as in (45) and the related port-controlled Hamiltonian system $H_{\hat{\Sigma}}$ as in (42). Suppose that $\dot{x} = f(x)$ is asymptotically stable and that Σ has the observability function $L_o(x)$ and the controllability function $L_c(x)$. Consider the reverse-time system of the p-subsystem of the related port-controlled Hamiltonian system*

$$\begin{cases} \dot{p} &= A^T(\phi(p))p + C^T(\phi(p))u_a \\ y_a &= g^T(\phi(p))p \end{cases} . \qquad (58)$$

Let $x = \phi(p)$ denote the inverse mapping of $p = \frac{\partial L_c}{\partial x}^T(x)$ and suppose that this system has the observability function $\tilde{L}_o(p)$. Then $\tilde{L}_o(p)$ is given by a Legendre transformation

$$\tilde{L}_o(p) = -L_c(x) + p^T x. \qquad (59)$$

Let $x = \phi(p)$ denote the inverse mapping of $p = \frac{\partial L_o}{\partial x}^T(x)$ and suppose the system (58) has a controllability function $\tilde{L}_c(p)$. Then $\tilde{L}_c(p)$ is given by a Legendre transformation as well

$$\tilde{L}_c(p) = -L_o(x) + p^T x. \qquad (60)$$

Legendre transformations give *the duality in the sense of Young* (Arnold, 1989) and this duality is intrinsically coordinate dependent. Thus, as in the linear case, we have duality in the sense of Young between inputs and outputs which appears to be closely related to port-controlled Hamiltonian adjoint systems.

ACKNOWLEDGMENT

The first author would like to thank Professor Arjan van der Schaft of the University of Twente for his insightful suggestions.

5. REFERENCES

Arnold, V. I. (1989). *Mathematical Methods of Classical Mechanics*. second edition ed.. Springer-Verlag. New York.

Ball, J. A. and A. J. van der Schaft (1996). *j*-inner-outer factorization, *j*-spectral factorization, and robust control for nonlinear systems. *IEEE Trans. Autom. Contr.* **AC-41**(3), 379–392.

Batt, J. (1970). Nonlinear compact mappings and their adjoints. *Math. Ann.* **189**, 5–25.

Crouch, P. E. and A. J. van der Schaft (1987). *Variational and Hamiltonian Control Systems*. Vol. 101 of *Lecture Notes on Control and Information Science*. Springer-Verlag. Berlin.

Fliess, M. (1974). Matrices de hankel. *J. Math. Pures. Appl.* pp. 197–222.

Fujimoto, K. and T. Sugie (1998). Canonical transformation and stabilization of generalized hamiltonian systems. Submitted (1999); Preliminary version is in *Proc. 4th IFAC Symp. NOLCOS '98* pp. 544–549.

Gray, W. S. and J. M. A. Scherpen (1998). Hankel operators and gramians for nonlinear systems. *Proc. 37th IEEE Conf. on Decision and Control* pp. 3349–3353.

Gray, W. S. and J. M. A. Scherpen (1999). (siam) hankel operators and gramians for nonlinear systems. Submitted.

Isidori, A. (1995). *Nonlinear Control Systems*. third ed.. Springer-Verlag. Berlin.

Maschke, B. M. J. and A. J. van der Schaft (1992). Port-controlled hamiltonian systems: modelling origins and system-theoretic properties. *IFAC Symp. NOLCOS* pp. 282–288.

Scherpen, J. M. A. (1993). Balancing for nonlinear systems. *Systems & Control Letters* **21**, 143–153.

Scherpen, J. M. A. and A. J. van der Schaft (1994). Normalized coprime factorization and balancing for unstable nonlinear systems. *Int. J. Control* **60**(6), 1193–1222.

Scherpen, J. M. A. and W. S. Gray (1999). On singular value functions and hankel operators for nonlinear systems. *Proc. ACC'99*.

Zhou, K., J. C. Doyle and K. Glover (1996). *Robust and Optimal Control*. Prentice-Hall, Inc.. Upper Saddle River, N.J.

STABILIZATION OF INVARIANT SETS IN NONLINEAR SYSTEMS: CHETAEV'S BUNDLES AND SPEED-GRADIENT

Anton Shiriaev†‡, Alexander Fradkov †

†Institute for Problems of Mechanical Engineering, 61, Bolshoy, V.O., 199178, St. Petersburg, Russia, e-mail: alf@ccs.ipme.ru

‡The Maersk Mc-Kinney Moller Institute for Production Technology, University of Southern Denmark, Odense University Campusvej 55, DK - 5230 Odense M, Denmark

Abstract: The current results related to stabilization of sets based on Speed-Gradient method and the notion of V-detectability are overviewed. Extensions to nonaffine and cascaded systems are presented. *Copyright ©2000 IFAC*

Keywords: Nonlinear control, Stabilization of sets.

1 INTRODUCTION

The first statement of the stability of sets or partial stability problem was done by Lyapunov (1892), who however investigated only the special case of stability of a point. The Lyapunov-like conditions of stability with respect to a part of variables were given by Rumyantsev (1957). The problems of partial stability and stabilization were intensively studied in Russia, see (Rumyantsev and Oziraner, 1987; Vorotnikov, 1993; 1998). Most results exploit (as in classical stability case) properties of known scalar smooth function V - Lyapunov function. Under mild assumptions the partial stability (stability of invariant set) of the system

$$\dot{x} = f(x,t), \qquad f(t,0) = 0, \quad \forall t \geq 0, \tag{1.1}$$

is equivalent to existence of appropriate V.

For example, according to (Rumyantsev and Oziraner, 1987), Theorem 5.1 if the function $V(t,x)$ is y-positive definite, i. e. there exists a time-invariant continuous function $W_1(y)$ with

$$V(x,t) \geq W_1(y) > 0, \quad \forall x = (z,y), \ y \neq 0, \tag{1.2}$$

and the time derivative of V along the solution of the system (1.1) is nonpositive:

$$\dot{V}(t,x(t)) \leq 0. \tag{1.3}$$

then the solution $x = 0$ is y-stable, i. e. $\forall \varepsilon > 0$, $t_0 \geq$ there exists $\delta(\varepsilon, t_0) > 0$ such that if $|x_0| \leq \delta$ then $|y(t,t_0,x_0)| < \varepsilon$ for all $t \geq t_0$. If, in addition, the derivative (1.3) of V is y-negative definite function, i. e. $\exists W_2(y)$:

$$\dot{V}(x,t) \leq W_2(y) < 0, \quad \forall x = (z,y), \ y \neq 0, \tag{1.4}$$

then

$$\lim_{t \to +\infty} |y(t,t_0,x_0)| = 0.$$

However, finding Lyapunov function V is extremely difficult in general. One of a few powerful methods of constructing V-functions: is known as "Chetaev's method of Bundles of First Integrals", see (Chetaev, 1946; 1962; Rouche, *et al.*, 1977). In the Chetaev's approach the knowledge of k first integrals $V_i(t,x)$ for the system (1.1) is assumed and the Lyapunov function is taken in the form

$$V(t,x) = \sum_{i=1}^{k} \lambda_i V_i + \sum_{j=1}^{k} \mu_j V_j^2, \tag{1.5}$$

where λ_i, μ_j are constants to be chosen to satisfy the partial stability criteria. The method was initially proposed by N.G.Chetaev for study of stability. It was extended to the case of partial stability by Rumyantsev (1957).

In this paper the Chetaev's method is extended to partial stabilization (design) problems (Section 2).

To find a stabilizing feedback the Speed-Gradient method (Fradkov, 1979; 1990) is used based on its recent extensions to partial stabilization problem, see (Fradkov, *et al.*, 1997; Shiriaev, 1997; 1999; Fradkov and Pogromsky, 1998). In the Section 3 a number of stabilizability results for affine, non-affine and cascaded systems are presented which cover many applications to control of oscillatory mechanical systems.

2 CHETAEV'S METHOD AND SPPED-GRADIENT FOR STABILIZATION OF INVARIANT SETS

Consider the control system

$$\dot{x} = f(x,t) + g(x,t)u, \tag{2.6}$$

where u is a vector of control inputs. Let the problem be stabilization of a set described by the given values c_i of the known k-first integrals V_i of the unforced system (1.1). Assume that there exists a control law $u = u^*(x,t)$ such that for the function

$$V^*(x,t) = \sum_{i=1}^{k} \mu_i \left(V_i(x,t) - c_i\right)^2$$

with some $\mu_i > 0$ the relations (1.2), (1.4) hold, (where $\dot{V}$ is the derivative along trajectories of the closed loo system), then the $\{x : V^*(x,t) = 0\}$ is globally attractive set of the closed loop system. To establish existence of such a control law is often difficult and nontrivial problem. We suggest to apply the so called Speed-Gradient control law (Fradkov, 1979; 1990):

$$u = -\gamma \nabla_u \dot{V}^*(x,t). \tag{2.7}$$

This control law ensures that the value of V^* will not increase along trajectories of the closed loop system and, under mild assumptions, it implies that $V^*(x(t),t)$ will tend to constant value.

However the closed loop system (2.6), (2.7) does not fit into the settings of the partial stabilization problem. Indeed, the inequality (1.2) means that there exists globally defined transformation of state space $x \to \tilde{x}$ such that

$$V^*(\tilde{x},t) \geq \alpha\left(|\tilde{y}|\right), \quad \tilde{x} = (\tilde{z},\tilde{y}), \tag{2.8}$$

where α is $\mathcal{K}$-function; the derivative of V^* satisfies only the inequality (1.3), while to use the known partial stability results to establish attractivity of the set $\tilde{y} = 0$ it should satisfy the inequality (1.4). This obstacle gives rise to an important stabilization problem: *When the regulator (2.7) provides the stabilization of the set* $\{x : V^*(x,t) = 0\}$.

The main difficulty of the partial stabilization algorithm design is to find constructive conditions localizing the limit set of the closed loop system. Usually the desired limit set is contained in $\mathcal{D}_0 = \{x : V(x) = 0\}$, where $V \geq 0$ is a given goal function. Obviously, $\mathcal{D}_0$ does not necessarily coincide with $\mathcal{D}_1 = \{x : \frac{d}{dt}V(x) = 0\}$. In this case, standard arguments based on La Salle principle require the evaluation of the largest invariant set containing in $\mathcal{D}_1$. Therefore they do not provide explicit conditions ensuring that the trajectories of the closed loop system approach the set $\mathcal{D}_0$. Additionally, the existing methods of the partial stabilization (e. g. (Vorotnikov, 1998; Mahony, *et al.*, 1997) use mainly the feedback linearization technique that often leads to the local results and makes it difficult to meet the "small control" requirement.

The above obstacles were overcome in (Fradkov, *et al.*, 1997; Shiriaev and Fradkov, 1998; Shiriaev, 2000) based on the concept of V-detectability (Shiriaev, 1998), which extends the zero-state detectability, see (Byrnes, *et al.*, 1991), to the case of the partial stabilization.

Below we present a number of set stabilizability theorems summarizing the existing results.

3 THEOREMS ON STABILIZATION OF INVARIANT SETS

3.1 Stabilization of Sets for Affine Systems

Consider the nonlinear control system

$$\begin{aligned} \dot{x} &= f(x) + g(x)u, & (3.1)\\ y &= h(x), & (3.2) \end{aligned}$$

where $x \in X = R^n$ is the state, $u \in U = R^m$ is the control, $y \in Y = R^m$ is the measurable output; f, g, h are smooth vector fields.

Definition 3.1 *The system (3.1), (3.2) is passive if there exists a* C^0 *nonnegative storage function* $V : X \to R$, $V(0) = 0$, *such that for all piecewise continuous control vector functions* $u(s)$ *along the solution* $x(t) = x(t, x_0)$ *defined on the maximal interval* $[0, t_*)$ *the inequality*

$$V(x(t)) - V(x_0) \leq \int_0^t y(s)^T u(s) ds, \quad t \in [0, t_*), \tag{3.3}$$

holds. ∎

It is well known, see (Hill and Moylan, 1976), that in the case of a C^1 smooth function V the notion of passivity of the system (3.1), (3.2) can be expressed in infinitesimal form

$$\begin{aligned} L_f V(x) &\leq 0, \quad \forall x \in R^n & (3.4)\\ L_g V(x) &= h(x)^T, \quad \forall x \in R^n. & (3.5) \end{aligned}$$

We will use the notions of *zero-state detectability* and *V-detectability.*

Definition 3.2 *(Byrnes, et al., 1991) The system (3.1), (3.2) is said to be* locally zero-state detectable *if there exists a neighbourhood* $\mathcal{U}$ *of* 0 *such that for all* $x_0 \in \mathcal{U}$

$$h(x(t, x_0, 0)) = 0 \quad \forall t \geq 0 \Rightarrow \lim_{t \to +\infty} x(t, x_0, 0) = 0, \tag{3.6}$$

where $x(t, x_0, 0)$ *is a solution of (3.1) with* $u(t) \equiv 0$. *If* $\mathcal{U} = R^n$, *the system (3.1), (3.2) is* zero-state detectable. ∎

Definition 3.3 *(Shiriaev, 1998) The system (3.1), (3.2) is said to be locally V-detectable if there exists a positive constant* $c > 0$ *such that for all* $x_0 \in \mathcal{U} = \{x \in R^n : V(x) < c\}$

$$h(x(t, x_0, 0)) = 0 \ \forall t \geq 0 \Rightarrow \\ \lim_{t \to +\infty} V(x(t, x_0, 0)) = 0, \tag{3.7}$$

where $x(t, x_0, 0)$ *is a solution of (3.1) with* $u(t) \equiv 0$. *If* $\mathcal{U} = R^n$, *the system (3.1), (3.2) is V-detectable.* ∎

The notion of V-detectability is natural and extends the notion of zero-state detectability. Indeed, if the function V is positive definite then these notions are equivalent.

The following statement describes a basic stabilizability property of passive systems. Recall that a function V is said to be *proper* if the set $V_c = \{x \in R^n : V(x) \leq c\}$ is compact for all $c \geq 0$.

Theorem 3.4 *Suppose system (3.1), (3.2) is passive with a nonnegative storage function* V. *Let* $\phi : Y \to U$ *be any smooth function such that* $\phi(0) = 0$, $y^T\phi(y) > 0$ *for each nonzero* y, *and take as feedback control*

$$u = -\phi(y). \tag{3.8}$$

Then
1.) If the set $V_0 = \{x \in R^n : V(x) = 0\}$ *is compact and the system (3.1), (3.2) is locally V-detectable then* V_0 *is a locally asymptotically stable set of the closed loop system (3.1), (3.2), (3.8);*
2.) If the function V *is proper and the system (3.1), (3.2) is V-detectable then* V_0 *is a globally asymptotically stable set of the closed loop system (3.1), (3.2), (3.8).*

The assumption that the storage function V is proper can be weakened in diferent ways. E.g. for the system (3.1), (3.2), evolving on a smooth manifold, one can only assume that V is a proper function on this manifold. This situation takes place in Hamiltonian controlled systems with some periodic coordinates. In fact we need to guarantee forward completeness of the closed loop system (3.1), (3.2), (3.8) and the existence of a compact ω-limit set for any solution, e.g. as follows.

Theorem 3.5 *Let the system (3.1), (3.2) be passive with a* C^r, $r \geq 1$, *smooth nonnegative storage function* V. *Suppose for any* $c > 0$ *the functions* $f(x)$, $g(x)$, $h(x)$, $\partial h(x)/\partial x$ *are bounded on the set* $V_c = \{x \in R^n : V(x) \leq c\}$, *and there exists a constant* δ, $\delta = \delta(c) > 0$, *such that any connected component of the set*

$$D_c = V_c \bigcap \{x \in X : |h(x)| \leq \delta\} \tag{3.9}$$

is compact. Suppose $\phi : Y \to U$ *is a smooth function with*

$$C_1|y|^2 \leq y^T\phi(y)| \leq C_2|y|^2 \quad \forall y,$$

where $C_2 \geq C_1 > 0$. *Take feedback control in the form (3.8). Then*
1). If the system (3.1), (3.2) is locally V-detectable then the set $V_0 = \{x \in R^n : V(x) = 0\}$ *is a locally asymptotically stable set of the closed loop system (3.1), (3.2), (3.8);*
2.) If the system (3.1), (3.2) is V-detectable then then V_0 *is a globally asymptotically stable set of the closed loop system (3.1), (3.2), (3.8).*

The condition of (local) V-detectability involved in theorem 3.4 and theorem 3.5 is nonconstructive and requires additional investigation. In the next statements we suggest three different sufficient conditions which imply V-detectability of (3.1), (3.2). Given a C^r, $r \geq 1$, smooth function $V(x)$, introduce the set

$$S = \{x \in R^n : L_{f_a}^m L_\tau V(x) = 0, \\ \text{for all } \tau \in D,\ 0 \leq m \leq r-1\}, \tag{3.10}$$

where D is a distribution which is defined by

$$D = \text{span}\{ad_{f_a}^k g_i, 0 \leq k \leq n-1, 1 \leq i \leq m\} \tag{3.11}$$

where $\text{span}\{v_1, v_2, \ldots\}$ is a linear space generated by columns of matrices v_1, v_2, ..., and g_i, $1 \leq i \leq m$ are the vector components of the smooth vector field $g_a(x)$.

Proposition 3.6 *Suppose that the system (3.1), (3.2) is passive with a* C^r, $r \geq 1$, *smooth proper nonnegative storage function* V. *If the set*

$$S \setminus \{x \in R^n : V(x) = 0\} \tag{3.12}$$

does not contain any whole trajectory of the unforced system (3.1) then the system (3.1), (3.2) is V-detectable.

To formulate another criteria of V-detectability let us assume that the C^r-smooth scalar function $V(x)$, $r \geq 1$, admits the factorization

$$V(x) = \frac{1}{2}|w(x)|^2, \qquad (3.13)$$

where $w : R^n \to R^l$, $l \leq n$, is a C^r-smooth vector function. Introduce the distribution $\hat{S}(x)$ as follows

$$\hat{S}(x) = \text{span}\{L_{f_a}^k L_{g_a} w(x),\ k = 0, 1, \ldots, r-1\}. \qquad (3.14)$$

It is easy to see that the dimension of the distribution $\hat{S}(x)$ cannot be greater than l for any point $x \in R^n$. Denote the set

$$\Lambda_1 = \{x \in R^n : (L_{f_a} w(x))^T L_{f_a}^k L_{g_a} w(x) = 0, \quad k = 0, 1, \ldots, r-1\}, \qquad (3.15)$$

and consider the set P_1 where the distribution $\hat{S}(x)$ has the maximal dimension, i. e.

$$P_1 = \{x \in \Lambda_1 : \dim \hat{S}(x) = l\}. \qquad (3.16)$$

Let the set Ω_1 be defined as follows

$\Omega_1 = \{x_0 \in X = R^n \setminus (P_1 \cup \{x \in R^n : V(x) = 0\}) :$ the whole trajectory $x = x(t, x_0)$ of (3.1) with $u = 0$ lies in $X\}$.

Proposition 3.7 *Suppose the system (3.1), (3.2) is passive with a proper C^r, $r \geq 1$, smooth storage function V, which can be presented in the form (3.13). If the set Ω_1 is empty then (3.1), (3.2) is V-detectable.*

Denote the sets

$$\Lambda_2 = \{x \in R^n : w(x)^T L_{f_a}^k L_{g_a} w(x) = 0, \quad k = 0, 1, \ldots, r-1\},$$

$$P_2 = \{x \in \Lambda_2 : \dim \hat{S}(x) = l\},$$

$\Omega_2 = \{x_0 \in X = R^n \setminus (P_2 \cup \{x \in R^n : V(x) = 0\}) :$ the whole trajectory $x = x(t, x_0)$ of (3.1) with $u = 0$ lies in $X\}$.

Proposition 3.8 *Suppose the system (3.1), (3.2) is passive with a proper C^r, $r \geq 1$, smooth storage function V which can be presented in the form (3.13). If the set Ω_2 is empty then (3.1), (3.2) is V-detectable.*

Propositions 3.6–3.8 will also remain true if one only assumes that any solution $x(t) = x(t, x_0, 0)$ of the unforced system (3.1) subjected to the constraint $y(t) = h(x(t)) = 0$ is well defined on $[0, +\infty)$ and has a nonempty compact ω-limit set. ■

3.2 Stabilization of Sets for Nonaffine Systems

Consider non-affine nonlinear system

$$\dot{x} = f(x, u), \qquad (3.17)$$
$$y = h(x, u). \qquad (3.18)$$

Here $x \in X = R^n$ is the state, $u \in U = R^m$ is the control, $y \in Y = R^m$ is the measurable output. The notion of passivity of non-affine in control nonlinear system (3.17), (3.18) with a C^0 smooth nonnegative storage function V is the same as in the case of affine in control system, see definition 3.1. But the infinitesimal variant of passivity with a C^1 smooth nonnegative storage function differs from the relations (3.4), (3.5). Let us formulate an almost obvious necessary condition for passivity of (3.17), (3.18). Given a C^1 smooth function $V(x)$, denote

$$f_a(x) = f(x, 0), \quad g_a(x) = \tfrac{\partial f(x,0)}{\partial u},$$
$$\mathcal{S} = \{x \in R^n : L_{f_a} V(x) = 0\}. \qquad (3.19)$$

Proposition 3.9 *(Lin, 1996) Let the system (3.17), (3.18) be passive with a C^1 smooth storage function V. Then the following relations*

$$L_{f_a} V(x) \leq 0, \quad \forall x \in R^n, \qquad (3.20)$$
$$L_{g_a} V(x) = h(x, 0)^T, \quad \forall x \in \mathcal{S} \qquad (3.21)$$

hold.

Definition 3.10 The system (3.17), (3.18) is said to be *affine-passive* with storage function V if the the relations (3.20), (3.21) hold.

Due to smoothness of f the system (3.17) can be globally rewritten in a new 'linearized' form. Such transformation is based on Hadamard lemma, see (Hartman, 1982) Lemma 3.1, p. 97.

$$f(x, u) = f_a(x) + g_a(x)u + \sum_{j=1}^{m} u_j(R_j(x, u)u), \qquad (3.22)$$

where u_j are the components of the vector u and R_j, $j = 1, \ldots, m$ are smooth. Denote $\rho(x)$ any scalar function satisfying the inequality

$$\max_{|u| \leq 1} |R_j(x, u)| \leq \rho(x), \quad j = 1, \ldots, m, \quad \forall x, \qquad (3.23)$$

The stabilizability conditions derived in this Section for the system (3.17), (3.18) essentially rely on properties of its affine part. Introduce the auxiliary affine system

$$\dot{x} = f_a(x) + g_a(x)u, \qquad (3.24)$$
$$y = h_a(x) = h(x, 0). \qquad (3.25)$$

Theorem 3.12 *Let the system (3.17), (3.18) be affine-passive with a C^r, $r \geq 1$, smooth storage function V. Let $\Phi : Y \to U$ be any smooth function such that*

$$y^T\Phi(y) \geq \psi(y)||\Phi(y)||^2 > 0,\ \forall y \neq 0,\ ||\Phi(y)|| \leq 1, \tag{3.26}$$

where $\psi(y)$ is some scalar function with $\psi(y) \geq \psi_0 > 0$, $\forall y$. Suppose $\alpha(x)$ is an arbitrary scalar smooth function such that $0 < \alpha(x) \leq \hat{\alpha}(x)$, where

$$\hat{\alpha}(x) = \frac{\beta}{m\left(1 + [||\partial V(x)/\partial x||\rho(x)]^2\right)}, \tag{3.27}$$

and β is a positive number, $0 < \beta < \min\{\psi_0, m\}$. Take the regulator

$$u = -\alpha(x)\Phi([L_{g_a}V(x)]^T). \tag{3.28}$$

Then
1.) If the set $V_0 = \{x \in R^n : V(x) = 0\}$ is compact and the system (3.24), (3.25) is locally V-detectable then V_0 is a locally asymptotically stable set of the closed loop system (3.17), (3.18), (3.28);
2.) If the function V is proper and the system (3.24), (3.25) is V-detectable then V_0 is a globally asymptotically stable set of the closed loop system (3.17), (3.18), (3.28).

3.3 Stabilization of Sets for Cascaded Systems

Consider a cascaded nonlinear system

$$\begin{aligned} \dot{x} &= f(x) + g(x)v, & (3.29)\\ \dot{v} &= a(x,v) + b(x,v)u, & (3.30)\\ y &= h(x), & (3.31) \end{aligned}$$

where $x \in \mathbf{R}^n$, $v \in \mathbf{R}^m$ are state vectors of subsystems, $u \in \mathbf{R}^m$ is control input vector, $y \in \mathbf{R}^m$ is output vector; $f(x)$, $g(x)$, $a(x,v)$, $b(x,v)$ are smooth vector fields of appropriate dimensions.

It is assumed that in the state space of the system (3.29) a goal set $\mathcal{S}$ is defined and that this set $\mathcal{S}$ is contained in the set of zeros of known smooth scalar function V, i. e. $\mathcal{S} \subset \{x : V(x) = 0\}$. Moreover, it is assumed that the subsystem (3.29), (3.31) is passive (or passifiable by some smooth feedback) with the storage function V. The problem is to find output feedback regulator and sufficient conditions which guarantee the control goal

$$\lim_{t\to+\infty} V(x(t)) = 0. \tag{3.32}$$

First consider a special case of problem, namely, suppose that the system of interest has the simplified form

$$\begin{aligned} \dot{x} &= f(x) + g(x)v, & (3.33)\\ \dot{v} &= u, & (3.34)\\ y &= h(x), & (3.35) \end{aligned}$$

that is the first system in cascade is the block of integrators.

Theorem 3.13 *Suppose that the system (3.33), (3.35) is passive with C^r, $r \geq 1$, smooth proper storage function V. If the system (3.33), (3.35) is V-detectable then for any solution of the closed loop system (3.33), (3.34), (3.35) with the regulator*

$$u = -Pv - y, \tag{3.36}$$

where P is an arbitrary positive definite $m \times m$ matrix, $P = P^* > 0$, the goal relation (3.32) is valid. Moreover,

$$\lim_{t\to+\infty} |v(t)|^2 = 0. \tag{3.37}$$

Now consider the general case.

Theorem 3.14 *Consider the system (3.29), (3.30), (3.31) under the following assumptions:*
1). the 2nd subsystem is passifiable, i. e. there exists smooth locally bounded function $\alpha(x)$ such that

$$\begin{aligned} &\tfrac{\partial V}{\partial x}[f(x) + g(x)\alpha(x)] \leq 0,\\ &L_gV(x) = h(x)^T, \quad \forall x \in \mathbf{R}^n, \end{aligned} \tag{3.38}$$

where $V(x)$ is some nonnegative proper function;
2). $\det b(v,x) \neq 0\ \forall v, x$;
3). the system

$$\dot{x} = (f(x) + g(x)\alpha(x)) + g(x)v, \quad y = h(x) \tag{3.39}$$

is V-detectable.
Then for any solution $[x(t), v(t)]$ of the closed loop system (3.29), (3.30), (3.31) with the regulator

$$\begin{aligned} u = b(v,x)^{-1}&\left[-P(v - \alpha(x)) + \right.\\ &\left.\tfrac{\partial \alpha}{\partial x}(f(x) + g(x)v) - y - a(v,x)\right] \end{aligned} \tag{3.40}$$

where P is an arbitrary positive definite $m \times m$ matrix, the goal relation (3.32) is valid. Moreover, $\lim_{t\to+\infty} |v(t) - \alpha(x(t))|^2 = 0$.

4 CONCLUSIONS

The above theorems constitute a kind of toolbox for partial stabilization problems. They unify and extend some previous results which were applied to control of oscillatory behavior of various mechanical systems. E.g. the conitions of achievment of given energy level were obtained for various pendular systems (Fradkov and Pogromsky, 1998); (Shiriaev et al., 1998); (Ludvigsen et al., 1999); (Shiriaev and Fradkov, 1999); (Shiriaev et al., 2000).

The work was supported in part by RFBR, project 99-01-00672.

REFERENCES

Byrnes, C.I., A. Isidori and J.C. Willems (1991). Passivity, feedback equivalence, and the global stabilization of minimum phase nonlinear systems. *IEEE Trans. Autom. Contr.*, **AC-36**, 1228–1240.

Chetaev, N.G. (1946). *Stability of Motion*, Gostekhizdat, Moscow, Russia (in Russian).

Chetaev, N.G. (1962). *Stability of Motion: Papers on Analytic Mechanics*, Izd. Akad Nauk SSSR, Moscow, Russia (in Russian).

Fradkov, A.L. (1979). Speed-gradient scheme and its applications in adaptive control, *Autom. Remote Control*, **40**(9), 1333–1342.

Fradkov, A.L. (1990). *Adaptive control of complex systems*. Nauka, Moscow, Russia (in Russian).

Fradkov, A.L., I.A. Makarov, A.S. Shiriaev, and O.P. Tomchina. (1997). Control of oscillations in Hamiltonian systems. In *Proceedings of the 4th European Control Conference*, Brussels, Belgium.

Fradkov, A.L., and A.Yu. Pogromsky (1998). *Introduction to Control of Oscillations and Chaos*. World Scientific, Singapore.

Hartman, P. (1982). *Ordinary differential equations*. 2nd edition, Birkhäuser.

Hill, D., and P. Moylan. (1976) The stability of nonlinear dissipative systems. *IEEE Transactions on Automatic Control*, **21**, 708–711.

Lin, W. (1996). Global asymptotic stabilization of general nonlinear systems with stable free dynamics via passivity and bounded feedback. *Automatica*, **32**(16), 915–924.

Ludvigsen, H., A.S. Shiriaev, and O. Egeland. (1999) Stabilization of stable manifold of upright position of spherical pendulum. In *American Control Conference*, 4039–4044, San Diego, USA.

Lyapunov, A.M. (1892). The general problem of the stability of motion. *Kharkov Mathematical Society*. English translation: *International Journal of Control*, **55**, 531–775, 1992.

Mahony, R., I. Mareels, G. Bastin, and G. Campion. (1997). Output stabilization of square nonlinear systems. *Automatica*, **33**(8), 1571–1577, 1997.

Rouche, N., P. Habets, and M. Laloy. (1977) *Stability Theory by Lyapunov's Direct Method*, Springer, New York.

Rumyantsev, V.V., and A.S. Oziraner. (1987). *Partial Stability and Stabilization*, IL, Moscow, Russia (in Russian).

Rumyantsev, V.V. (1957) Stability of motion with respect to a part of variables. *Vestnik MGU*, (4), 9–16, Moscow, Russia (in Russian).

Shiriaev, A.S. (1998). The notion of V-detectability and stabilization of invariant sets of nonlinear systems. *Proc. of 37th CDC*, 2509–2514, Tampa, IEEE. To appear in: *Systems & Control Letters*, 2000.

Shiriaev, A.S. (2000). Stabilization of compact sets for passive affine nonlinear systems. To appear in: *Automatica*, 2000.

Shiriaev, A.S., and A.L. Fradkov. (1998). Stabilization of invariant manifolds for nonaffine nonlinear systems. In *Proceedings of the 4th IFAC Symposium on Nonlinear Control Systems Design*, **1**, 215–220, Enschede, The Netherlands.

Shiriaev, A.S., and A.L. Fradkov. (1999). Stabilization of invariant sets of cascaded nonlinear systems. *Proc. CDC'99*, IEEE.

Shiriaev, A.S., H. Ludvigsen, and O. Egeland. (1999). Swinging up of the spherical pendulum. In: *Preprints of the 14th IFAC World Congress*, **E**, 65–70.

Shiriaev, A.S., A. Pogromsky, H. Ludvigsen, and O. Egeland. (2000). On global properties of passivity based control of an inverted pendulum. To appear in: *International Journal of Robust & Nonlinear Control.*

Vorotnikov, V.I. (1993). Stability and stabilization of motion: research approaches, results, distinctive characteristics. *Automation and Remote Control*, (3), 3–62.

Vorotnikov, V.I. (1998). *Partial Stability and Control*, Birkhauser Verlag.

ASYMPTOTIC STABILIZATION OF EULER-POINCARÉ MECHANICAL SYSTEMS

Anthony M. Bloch *,[1] **Dong Eui Chang** ***,[3]
Naomi E. Leonard **,[2] **Jerrold E. Marsden** ***,[3]
Craig Woolsey **,[2]

* *Department of Mathematics, University of Michigan, Ann Arbor, MI 48109*
** *Department of Mechanical and Aerospace Engineering, Princeton University, Princeton, NJ 08544*
*** *Control and Dynamical Systems, California Institute of Technology 107-81, Pasadena, CA 91125*

Abstract: Stabilization of mechanical control systems by the method of controlled Lagrangians and matching is used to analyze asymptotic stabilization of systems whose underlying dynamics are governed by the Euler-Poincaré equations. In particular, we analyze asymptotic stabilization of a satellite. *Copyright© 2000 IFAC*

Keywords: Feedback stabilization, Lyapunov methods, Nonlinear control, Dissipation

1. INTRODUCTION

This paper develops a constructive approach to the determination of asymptotically stabilizing control laws for a class of Lagrangian mechanical systems with symmetry – systems described by the Euler-Poincaré equations. This work complements and extends the class of systems discussed in Bloch, Leonard and Marsden [1997, 1998, 1999a,b, 2000a]. Here we concentrate on the details of asymptotic stability. The paper is complementary to Bloch, Leonard and Marsden [2000b] and uses some of the ideas from Bloch, Chang, Leonard and Marsden [2000] and Woolsey and Leonard [1999].

The specific case we consider is that in which the configuration space is $Q = H \times G$, where H is a Lie group and G is an *Abelian* Lie group. We also assume that the Lagrangian $L : TQ \to \mathrm{R}$ is left invariant under both G and H, so the G variables are cyclic and the controls act only on these cyclic variables.

As in our previous analysis, the guiding principle behind our methodology is to begin by considering a class of control laws that yield closed-loop dynamics which remain in Lagrangian form. The goal with this first step is to achieve stabilization within the class of conservative systems. Secondly, we append controls that are dissipative in nature to turn the conservative stabilization into asymptotic stabilization.

2. EULER–POINCARÉ MATCHING AND STABILIZATION.

In this section we recall the Euler-Poincaré matching theorem. This will be illustrated in §4 by the spacecraft with rotors.

The spacecraft example has two symmetry groups associated with it, as do many other examples.

[1] Research partially supported by NSF grant, DMS-9803181, AFOSR grant F49620-96-1-0100, and an NSF group infrastructure grant at the University of Michigan
[2] Research partially supported by NSF grant BES-9502477 and ONR grant N00014-98-1-0649
[3] Research partially supported by AFOSR grant F49620-95-1-0419

One group, which in this case is the nonabelian group SO(3), is associated with the rotational symmetry of the overall problem and another group, an Abelian group, is the product of several copies of S^1 associated to the rotors, which are also the control directions. In this section, we use this setting to get a concrete and readily implementable Euler–Poincaré matching theorem.

2.1 *Euler–Poincaré Matching.*

Let $L : T(H \times G) \to \mathrm{R}$ denote a given left invariant Lagrangian and $l : \mathrm{h} \times TG \to \mathrm{R}$ be the restriction of L to the Lie algebra of H. For a curve $h(t) \in H$ let $\eta(t) = T_{h(t)}L_{h(t)^{-1}}\dot{h}$, which we also write as $\eta(t) = h(t)^{-1}\dot{h}(t)$. We consider Lagrangians that are purely kinetic energy Lagrangians, and correspondingly, the (reduced) Lagrangian takes the form

$$l(\eta^\alpha, \dot{\theta}^a) = \frac{1}{2}g_{\alpha\beta}\eta^\alpha\eta^\beta + g_{\alpha a}\eta^\alpha\dot{\theta}^a + \frac{1}{2}g_{ab}\dot{\theta}^a\dot{\theta}^b. \tag{2.1}$$

Here η^α are the variables in h and θ^a are the control variables. Note that $g_{\alpha\beta}$, $g_{\alpha a}$ and g_{ab} are all constant (fixed) matrices.

The cyclic variables θ^a in G give rise to the conserved quantity

$$J_a = \frac{\partial l}{\partial \dot{\theta}^a} = g_{a\alpha}\eta^\alpha + g_{ab}\dot{\theta}^b. \tag{2.2}$$

The equations of motion for the control system where the controls u_a act in the θ^a directions are the ***controlled Euler-Poincaré equations***:

$$\frac{d}{dt}\frac{\partial l}{\partial \eta^\alpha} = c^\beta_{\alpha\delta}\eta^\delta\frac{\partial l}{\partial \eta^\beta} \tag{2.3}$$

$$\frac{d}{dt}\frac{\partial l}{\partial \dot{\theta}^a} = u_a \tag{2.4}$$

where $c^\beta_{\alpha\delta}$ are the structure constants of the Lie algebra h.

We choose the ***controlled Lagrangian*** to be

$$\begin{aligned} l_{\tau,\sigma,\rho} &= l(\eta^\alpha, \dot{\theta}^a + \tau^a_\alpha\eta^\alpha) + \frac{1}{2}\sigma_{ab}\tau^a_\alpha\tau^b_\beta\eta^\alpha\eta^\beta \\ &+ \frac{1}{2}(\rho_{ab} - g_{ab})(\dot{\theta}^a + g^{ac}g_{c\alpha}\eta^\alpha + \tau^a_\alpha\eta^\alpha) \\ &\cdot (\dot{\theta}^b + g^{bc}g_{c\beta}\eta^\beta + \tau^b_\beta\eta^\beta). \end{aligned} \tag{2.5}$$

To preserve symmetry τ^a_α, σ_{ab} and ρ_{ab} are constant matrices.

From (2.5) we find that the ***controlled conserved quantity*** is given by

$$\tilde{J}_a = \frac{\partial l_{\tau,\sigma,\rho}}{\partial \dot{\theta}^a} = \rho_{ab}(\dot{\theta}^b + g^{bc}g_{c\alpha}\eta^\alpha + \tau^b_\alpha\eta^\alpha). \tag{2.6}$$

The controlled Lagrangian prescribes the closed-loop system, i.e., the closed-loop dynamics are the Euler-Poincaré equations corresponding to $l_{\tau,\sigma,\rho}$:

$$\frac{d}{dt}\frac{\partial l_{\tau,\sigma,\rho}}{\partial \eta^\alpha} = c^\beta_{\alpha\delta}\eta^\delta\frac{\partial l_{\tau,\sigma,\rho}}{\partial \eta^\beta} \tag{2.7}$$

$$\frac{d}{dt}\frac{\partial l_{\tau,\sigma,\rho}}{\partial \dot{\theta}^a} = 0. \tag{2.8}$$

To effect this closed-loop system, the control inputs u_a must be chosen so that (2.4) and (2.8) are equivalent. Additionally, the controlled Lagrangian must satisfy matching conditions, i.e., it must be chosen so that (2.3) and (2.7) are equivalent.

We make the following assumptions:

Assumption EP-1. $\tau^a_\alpha = -\sigma^{ab}g_{b\alpha}$.

Assumption EP-2. $\sigma^{ab} + \rho^{ab} = g^{ab}$.

The following is proved in Bloch, Leonard and Marsden [2000b].

Theorem 1. Under the assumptions EP-1 and EP-2, the Euler-Poincaré equations for the controlled Lagrangian (2.7)-(2.8) coincide with the controlled Euler-Poincaré equations (2.3)-(2.4).

The control law u_a can be determined by comparing (2.4) to the controlled conservation law

$$\frac{d}{dt}\frac{\partial l_{\tau,\sigma,\rho}}{\partial \dot{\theta}^a} = \frac{d}{dt}\tilde{J}_a = 0 \tag{2.9}$$

where $\tilde{J}_a$ is given by (2.6). We find

$$u_a = g_{ab}\sigma^{bc}g_{c\alpha}\dot{\eta}^\alpha. \tag{2.10}$$

To get a control law that is a function of velocities rather than accelerations we can substitute for accelerations from the Euler-Poincaré equations. This yields

$$\begin{aligned} u_a &= k^\alpha_a\frac{d}{dt}\frac{\partial l}{\partial \eta^\alpha} = k^\alpha_a c^\psi_{\alpha\delta}\eta^\delta\frac{\partial l}{\partial \eta^\psi} \\ &= k^\alpha_a c^\psi_{\alpha\delta}\eta^\delta(g_{\psi\beta}\eta^\beta + g_{\psi b}\dot{\theta}^b), \end{aligned} \tag{2.11}$$

where k^α_a are control gains defined by

$$k^\alpha_a = D_{ab}\sigma^{bc}g_{c\beta}B^{\alpha\beta}, \tag{2.12}$$

$$B_{\alpha\beta} = g_{\alpha\beta} - g_{\alpha b}g^{ab}g_{a\beta}, \tag{2.13}$$

$$D^{ba} = g^{ba} + \sigma^{bc}g_{c\beta}B^{\alpha\beta}g_{\alpha e}g^{ae}. \tag{2.14}$$

2.2 *Euler–Poincaré Stabilization*

We now use the energy-Casimir method to determine stability (see e.g. Marsden and Ratiu [1994]). Recall that for mechanical systems, an eigenvalue analysis alone is not sufficient for determining stability.

Define l_0 on h by

$$l_0(\eta^\alpha) = \frac{1}{2}g_{\alpha\beta}\eta^\alpha\eta^\beta. \tag{2.15}$$

A (relative) equilibrium η_e for the corresponding dynamical equations satisfies the equation

$$c_{\alpha\delta}^{\beta}\eta^{\delta}\frac{\partial l_0}{\partial\eta^{\beta}} = 0. \qquad (2.16)$$

Now suppose we have a collection of Casimir functions $C^1(M_\alpha), \cdots, C^m(M_\alpha)$ where

$$M_\alpha = \frac{\partial l_0}{\partial\eta^\alpha} = g_{\alpha\beta}\eta^\beta.$$

Now set

$$E_\Phi = l_0 + \Phi(C^1, \cdots C^m). \qquad (2.17)$$

We require that the Casimir functions be chosen so that the first variation of E_Φ vanishes at equilibrium, i.e.,

$$\delta(E_\Phi)|_{\eta_e} = g_{\alpha\beta}\delta\eta^\beta\left(\eta^\alpha + \sum_{k=1}^{m}(D_k\Phi)\frac{\partial C^k}{\partial M_\alpha}\right)\Bigg|_e = 0. \qquad (2.18)$$

The second variation at equilibrium is given by

$$\delta^2(E_\Phi)|_{\eta_e} = (g_{\alpha\beta} + g_{\alpha\mu}H^{\mu\nu}g_{\nu\beta})\,\delta\eta^\alpha\delta\eta^\beta, \quad (2.19)$$

$$H^{\mu\nu} = \left(\sum_{k,j=1}^{m}(D_{kj}\Phi)\frac{\partial C^k}{\partial M_\mu}\frac{\partial C^j}{\partial M_\nu} + \sum_{k=1}^{m}(D_k\Phi)\frac{\partial^2 C^k}{\partial M_\mu\partial M_\nu}\right)\Bigg|_e. \qquad (2.20)$$

Now consider the full uncontrolled Lagrangian l. Using (2.3), the full system will still have η_e as an equilibrium together with $\dot\theta_e^a$ provided

$$c_{\alpha\delta}^{\beta}\eta_e^\delta\left(g_{\beta\delta}\eta_e^\delta + g_{\beta a}\dot\theta_e^a\right) = 0. \qquad (2.21)$$

This is satisfied if $c_{\alpha\delta}^{\beta}\eta_e^\delta g_{\beta a}\dot\theta_e^a = 0$ and in particular if $\dot\theta_e^a = 0$.

This implies, from our matching conditions, that $l_{\tau,\sigma,\rho}$ also has this equilibrium. Set

$$\begin{aligned}\tilde M_\alpha = \frac{\partial l_{\tau,\sigma,\rho}}{\partial\eta^\alpha} &= \frac{\delta l}{\delta\eta^\alpha} = g_{\alpha\beta}\eta^\beta + g_{\alpha a}\dot\theta^a\\ &= G_{\alpha\beta}\eta^\beta + g_{\alpha a}\rho^{ab}\tilde J_b \end{aligned} \qquad (2.22)$$

using (2.6) and where we define

$$G_{\alpha\beta} = g_{\alpha\beta} - g_{a\alpha}\rho^{ab}g_{b\beta}.$$

For stability of the controlled system we use

$$E_{\tilde\Phi} = l_{\tau,\sigma,\rho} + \tilde\Phi\left(C^k(\tilde M_\alpha), \tilde J^a\right) \qquad (2.23)$$

where $\tilde\Phi$ is a smooth function.

We next compute the first and second variations of $E_{\tilde\Phi}$. Using again the conserved quantities $\tilde J_a$ and assumption EP-2, we get

$$\begin{aligned}l_{\tau,\sigma,\rho} &= \frac{1}{2}\left(g_{\alpha\beta} + g_{a\alpha}\left(\sigma^{ab} - g^{ab}\right)g_{b\beta}\right)\eta^\alpha\eta^\beta + \frac{1}{2}\rho^{ab}\tilde J_a\tilde J_b\\ &\equiv \frac{1}{2}G_{\alpha\beta}\eta^\alpha\eta^\beta + \frac{1}{2}\rho^{ab}\tilde J_a\tilde J_b. \end{aligned} \qquad (2.24)$$

The first variation is

$$\begin{aligned}\delta E_{\tilde\Phi} &= G_{\alpha\beta}\delta\eta^\beta\left(\eta^\alpha + \sum_{k=1}^{m}\left(D_k\tilde\Phi\right)\frac{\partial C^k}{\partial M_\alpha}\right)\Bigg|_e\\ &+ \delta\tilde J_a\left(\rho^{ab}\tilde J_b + D_{m+a}\tilde\Phi + \sum_{k=1}^{m}\left(D_k\tilde\Phi\right)\frac{\partial C^k}{\partial M_\alpha}g_{\alpha b}\rho^{ab}\right)\Bigg|_e = 0. \end{aligned} \qquad (2.25)$$

Thus, we require

$$\left(\sum_{k=1}^{m}\left(D_k\tilde\Phi\right)\frac{\partial C^k}{\partial\tilde M_\alpha}\right)\Bigg|_e = -\eta_e^\alpha, \qquad (2.26)$$

$$D_{m+a}\tilde\Phi|_e = \left(-\rho^{ab}\tilde J_b + \rho^{ab}g_{\alpha b}\eta^\alpha\right)\Big|_e = -\dot\theta^a|_e \qquad (2.27)$$

Similarly, we can compute the second variation. Consider the case (apparently sufficient for applications) where

$$\tilde\Phi\left(C^1, \cdots, C^m, \tilde J^a\right) \equiv \Phi\left(C^1, \cdots, C^m\right) + \Psi\left(\tilde J^a\right). \qquad (2.28)$$

Accordingly, (2.27) becomes

$$D_a\Psi|_e = -\dot\theta^a|_e. \qquad (2.29)$$

Now define

$$\tilde H^{\alpha\beta} = \left(\sum_{k,j=1}^{m}D_{kj}\Phi\frac{\partial C^k}{\partial\tilde M_\alpha}\frac{\partial C^j}{\partial\tilde M_\beta} + \sum_{k=1}^{m}D_k\Phi\frac{\partial^2 C^k}{\partial\tilde M_\alpha\partial\tilde M_\beta}\right)\Bigg|_e. \qquad (2.30)$$

Then, the second variation is given by

$$\delta^2 E_{\tilde\Phi}|_e \equiv \delta^2 E_{\Phi,\Psi}|_e \qquad (2.31)$$

$$\begin{aligned} &= N_{\alpha\beta}\delta\eta^\alpha\delta\eta^\beta + 2P_\alpha^a\delta\eta^\alpha\delta\tilde J_a + R^{ab}\delta\tilde J_a\delta\tilde J_b\\ &= N_{\alpha\beta}\delta\xi^\alpha\delta\xi^\beta + \left(R^{ab} - N^{\gamma\delta}P_\gamma^a P_\delta^b\right)\delta\tilde J_a\delta\tilde J_b, \end{aligned} \qquad (2.32)$$

where

$$N_{\alpha\beta} = G_{\alpha\beta} + G_{\alpha\gamma}\tilde H^{\gamma\delta}G_{\delta\beta},$$

$$P_\alpha^a = G_{\alpha\gamma}\tilde H^{\gamma\delta}g_{\delta b}\rho^{ab},$$

$$R^{ab} = \rho^{ab} + \left(\frac{\partial^2\Psi}{\partial\tilde J_a\partial\tilde J_b}\right)\Bigg|_e + \tilde H^{\alpha\beta}g_{\alpha c}\rho^{ac}g_{\beta d}\rho^{bd},$$

$$\delta\xi^\alpha = \delta\eta^\alpha + N^{\alpha\beta}P_\beta^b\delta\tilde J_b.$$

Definiteness of this quantity at the given equilibrium implies nonlinear stability. Using the freedom in choosing $\left(\frac{\partial^2\Psi}{\partial\tilde J_a\partial\tilde J_b}\right)\Big|_e$ we can make the second term on the right hand side of (2.32) have whatever definiteness we require. Then, stability will be guaranteed if we can choose ρ_{ab} such that $N_{\alpha\beta}$ is definite (under the restrictions that (2.26) and (2.27) are satisfied).

Theorem 2. Let η_e be an equilibrium for the uncontrolled dynamics given by l_0 (2.15). Suppose that $\dot{\theta}_e$ satisfies (2.21). Then, $(\eta_e, \dot{\theta}_e)$ is an equilibrium for the controlled system described by $l_{\tau,\sigma,\rho}$ (2.24). This equilibrium is Lyapunov stable for the controlled dynamics if ρ_{ab} and $\Phi(C_1, \ldots, C_m)$ can be found so that (2.26) is satisfied and $G_{\alpha\beta} + G_{\alpha\gamma}\tilde{H}^{\gamma\delta}G_{\delta\beta}$ is definite. Here $\tilde{\Phi}$ is assumed to be of the form (2.28).

If the equilibrium is spectrally unstable for the uncontrolled dynamics, one cannot make $g_{\alpha\beta} + g_{\alpha\gamma}H^{\gamma\delta}g_{\delta\beta}$ of (2.19) definite. In the controlled setting, however, we can require that $G_{\alpha\beta} + G_{\alpha\gamma}\tilde{H}^{\gamma\delta}G_{\delta\beta}$ be definite. The tensor $G_{\alpha\beta}$ is the horizontal part of the metric for the controlled system, i.e., the "controlled inertia" associated with the group H variables. Since $G_{\alpha\beta} = g_{\alpha\beta} - g_{a\alpha}\rho^{ab}g_{b\beta}$, it is clear how the control gain ρ_{ab} enters in to provide stabilization, i.e., by modifying the inertia.

3. DISSIPATION AND ASYMPTOTIC STABILIZATION

To obtain asymptotic stability, we introduce an additional term in the control law to simulate dissipation as follows

$$\frac{d}{dt}\frac{\partial l_{\tau,\sigma,\rho}}{\partial \eta^\alpha} = c^\beta_{\alpha\delta}\eta^\delta \frac{\partial l_{\tau,\sigma,\rho}}{\partial \eta^\beta} \tag{3.1}$$

$$\frac{d}{dt}\frac{\partial l_{\tau,\sigma,\rho}}{\partial \dot{\theta}^a} = \dot{\tilde{J}}_a = u_a^{\text{diss}} \tag{3.2}$$

In this case the complete control law takes the form

$$\begin{aligned} u_a &= u_a^{\text{cons}} + D_{ab}\rho^{bc}u_c^{\text{diss}} \\ &= u_a^{\text{cons}} + (g_{ab} - k_a^\alpha g_{\alpha b})\rho^{bc}u_c^{\text{diss}} \end{aligned} \tag{3.3}$$

where

$$u_a^{\text{cons}} = k_a^\alpha c^\psi_{\alpha\delta}\eta^\delta(g_{\psi\beta}\eta^\beta + g_{\psi b}\dot{\theta}^b)$$

and the relationship between ρ^{ab} and k_a^α is given by Assumption EP-2 and (2.12).

Assume that we have found Casimir functions $C^k(\tilde{M}_\alpha)$ and a function $\tilde{\Phi}(C^k, \tilde{J}_a) = \Phi(C^k) + \Psi(\tilde{J}_a)$ such that the Lyapunov function

$$E_{\tilde{\Phi}} = l_{\tau,\sigma,\rho} + \tilde{\Phi}(C^k, \tilde{J}_a) = l_{\tau,\sigma,\rho} + \Phi(C^k) + \Psi(\tilde{J}_a)$$

yields (Lyapunov) stability of the (relative) equilibrium, $(\eta_e^\alpha, \dot{\theta}_e^a)$. Then,

$$\begin{aligned} \frac{d}{dt}E_{\tilde{\Phi}} &= \frac{d}{dt}l_{\tau,\sigma,\rho} + \frac{\partial \Phi}{\partial C^k}\dot{C}^k + \frac{\partial \Psi}{\partial \tilde{J}_a}\dot{\tilde{J}}_a \\ &= \left(\dot{\theta}^a + \frac{\partial \Psi}{\partial \tilde{J}_a}\right)u_a^{\text{diss}} \end{aligned} \tag{3.4}$$

where we have used $\frac{d}{dt}l_{\tau,\sigma,\rho} = \dot{\theta}^a u_a^{\text{diss}}$, and since actuation is internal,

$$C^k = \text{constant.} \tag{3.5}$$

Assume that $E_{\tilde{\Phi}}$ has a local maximum at the equilibrium. Choose

$$u_a^{\text{diss}} = c_{ab}\left(\dot{\theta}^b + \frac{\partial \Psi}{\partial \tilde{J}_b}\right), \tag{3.6}$$

where c_{ab} is a positive definite matrix. Then,

$$\frac{d}{dt}E_{\tilde{\Phi}} = c_{ab}\left(\dot{\theta}^a + \frac{\partial \Psi}{\partial \tilde{J}_a}\right)\left(\dot{\theta}^b + \frac{\partial \Psi}{\partial \tilde{J}_b}\right) \geq 0. \tag{3.7}$$

In the case that the equilibrium of interest is such that $\dot{\theta}^a|_e = 0$, we can take Ψ as

$$\Psi(\tilde{J}) = \frac{1}{2}\epsilon^{bc}\tilde{J}_b\tilde{J}_c,$$

where ϵ^{bc} is a sign definite symmetric matrix. Then, (3.7) becomes

$$\frac{d}{dt}E_{\tilde{\Phi}} = c_{ab}\left(\dot{\theta}^a + \epsilon^{ac}\tilde{J}_c\right)\left(\dot{\theta}^b + \epsilon^{bd}\tilde{J}_d\right) \geq 0. \tag{3.8}$$

To obtain asymptotic stability we use the LaSalle invariance principle and the details of the specific system, as illustrated below.

4. GYROSCOPIC STABILIZATION WITH ROTORS

We show how the preceding results on Euler–Poincaré matching and stabilization apply to an important class of examples, namely rigid bodies carrying internal rotors. We can treat many systems, such as the spacecraft with internal rotors, the underwater vehicle with internal rotors and the heavy top with rotors, but we confine ourselves to the spacecraft here due to limited space.

Following Krishnaprasad [1985] and Bloch, Krishnaprasad, Marsden and Sánchez de Alvarez [1992], consider a rigid body with a rotor aligned along the third principal axis of the body as in Figure 4.1. The rotor spins under the influence of a torque u acting on the rotor. The configuration space is $Q = SO(3) \times S^1$, with the first factor $H = SO(3)$ being the spacecraft attitude and the second factor $G = S^1$ being the rotor angle. The Lagrangian is total kinetic energy of the system, (rigid carrier plus rotor), with no potential energy.

The reduced Lagrangian on so(3) $\times TS^1$ is

$$l(\Omega, \dot{\phi}) = \frac{1}{2}(\lambda_1\Omega_1^2 + \lambda_2\Omega_2^2 + I_3\Omega_3^2 + J_3(\Omega_3 + \dot{\phi})^2) \tag{4.1}$$

where $\Omega = (\Omega_1, \Omega_2, \Omega_3)$ is the body angular velocity vector of the carrier, ϕ is the relative angle of the rotor, $I_1 > I_2 > I_3$ are the rigid body moments of inertia, $J_1 = J_2$ and J_3 are the rotor moments of inertia and $\lambda_i = I_i + J_i$.

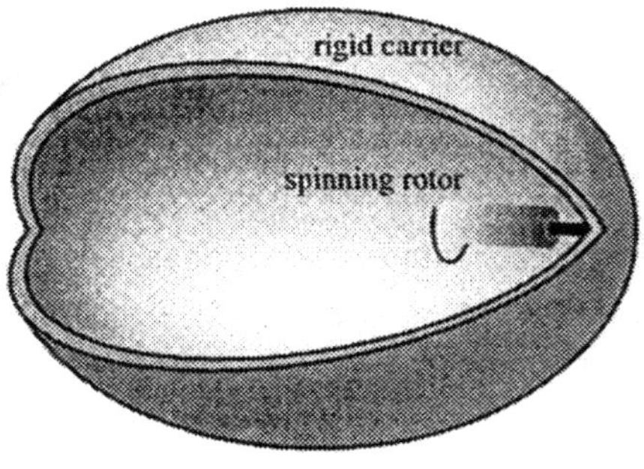

Fig. 4.1. The spacecraft with a rotor attached along the long axis.

The body angular momenta are determined by the Legendre transform to be

$$\Pi_1 = \tilde{M}_1 = \lambda_1 \Omega_1, \quad \Pi_2 = \tilde{M}_2 = \lambda_2 \Omega_2,$$
$$\Pi_3 = \tilde{M}_3 = \lambda_3 \Omega_3 + J_3 \dot{\phi}.$$

The momentum conjugate to ϕ is

$$\frac{\partial l}{\partial \dot{\phi}} = l_3 = J_3(\Omega_3 + \dot{\phi}).$$

The equations of motion with a control torque u acting on the rotor are

$$\lambda_1 \dot{\Omega}_1 = \lambda_2 \Omega_2 \Omega_3 - (\lambda_3 \Omega_3 + J_3 \dot{\phi}) \Omega_2 \quad (4.2)$$
$$\lambda_2 \dot{\Omega}_2 = -\lambda_1 \Omega_1 \Omega_3 + (\lambda_3 \Omega_3 + J_3 \dot{\phi}) \Omega_1 \quad (4.3)$$
$$\lambda_3 \dot{\Omega}_3 + J_3 \ddot{\phi} = (\lambda_1 - \lambda_2) \Omega_1 \Omega_2. \quad (4.4)$$
$$\dot{l}_3 = u. \quad (4.5)$$

Next, we form the controlled Lagrangian and apply the Euler-Poincaré matching theorem. Since the Abelian group $G = S^1$ is one-dimensional, g_{ab}, σ_{ab} and ρ_{ab} are all scalars. From (4.1), $g_{ab} = J_3$. Let $\sigma_{ab} = \sigma J_3$ and $\rho_{ab} = \rho J_3$ where σ and ρ are dimensionless scalars. For matching, choose τ^a_α according to Assumption EP-1, i.e.,

$$(\, \tau^\phi_{\Omega_1} \; \tau^\phi_{\Omega_2} \; \tau^\phi_{\Omega_3} \,) = -\frac{1}{\sigma J_3}(\, 0 \; 0 \; J_3 \,). \quad (4.6)$$

To satisfy Assumption EP-2, ρ should satisfy

$$\frac{1}{\sigma J_3} + \frac{1}{\rho J_3} = \frac{1}{J_3} \quad \text{i.e.,} \quad \rho = \frac{\sigma}{\sigma - 1}. \quad (4.7)$$

Substituting into equation (2.5) with these choices, the controlled Lagrangian is given by

$$l_{\tau,\sigma,\rho} = \frac{1}{2}\left(\lambda_1 \Omega_1^2 + \lambda_2 \Omega_2^2 + I_3 \Omega_3^2 + \frac{1}{\sigma} J_3 \Omega_3^2 \right) + \frac{1}{2}\left(\frac{\sigma}{\sigma - 1} J_3 \left(\Omega_3 + \dot{\phi} - \frac{1}{\sigma}\Omega_3 \right)^2 \right) \quad (4.8)$$

where σ is a free variable and matching is ensured by Theorem 1. The controlled conserved quantity is

$$\tilde{l}_3 = \tilde{J} = \frac{\partial l_{\tau,\sigma,\rho}}{\partial \dot{\phi}} = J_3 \Omega_3 + \rho J_3 \dot{\phi}.$$

We use the formula (2.11) to get the control law

$$u = u^{\text{cons}} = k(\lambda_1 - \lambda_2)\Omega_1 \Omega_2.$$

We note that

$$\frac{1}{\sigma} = \frac{k}{1-k}\frac{I_3}{J_3}, \quad \frac{1}{\rho} = \frac{(1-k)J_3 - kI_3}{(1-k)J_3} \quad (4.9)$$

Consider $(\Omega_1, \Omega_2, \Omega_3, \dot{\phi}) = (0, \bar{\Omega}, 0, 0)$, $\bar{\Omega} \neq 0$, the equilibrium corresponding to steady rotation about the intermediate axis (unstable for the uncontrolled spacecraft). In contrast to earlier work, we carry out our analysis on the Lagrangian side and we do not restrict the stability analysis to the zero level set of the conserved momentum.

The Casimir for this problem is the total angular momentum of the body plus rotor system. We let

$$C = \frac{1}{2}\left((\lambda_1 \Omega_1)^2 + (\lambda_2 \Omega_2)^2 + \left(\left(I_3 + \frac{J_3}{\sigma} \right) \Omega_3 + \frac{\sigma - 1}{\sigma} \tilde{l}_3 \right)^2 \right). \quad (4.10)$$

The Lyapunov function becomes

$$E_{\tilde{\Phi}} = \frac{1}{2}\left(\lambda_1 \Omega_1^2 + \lambda_2 \Omega_2^2 + \left(I_3 + \frac{J_3}{\sigma} \right) \Omega_3^2 + \frac{1}{\rho J_3} \tilde{l}_3^2 \right) + \Phi(C) + \Psi(\tilde{l}_3). \quad (4.11)$$

To satisfy (2.26) of Theorem 2, we require

$$\Phi'|_e \lambda_2 \bar{\Omega}_2 = -\bar{\Omega}_2, \quad \text{i.e.,} \quad \Phi'|_e = -\frac{1}{\lambda_2}.$$

We show that $N_{\alpha\beta} = G_{\alpha\beta} + G_{\alpha\gamma} \tilde{H}^{\gamma\delta} G_{\delta\beta}$ can be made definite. The matrix $G_{\alpha\beta} = g_{\alpha\beta} - g_{a\alpha} \rho^{ab} g_{b\beta}$ is diagonal: $\text{diag}(\lambda_1, \lambda_2, I_3 + J_3/\sigma)$. The matrix $\tilde{H}^{\alpha\beta}$ defined by (2.30) is computed to be

$$\tilde{H} = \text{diag}\left(\Phi'|_e, \Phi'|_e + \Phi''|_e \lambda_2^2 \bar{\Omega}^2, \Phi'|_e \right) = \text{diag}\left(-\frac{1}{\lambda_2}, -\frac{1}{\lambda_2} + \Phi''|_e \lambda_2^2 \bar{\Omega}^2, -\frac{1}{\lambda_2} \right).$$

$N_{\alpha\beta}$ is computed to be diagonal with diagonal elements

$$\left(\lambda_1 - \frac{\lambda_1^2}{\lambda_2}, \; \Phi''|_e \lambda_2^4 \bar{\Omega}^2, \; I_3 + \frac{J_3}{\sigma} - \frac{1}{\lambda_2}\left(I_3 + \frac{J_3}{\sigma} \right)^2 \right).$$

Since the first diagonal element is negative, we choose $\Phi''|_e$ to be negative also and require that

$$I_3 + \frac{J_3}{\sigma} - \frac{1}{\lambda_2}\left(I_3 + \frac{J_3}{\sigma} \right)^2 < 0$$

This condition holds if $k > 1 - I_3/\lambda_2$ and $E_{\tilde{\Phi}}$ has a local maximum at the equilibrium of interest. So, by Theorem 2 we have proved

Proposition 3. For $k > 1 - I_3/\lambda_2$, the equilibrium $(0, \bar{\Omega}, 0, 0)$ is nonlinearly stable for the feedback controlled system.

5. ASYMPTOTIC STABILITY OF SPACECRAFT WITH A ROTOR

We consider one approach to asymptotic stabilization; others are examined in a future publication.

Dissipation is introduced so that $E_{\tilde{\Phi}}$ becomes a Lyapunov function for the closed-loop system with the complete control law which is computed below. Any given initial condition uniquely determines the equilibrium that can be asymptotically stabilized; the conservation of the Casimir function (4.10) gives the magnitude as $\|\lambda_2\bar{\Omega}\|^2 = \|\Pi\|^2 = \|\Pi(0)\|^2$ and the stability implies that the sign of $\bar{\Omega}$ is the same as that of $\Omega_2(0)$ since the flow stays near $(0, \bar{\Omega}, 0, 0)$.

In the expression for $E_{\tilde{\Phi}}$, Ψ is given by

$$\Psi(\tilde{l}_3) = \frac{1}{2\epsilon J_3}\tilde{l}_3^2$$

with $\epsilon < 0$ and $|\epsilon| \ll 1$. By (3.6)

$$u^{\text{diss}} = c\left(\dot{\phi} + \frac{1}{\epsilon J_3}\tilde{l}_3\right) = c\left(\frac{1}{\epsilon}\Omega_3 + \left(1 + \frac{\rho}{\epsilon}\right)\dot{\phi}\right)$$

with $c > 0$, and the complete control law is

$$u = k(\lambda_1 - \lambda_2)\Omega_1\Omega_2 + (1-k)\frac{1}{\rho}u^{\text{diss}},$$

where ρ is given by (4.9). Suppose that the flow $(\Omega_1(t), \Omega_2(t), \Omega_3(t), \dot{\phi}(t))$ satisfies $\dot{E}_{\tilde{\Phi}} = 0$, equivalently $u^{\text{diss}} = 0$. Then, since $\dot{\tilde{l}}_3 = u^{\text{diss}}$, $\tilde{l}_3$ is constant. This implies that

$$\dot{\phi}(t) = \dot{\phi}(0) = \text{constant},$$
$$\Omega_3(t) = \Omega_3(0) = \text{constant}.$$

Substituting these into (4.4), we get

$$\Omega_1\Omega_2 = 0. \tag{5.1}$$

Since $\Omega_2(t)$ stays near $\bar{\Omega} \neq 0$ by stability, (5.1) implies that

$$\Omega_1(t) = 0$$

for all t. Substitution of this into (4.3) gives

$$\Omega_2(t) = \Omega_2(0) = \text{constant}. \tag{5.2}$$

Substitute these two into (4.2) and we get

$$\left((\lambda_2 - \lambda_3)\Omega_3 - J_3\dot{\phi}\right)\Omega_2(0) = 0$$

or

$$(\lambda_2 - \lambda_3)\Omega_3 - J_3\dot{\phi} = 0 \tag{5.3}$$

since $\Omega_2(0) \neq 0$ by stability. We also have $u^{\text{diss}} = 0$, which is given by

$$\Omega_3 + (\epsilon + \rho)\dot{\phi} = 0. \tag{5.4}$$

All we require on ϵ is that it should be negative and satisfy some inequality to guarantee the (Lyapunov) stability. We can find ϵ satisfying

$$(\lambda_2 - \lambda_3)(\epsilon + \rho) + J_3 \neq 0 \tag{5.5}$$

such that the two equations in (5.3) and (5.4) are independent. Then $\Omega_3 = \dot{\phi} = 0$. Thus, the only possible flow satisfying $u^{\text{diss}} = 0$ is

$$\Omega_1(t) = \Omega_3(t) = \dot{\phi}(t) = 0, \qquad \Omega_2(t) = \Omega_2(0).$$

This implies that

$$|\lambda_2\Omega_2(0)|^2 = |\lambda_1\Omega_1|^2 + |\lambda_2\Omega_2|^2 + |\lambda_3\Omega_3 + J_3\dot{\phi}|^2$$
$$= \|\Pi\|^2 = |\lambda_2\bar{\Omega}|^2$$

and so $\Omega_2(0) = \bar{\Omega}$ by stability. Thus, the only possible flow satisfying $u^{\text{diss}} = 0$ is the equilibrium. By the LaSalle invariance principle, it is asymptotically stable.

References.

Bloch, A.M., D.E. Chang, N.E Leonard and J.E. Marsden [2000] Controlled Lagrangians and the stabilization of mechanical systems II; Potential shaping and tracking, in preparation.

Bloch, A.M., P.S. Krishnaprasad, J.E. Marsden and G. Sánchez de Alvarez [1992] Stabilization of rigid body dynamics by internal and external torques. *Automatica* **28**, 745–756.

Bloch, A.M., N.E. Leonard and J.E. Marsden [1997] Stabilization of mechanical systems using controlled Lagrangians, *Proc. CDC* **36**, 2356–2361.

Bloch, A.M., N.E. Leonard and J.E. Marsden [1998] Matching and stabilization by the method of controlled Lagrangians, *Proc. CDC* **37**, 1446–1451.

Bloch, A.M., N.E. Leonard and J.E. Marsden [1999a] Stabilization of the pendulum on a rotor arm by the method of controlled Lagrangians, *Proc. ICRA*, IEEE, 500- 505.

Bloch, A.M., N.E. Leonard and J.E. Marsden [1999b] Potential shaping and the method of controlled Lagrangians *Proc. CDC* **38**, 1653–1657.

Bloch, A.M., N.E. Leonard and J.E. Marsden [2000a] Controlled Lagrangians and the stabilization of mechanical systems I: The first matching theorem, *IEEE Trans. Auto. Control*, to appear.

Bloch, A.M., N.E. Leonard and J.E. Marsden [2000b] Controlled Lagrangians and the stabilization of Euler-Poincaré mechanical systems, Preprint.

Krishnaprasad, P.S. [1985] Lie-Poisson structures, dual-spin spacecraft and asymptotic stability, *Nonl. Anal. Th. Meth. and Appl.* **9**, 1011–1035.

Marsden, J.E. and T.S. Ratiu [1994] *Symmetry and Mechanics*. Texts in Applied Mathematics, **17**, Springer-Verlag. Second Edition, 1999.

C.A. Woolsey and N.E. Leonard [1999], Underwater vehicle stabilization by internal rotors, *Proc. ACC* 3417-3421.

CONTROLLED LAGRANGIANS, SYMMETRIES AND CONDITIONS FOR STRONG MATCHING

Johan Hamberg

Department of Guidance and Control
Swedish Defense Research Establishment
E-mail: johan.hamberg@sto.foa.se

Abstract: The method of controlled Lagrangians consists in constructing state feedback laws for underactuated Lagrangian systems such that the closed loop equations are Lagrangian. In this paper the general matching problem is solved and the matching conditions are given a geometrical formulation. This geometric formulation is used for simplifying and extending the method of Auckly, Kapitanski and White. In the original symmetric context of Bloch, Leonard and Marsden, a new notion of strong matching is introduced, for which necessary and sufficient conditions are given. *Copyright ©2000 IFAC*

Keywords: Differential geometric methods, Stabilization methods, Energy control, Nonlinear systems theory

1. BACKGROUND

1.1 *General*

The method of controlled Lagrangians consists in constructing state feedback laws for underactuated Lagrangian systems such that the closed loop equations are Lagrangian. The new Lagrangian is said to be *matching.* This method was introduced as a means of stabilizing unstable equilibrium points of underactuated Lagrangian systems by Bloch, Leonard and Marsden (1997*b*) (generalizing an example in Bloch *et al.*, 1997*a*) in a special context. In this context the system is invariant under a symmetry Lie group acting on the configuration manifold.

Sufficient conditions of increasing generality have been given for matching by Bloch, Leonard and Marsden (1998*a*; 1998*b*; 1999). Necessary and sufficient conditions for matching in a general context have been given independently by Auckly, Kapitanski and White (1998) and by the author (Hamberg, 1999). In this paper the matching problem is solved for general Lagrangians and the matching conditions are given a geometrical formulation. This geometric formulation is then used for simplifying and extending the method of (Auckly *et al.*, 1998). In the original symmetric context of Bloch, Leonard and Marsden, a new notion of strong matching is introduced, for which necessary and sufficient conditions are given.

1.2 *Underactuated Lagrangian Systems*

The controlled Lagrangian method has been introduced in order to control underactuated mechanical systems. The concept of an Underactuated Lagrangian System is formalized as follows

Definition 1. An *Underactuated Lagrangian System* is a triple $(\mathcal{Q}, L, \Lambda)$, where $\mathcal{Q}$ is a configuration manifold, $L : \mathcal{TQ} \to \mathbb{R}$ is a regular Lagrangian defined on the tangent bundle $\mathcal{TQ}$ of $\mathcal{Q}$, and Λ is a subbundle of $\mathcal{TQ}$, called the *bundle of unactuated directions.* (The Lagrangian is said to be *regular* if $G_{ij} = L''_{\dot{q}^i\dot{q}^j}$ is nonsingular everywhere.)

The triple $(\mathcal{Q},\ L,\ \Lambda)$ defines a control system on $\mathcal{TQ}$, for which the dynamics is given by Lagrange's equations,

$$\frac{d}{dt}\left(\frac{\partial L}{\partial \dot{q}^i}\right) - \frac{\partial L}{\partial q^i} = Q_i \qquad (1)$$

where $q^i, (i = 1 \dots n)$ denote coordinates on $\mathcal{Q}$.

More precisely, Lagrange's equations is the condition that the state space trajectory $\mathbb{R} \to \mathcal{TQ}$ is the tangent lift χ_* of a motion $\chi : \mathbb{R} \to \mathcal{Q}$, and that it satisfies $\frac{d}{dt}(L'_{\dot{q}^i} \circ \chi_*) - L'_{q^i} \circ \chi_* = Q_i$.

The Q_i are controls, and it is demanded that these take values in the cotangent subbundle $\Lambda^\perp$ which is the annihilator of Λ, explicitly: $\Lambda^i_A\ Q_i = 0$, where Λ_A, $(A = 1 \dots m)$, is a local basis for Λ. Such Q_i are called *admissible.* The Λ^i_A may be thought of as components of the injective bundle mapping $i_\Lambda : \Lambda \hookrightarrow \mathcal{TQ}$.

1.3 *The Controlled Lagrangian Strategy*

The *method of controlled Lagrangians* consists in finding an admissible state feedback law (*i.e.* a bundle mapping $\mathcal{TQ} \to \Lambda^\perp \hookrightarrow \mathcal{T}^*\mathcal{Q}$) determining Q_i, the "generalized forces from the actuators", such that the closed loop dynamics is still in Lagrangian form, *viz.*

$$\frac{d}{dt}\left(\frac{\partial \tilde{L}}{\partial \dot{q}^i}\right) - \frac{\partial \tilde{L}}{\partial q^i} = \tilde{Q}_i \qquad (2)$$

The function $\tilde{L}$ is called the *controlled Lagrangian.*

Clearly any admissible Q_i and any $\tilde{L}$ can be chosen, provided that there are no conditions on the $\tilde{Q}_i$. Conditions on the $\tilde{Q}_i$ allowed imply conditions on the possible Q_i and $\tilde{L}$.

Several sorts of conditions on the $\tilde{Q}_i$ can be considered, for instance

- *matching,* which is the condition $\tilde{Q}_i = 0$,
- *generalized matching* which is the condition that $\tilde{Q}_i(q,\dot{q})\dot{q}^i \leq 0$ identically,
- *strong matching,* which will be defined later in this paper.

2. THE MATCHING CONDITION

2.1 *Matching for general Lagrangians*

Written out in full, Lagrange's equations read

$$L''_{\dot{q}^i\dot{q}^j}\ddot{q}^j + L''_{\dot{q}^i q^j}\dot{q}^j - L'_{q^i} = Q_i \qquad (3)$$

Due to the regularity assumption, (1) can be written in the form $\ddot{q}^i + G^{ij}B_j = G^{ij}Q_j$, where $B_j = L''_{\dot{q}^j q^k}\dot{q}^k - L'_{q^j}$ and analogously (2) is written as $\ddot{q}^i + \tilde{G}^{ij}\tilde{B}_j = \tilde{G}^{ij}\tilde{Q}_j$. Their difference gives the following condition

$$Q_j = H^j_i\tilde{Q}_j + B_i - H^j_i\ \tilde{B}_j \qquad (4)$$

where $H^j_i = G_{ik}\tilde{G}^{kj}$.

From (4), the following two propositions immediately follow.

Proposition 1. A necessary and sufficient condition for matching (i.e. that $\tilde{Q}_i = 0$ corresponds to admissible Q_i) is that $\tilde{L}$ satisfies the system of partial differential equations

$$\Lambda^i_A\left(H^j_i\ \tilde{B}_j - B_i\right) = 0 \qquad (5)$$

Proposition 2. Suppose that (5) is satisfied. Then those $\tilde{Q}_i$ that correspond to admissible Q_i are precisely the ones that satisfy

$$\Lambda^i_A H^j_i \tilde{Q}_j = 0 \qquad (6)$$

2.2 *The case of unactuated systems*

For unactuated systems, *i.e.* the case when $\Lambda = \mathcal{TQ}$, the matching condition (5) simply reads $H^j_i\ \tilde{B}_j - B_i = 0$, which can be rewritten as $\tilde{L}''_{\dot{q}^j q^k}\dot{q}^k - \tilde{L}'_{q^j} = \tilde{L}''_{\dot{q}^j\dot{q}^k}G^{kl}B_l$, which is a *linear* system of PDEs for those $\tilde{L}$ that are equivalent to L . Some examples of this:

Example 1. For the unactuated Lagrangian system $\left(\mathbb{R}^n, \frac{1}{2}\delta_{ij}\dot{q}^i\dot{q}^j, \mathcal{T}\mathbb{R}^n\right)$it is straight forward to solve the above PDEs (in the class of real analytic regular Lagrangians $\tilde{L}$). The matching condition is then equivalent to $\tilde{L} = f(\dot{q}) + S'_{q^i}(q)\,\dot{q}^i$, with f and S arbitrary, but such that f is a regular Lagrangian.

Example 2. For a general (curved) Riemannian metric, the solution of the corresponding problem can be expressed in terms of the family of completely symmetric covariantly constant tensor fields. In the generic case these are all generated by the metric, and the general solution is then given by $\tilde{L} = f\left(\frac{1}{2}g_{ij}(q)\,\dot{q}^i\dot{q}^j\right) + S'_{q^i}(q)\,\dot{q}^i$, with f and S arbitrary, apart from the regularity assumption. A similar result holds for a general (regular) Lagrangian L, homogenous in the $\dot{q}$.

2.3 *The case of classical Lagrangians*

It is now assumed that both G_{ij} and $\tilde{G}_{ij}$ are independent of $\dot{q}$, so G_{ij}, $\tilde{G}_{ij}$ and H^j_i are components of tensor fields on $\mathcal{Q}$, and lower case notation, g_{ij}, $\tilde{g}_{ij}$ and h^j_i, is used to distinguish this case.

According to (6), $(\mathcal{Q}, \tilde{L}, \tilde{\Lambda})$ is a new underactuated Lagrangian system, where $\tilde{\Lambda}$ is the tangent subbundle spanned by $\tilde{\Lambda}_A^j = \Lambda_A^i h_i^j$. Such $(\mathcal{Q}, \tilde{L}, \tilde{\Lambda})$ and $(\mathcal{Q}, L, \Lambda)$ are said to be *CL-equivalent.* (It is clear that this is an equivalence relation, since it is the condition that $\tilde{\Lambda}_A^j \tilde{Q}_j = 0 \Leftrightarrow \Lambda_A^i Q_i = 0$.) For CL-equivalent systems, there is a natural isomorphism $\lambda : \Lambda \to \tilde{\Lambda}$ given by $\tilde{\Lambda}_A^j = \Lambda_A^i h_i^j$.

Since $g_{ij} = L''_{\dot{q}^i \dot{q}^j}$ is independent of $\dot{q}$, L is of the classical form, $L = \frac{1}{2} g_{ij}(q)\dot{q}^i \dot{q}^j + A_i(q)\dot{q}^i - V(q)$ and similarly for $\tilde{L}$. When this is substituted in (5), the latter becomes a set of second degree polynomial identities in $\dot{q}$. After separation into homogeneous terms, the following explicit conditions follow.

$$\begin{aligned} &\Lambda_A^i g_{ij} \left(\tilde{\Gamma}_{kl}^j - \Gamma_{kl}^j\right) = 0 \\ &\tilde{\Lambda}_A^j \left(\tilde{A}_{j,k} - \tilde{A}_{k,j}\right) - \Lambda_A^j \left(A_{j,k} - A_{k,j}\right) = 0 \\ &\tilde{\Lambda}_A^j \tilde{V}_{,j} - \Lambda_A^j V_{,j} = 0 \end{aligned} \quad (7)$$

Here the Γ_{kl}^j and $\tilde{\Gamma}_{kl}^j$ denote the Christoffel symbols of the metrics g_{ij} and $\tilde{g}_{ij}$ respectively. In particular it is seen that if $A_j \equiv 0$, then $\tilde{A}_j \equiv 0$ is a possible choice, so in that case the matching condition of (Hamberg, 1999) and (Auckly *et al.*, 1998) are recovered.

2.4 *The case of fully actuated systems*

Consider a *fully actuated* system on $\mathbb{R}^n$ with a classical Lagrangian, L as above. By extending the configuration manifold to $\mathbb{R}^{n+1}$ the controlled Lagrangian strategy may be used for *output regulation, i.e.* the controller depends on the configuration and not on the velocity. An auxiliary variable q^0 is adjoined and and the following control is chosen: $Q_j = V_{,j} - W_{,j}$ $(j = 1 \ldots n)$, where $W(q^0, q^1, .., q^n)$ is a suitable positive definite function and $\ddot{q}^0 = -\frac{\partial W}{\partial q^0} - \dot{q}^0$. The closed loop equations are then Lagrangian with $\tilde{L} = L + (\frac{1}{2}\dot{q}^0)^2 + V - W$ and $\tilde{Q} = 0$, except $\tilde{Q}_0 = -\dot{q}^0$. If W is judiciously chosen, global asymptotic stability follows by LaSalle's theorem. This is in the spirit of (Ortega *et al.*, 1998).

3. GEOMETRICAL FORMULATION

In this section a more geometric formulation is given. The classical Lagrangian can be written in coordinate-free form

$$L = \frac{1}{2} g + \alpha - V \quad (8)$$

Here the Riemannian metric $g = g_{ij}(q)dq^i \otimes dq^j$ and the 1-form $\alpha = A_i(q)dq^i$ are naturally identified with polynomial functions on $\mathcal{TQ}$ and V is identified with its pullback to $\mathcal{TQ}$.

3.1 *Geometrical Interpretation of the Matching Condition*

The first condition of (7) can be written as $\nabla(\mathrm{i}_{\Lambda_A} g) - \tilde{\nabla}(\mathrm{i}_{\Lambda_A} g) = 0$, which in turn is equivalent to its symmetrization, $\mathrm{symm}(\nabla(\mathrm{i}_{\Lambda_A} g)) - \mathrm{symm}\left(\tilde{\nabla}(\mathrm{i}_{\Lambda_A} g)\right) = 0$, since the skew-symmetric parts of $\nabla(\mathrm{i}_{\Lambda_A} g)$ and $\tilde{\nabla}(\mathrm{i}_{\Lambda_A} g)$ coincide due to vanishing torsion – both are equal to the exterior derivative of the 1-form $\mathrm{i}_{\Lambda_A} g$. By the defining relation $\mathrm{i}_{\Lambda_A} g = \mathrm{i}_{\tilde{\Lambda}_A} \tilde{g}$ for $\tilde{\Lambda}_A$, the condition takes the form

$$\mathrm{symm}(\nabla(\mathrm{i}_{\Lambda_A} g)) - \mathrm{symm}\left(\tilde{\nabla}(\mathrm{i}_{\tilde{\Lambda}_A} \tilde{g})\right) = 0 \quad (9)$$

In (9) the expression for the Lie derivative of a metric is recognized, and the final form of the first equation becomes

$$\mathfrak{L}_{\tilde{\Lambda}_A} \tilde{g} - \mathfrak{L}_{\Lambda_A} g = 0 \quad (10)$$

The geometric reformulation of the remaining two equations is immediate. This gives

Proposition 3. The matching condition (7) for classical underactuated Lagrangian systems can be reexpressed as

$$\begin{aligned} &\mathfrak{L}_{\tilde{\Lambda}_A} \tilde{g} - \mathfrak{L}_{\Lambda_A} g = 0 \\ &\mathrm{i}_{\tilde{\Lambda}_A} d\tilde{\alpha} - \mathrm{i}_{\Lambda_A} d\alpha = 0 \\ &\mathfrak{L}_{\tilde{\Lambda}_A} \tilde{V} - \mathfrak{L}_{\Lambda_A} V = 0 \end{aligned} \quad (11)$$

where $\tilde{\Lambda}_A$ is given by $\mathrm{i}_{\tilde{\Lambda}_A} \tilde{g} - \mathrm{i}_{\Lambda_A} g = 0$.

In particular, if $\alpha \equiv \tilde{\alpha} \equiv 0$, this can be written as $\mathfrak{L}_{\tilde{\Lambda}_A} \tilde{L} - \mathfrak{L}_{\Lambda_A} L = 0$. For general α and $\tilde{\alpha}$, the following theorem follows, expressed in terms of the *fiber derivative*, $\mathfrak{F}_X L = X^i L'_{\dot{q}^i}$.

Theorem 4. Necessary and sufficient conditions for two classical underactuated Lagrangian systems $(\mathcal{Q}, \tilde{L}, \tilde{\Lambda})$ and $(\mathcal{Q}, L, \Lambda)$ to be CL-equivalent is that there are functions S_A, $(A = 1 \ldots m)$ on $\mathcal{Q}$ such that

$$\begin{aligned} &\mathfrak{F}_{\tilde{\Lambda}_A} \tilde{L} - \mathfrak{F}_{\Lambda_A} L = S_A \\ &\mathfrak{L}_{\tilde{\Lambda}_A} \tilde{L} - \mathfrak{L}_{\Lambda_A} L = dS_A \end{aligned} \quad (12)$$

In particular, if neither Lagrangian contains terms linear in $\dot{q}$, then $S_A \equiv 0$ in the condition (12).

Proof. *Necessity*: According to (11), $\mathfrak{L}_{\tilde{\Lambda}_A} \tilde{L} - \mathfrak{L}_{\Lambda_A} L$ equals $(\mathfrak{L}_{\tilde{\Lambda}_A} \tilde{\alpha} - \mathfrak{L}_{\Lambda_A} \alpha) - (\mathrm{i}_{\tilde{\Lambda}_A} d\tilde{\alpha} - \mathrm{i}_{\Lambda_A} d\alpha)$. Hence, Cartan's formula, $\mathfrak{L}_X = d\,\mathrm{i}_X + \mathrm{i}_X d$, gives that $\mathfrak{L}_{\tilde{\Lambda}_A} \tilde{L} - \mathfrak{L}_{\Lambda_A} L = dS_A$, where $S_A = \mathrm{i}_{\tilde{\Lambda}_A} \tilde{\alpha} - \mathrm{i}_{\Lambda_A} \alpha$, $(A = 1 \ldots m)$ are functions on $\mathcal{Q}$. Since $\mathfrak{F}_{\Lambda_A} L = \mathrm{i}_{\Lambda_A} g + \mathrm{i}_{\Lambda_A} \alpha$, the defining equation of S_A and the identity $\mathrm{i}_{\tilde{\Lambda}_A} \tilde{g} = \mathrm{i}_{\Lambda_A} g$ can be summarized as $\mathfrak{F}_{\tilde{\Lambda}_A} \tilde{L} - \mathfrak{F}_{\Lambda_A} L = S_A$.

Sufficiency: Assume (12). Separation of (12_1) into its homogenous parts, gives $\mathrm{i}_{\tilde{\Lambda}_A}\tilde{g} - \mathrm{i}_{\Lambda_A} g = 0$ and $S_A = \mathrm{i}_{\tilde{\Lambda}_A}\tilde{\alpha} - \mathrm{i}_{\Lambda_A}\alpha$. When this is substituted in (12_2), separation into homogenous parts gives (11). ■

3.2 *Lagrange's Equations in Geometric Form*

The results of the preceding paragraph indicate that the Lie derivative formulation of the matching condition is more than a coincidence. It will be shown that these Lie derivative conditions arise quite naturally, when the Lagrange's equations are expressed in terms of geometric quantities.

Let $\mathfrak{J}^{(k)}$ denote the space of k-jets $\mathrm{j}^{(k)}\chi$ of motions $\chi : \mathbb{R}_t \to \mathcal{Q}$, and $\pi : \mathfrak{J}^{(2)} \to \mathfrak{J}^{(1)}$ the natural projection. The *formal total time derivative* $\mathfrak{D}_t : C^\infty\left(\mathfrak{J}^{(1)}\right) \to C^\infty\left(\mathfrak{J}^{(2)}\right)$is characterized by the condition that $\mathfrak{D}_t(L) \circ \mathrm{j}^{(2)}\chi = \frac{d}{dt}\left(L \circ \mathrm{j}^{(1)}\chi\right)$ identically in L and χ.

Lemma 5. The expression $\mathfrak{E}_X(L) = \mathfrak{D}_t(\mathfrak{F}_X L) - \pi^\star \circ \mathfrak{L}_X L$ is C^∞-linear in the vector field X on $\mathcal{Q}$, and for a motion $\chi : \mathbb{R} \to \mathcal{Q}$, $\mathfrak{E}_{\partial/\partial q^i}(L) \circ \mathrm{j}^{(2)}\chi$ equals

$$\frac{d}{dt}(L'_{\dot{q}^i} \circ \mathrm{j}^{(1)}\chi) - L'_{q^i} \circ \mathrm{j}^{(1)}\chi \qquad (13)$$

Proof. $\mathfrak{E}_{fX}(.) - f\mathfrak{E}_X(.)$ is a derivation on the algebra $C^\infty\left(\mathfrak{J}^{(1)}\right)$ which clearly vanishes on Lagrangians of the zeroth and first degrees. It it therefore vanishes identically, and C^∞-linearity follows. The formula (13) follows directly by the above definitions. ■

By the lemma, Lagrange's equations may be written as

$$\mathfrak{D}_t(\mathfrak{F}_X L) - \pi^\star \circ \mathfrak{L}_X L = \pi^\star \circ \mathrm{i}_X Q \qquad (14)$$

where $Q = Q_i dq^i$. By subtraction of two copies of (14), it follows that

$$\pi^\star \circ \mathfrak{C} = \pi^\star \circ \left(\mathrm{i}_{\tilde{\Lambda}}\tilde{Q} - \mathrm{i}_\Lambda Q\right) + \mathfrak{D}_t \mathfrak{c} \qquad (15)$$

where $\mathfrak{C} = \mathfrak{L}_{\tilde{\Lambda}}\tilde{L} - \mathfrak{L}_\Lambda L$ and $\mathfrak{c} = \mathfrak{F}_{\tilde{\Lambda}}\tilde{L} - \mathfrak{F}_\Lambda L$, from which the following theorem is inferred.

Theorem 6. Necessary and sufficient conditions for two (general) underactuated Lagrangian systems $(\mathcal{Q}, \tilde{L}, \tilde{\Lambda})$ and $(\mathcal{Q}, L, \Lambda)$ to be CL-equivalent is that there are functions S_A, $(A = 1 \ldots m)$ on $\mathcal{Q}$ such that

$$\begin{array}{l} \mathfrak{F}_{\tilde{\Lambda}_A}\tilde{L} - \mathfrak{F}_{\Lambda_A} L = S_A \\ \mathfrak{L}_{\tilde{\Lambda}_A}\tilde{L} - \mathfrak{L}_{\Lambda_A} L = dS_A \end{array} \qquad (16)$$

Proof. *Sufficiency*: Assume (16). Then (15) reduces to $\mathrm{i}_{\tilde{\Lambda}_A}\tilde{Q} = \mathrm{i}_{\Lambda_A} Q$, so CL-equivalence follows.

Necessity: Assume that $(\mathcal{Q}, \tilde{L}, \tilde{\Lambda})$ and $(\mathcal{Q}, L, \Lambda)$ are CL-equivalent, so $\mathrm{i}_{\tilde{\Lambda}_A}\tilde{Q} = 0 \Leftrightarrow \mathrm{i}_{\Lambda_A} Q = 0$. Then (15) reduces to $\pi^\star \circ (\mathfrak{L}_{\tilde{\Lambda}_A}\tilde{L} - \mathfrak{L}_{\Lambda_A} L) = \mathfrak{D}_t\left(\mathfrak{F}_{\tilde{\Lambda}_A}\tilde{L} - \mathfrak{F}_{\Lambda_A} L\right)$, which in particular says that $\mathfrak{F}_{\tilde{\Lambda}_A}\tilde{L} - \mathfrak{F}_{\Lambda_A} L$ is independent of $\dot{q}$, i.e. $\mathfrak{F}_{\tilde{\Lambda}_A}\tilde{L} - \mathfrak{F}_{\Lambda_A} L = S_A$, hence $\pi^\star \circ (\mathfrak{L}_{\tilde{\Lambda}_A}\tilde{L} - \mathfrak{L}_{\Lambda_A} L) = \mathfrak{D}_t S_A = \pi^\star \circ dS_A$. ■

3.3 *Symmetry Vector Fields Parallel to* Λ

Consider two CL-equivalent systems $(\mathcal{Q}, \tilde{L}, \tilde{\Lambda})$ and $(\mathcal{Q}, L, \Lambda)$, both with classical Lagrangians without terms linear in $\dot{q}$. Suppose that the vector field X, parallel to Λ, is a symmetry generator, $\mathfrak{L}_X L = 0$. According to (14), $\mathfrak{F}_X L$ is then a constant of motion. It follows from (12_2) that $\lambda(X)$ generates a corresponding symmetry of $\tilde{L}$. Furthermore, according to (12_1) the constants of motion $\mathfrak{F}_X L$ and $\mathfrak{F}_{\lambda(X)}\tilde{L}$ coincide.

3.4 *The Method of Auckly, Kapitanski and White*

Auckly, Kapitanski and White (1998) have introduced a systematic method for simplifying the integration of (7). In the terminology of the present paper, this method can be described as

- first finding a candidate for $\tilde{\Lambda}$ and $\lambda : \Lambda \to \tilde{\Lambda}$,
- then constructing $\tilde{L}$ by means of λ.

This method becomes simple and natural in the present geometric formulation, and a more precise condition for the first step is obtained. The discussion is restricted to classical Lagrangians without terms linear in $\dot{q}$.

Lemma 7. Consider two systems $(\mathcal{Q}, \tilde{L}, \tilde{\Lambda})$ and $(\mathcal{Q}, L, \Lambda)$, both with classical Lagrangians without terms linear in $\dot{q}$. It then generally holds that

$$\mathfrak{K}_{AB} = -\mathfrak{L}_{\tilde{\Lambda}_B}\mathfrak{c}_A + \mathrm{i}_{\tilde{\Lambda}_A}\mathfrak{C}_B \qquad (17)$$

where $\mathfrak{K}_{AB} = \mathfrak{L}_{\tilde{\Lambda}_B}(\mathrm{i}_{\Lambda_A} g) + \mathrm{i}_{[\tilde{\Lambda}_B, \tilde{\Lambda}_A]}\tilde{g} - \mathrm{i}_{\tilde{\Lambda}_A}(\mathfrak{L}_{\Lambda_B} g)$, $\mathfrak{c}_A = \mathrm{i}_{\tilde{\Lambda}_A}\tilde{g} - \mathrm{i}_{\Lambda_A} g$ and $\mathfrak{C}_B = \mathfrak{L}_{\tilde{\Lambda}_B}\tilde{g} - \mathfrak{L}_{\Lambda_B} g$

Proof. Using $\mathfrak{L}_{\tilde{\Lambda}_B}\left(\tilde{\Lambda}_A\right) = \left[\tilde{\Lambda}_B, \tilde{\Lambda}_A\right]$, a computation gives $\mathfrak{L}_{\tilde{\Lambda}_B}(\mathrm{i}_{\tilde{\Lambda}_A}\tilde{g} - \mathrm{i}_{\Lambda_A} g) = \mathrm{i}_{[\tilde{\Lambda}_B, \tilde{\Lambda}_A]}\tilde{g} + \mathrm{i}_{\tilde{\Lambda}_A}(\mathfrak{L}_{\tilde{\Lambda}_B}\tilde{g}) - \mathfrak{L}_{\tilde{\Lambda}_B}(\mathrm{i}_{\Lambda_A} g) = \mathrm{i}_{[\tilde{\Lambda}_B, \tilde{\Lambda}_A]}\tilde{g} + \mathrm{i}_{\tilde{\Lambda}_A}(\mathfrak{L}_{\Lambda_B} g) - \mathfrak{L}_{\tilde{\Lambda}_B}(\mathrm{i}_{\Lambda_A} g) + \mathrm{i}_{\tilde{\Lambda}_A}(\mathfrak{L}_{\tilde{\Lambda}_B}\tilde{g} - \mathfrak{L}_{\Lambda_B} g)$. ■

For CL-equivalent systems, the RHS of (17) vanishes by Proposition 3. The idea is now to use the vanishing of the LHS, $\mathfrak{K}_{AB} = 0$, as an equation for λ, independent of $\tilde{g}$.

Theorem 8. Suppose the independent vector fields $\tilde{\Lambda}_A$, $(A = 1, .., m)$ satisfy the system of PDEs

$$\begin{aligned} \left[\tilde{\Lambda}_A, \tilde{\Lambda}_B\right] &= C^C_{AB}\tilde{\Lambda}_C \\ \mathfrak{L}_{\tilde{\Lambda}_B}\left(\mathrm{i}_{\Lambda_A} g\right) - C^C_{AB}\mathrm{i}_{\Lambda_C} g - \mathrm{i}_{\tilde{\Lambda}_A}\left(\mathfrak{L}_{\Lambda_B} g\right) &= 0 \end{aligned} \tag{18}$$

where C^C_{AB} are functions on $\mathcal{Q}$

then, for any solution $\tilde{L} = \frac{1}{2}\tilde{g} - \tilde{V}$ to the PDEs $\mathfrak{L}_{\tilde{\Lambda}_B}\tilde{L} = \mathfrak{L}_{\Lambda_B} L$ satisfying

the "initial condition" $\mathrm{i}_{\tilde{\Lambda}_A}\tilde{g} - \mathrm{i}_{\Lambda_A} g$ on a $n - m$ dimensional surface Γ transversal to $\tilde{\Lambda}$ (a Cauchy surface), then for some neighborhood $\mathcal{Q}_0$ of Γ, $(\mathcal{Q}_0, \tilde{L}, \tilde{\Lambda})$ and $(\mathcal{Q}_0, L, \Lambda)$ are CL-equivalent.

Proof. By assumption both $\mathfrak{L}_{\tilde{\Lambda}_B}\left(\mathrm{i}_{\Lambda_A} g\right) + \mathrm{i}_{[\tilde{\Lambda}_B, \tilde{\Lambda}_A]}\tilde{g} - \mathrm{i}_{\tilde{\Lambda}_A}\left(\mathfrak{L}_{\Lambda_B} g\right)$ and $\mathfrak{C}_B$ vanish, so (17) becomes $\mathfrak{L}_{\tilde{\Lambda}_B}\mathfrak{c}_A = 0$, with initial the condition $\mathfrak{c}_A = 0$ on Γ. Hence $\mathfrak{c}_A = 0$ identically, so (16) is satisfied. ■

Remark 1. The "λ-equation" of (Auckly *et al.*, 1998), derived by different methods, is equivalent to the (A, B)-symmetric part of (18_2).

Theorem 9. Suppose that the independent vector fields $\tilde{\Lambda}_A$, $(A = 1, .., m)$ satisfy the system of PDEs

$$\begin{aligned} \left[\tilde{\Lambda}_A, \tilde{\Lambda}_B\right] &= 0 \\ \mathfrak{L}_{\tilde{\Lambda}_B}\left(\mathrm{i}_{\Lambda_A} g\right) - \mathrm{i}_{\tilde{\Lambda}_A}\left(\mathfrak{L}_{\Lambda_B} g\right) &= 0 \\ \mathfrak{L}_{\tilde{\Lambda}_A}\mathfrak{L}_{\Lambda_B} L - \mathfrak{L}_{\tilde{\Lambda}_B}\mathfrak{L}_{\Lambda_A} L &= 0 \end{aligned} \tag{19}$$

then, for any initial values of $\tilde{L}$ satisfying the "initial condition" $\mathrm{i}_{\tilde{\Lambda}_A}\tilde{g} - \mathrm{i}_{\Lambda_A} g$ on the Cauchy surface Γ, there is a unique solution $\tilde{L} = \frac{1}{2}\tilde{g} - \tilde{V}$ of the PDEs $\mathfrak{L}_{\tilde{\Lambda}_B}\tilde{L} = \mathfrak{L}_{\Lambda_B} L$ on some neighborhood $\mathcal{Q}_0$ of Γ, and $(\mathcal{Q}_0, \tilde{L}, \tilde{\Lambda})$ and $(\mathcal{Q}_0, L, \Lambda)$ are CL-equivalent.

Proof. This is a corollary to the previous theorem together with the integrability conditions for the equation $\mathfrak{L}_{\tilde{\Lambda}_B}\tilde{L} = \mathfrak{L}_{\Lambda_B} L$. ■

4. SYSTEMS WITH SYMMETRY

4.1 *General*

In this section the above general results are applied to the symmetric special context of Bloch, Leonard and Marsden.

Definition 2. A *Special Lagrangian System* is an underactuated Lagrangian system $(\mathcal{Q}, L, \Lambda)$, such that $\mathcal{Q}$ is the total space of a principal fiber bundle $\pi : \mathcal{Q} \to \mathcal{B}$ with structure group $\mathcal{G}$, acting on $\mathcal{Q}$ on the left and such that

- Λ is integrable and the integral manifolds of Λ are bundle sections.
- $\mathcal{G}$ leaves invariant both L and Λ

The Λ-bundle is best thought of as (the horizontal distribution of) a flat connection on the principal fiber bundle. (This connection in general differs from the so called mechanical connection, formed by the kinetic-orthogonal complements of the fiber tangent planes.) The arguments in the following paragraph are standard in gauge theory (see e.g. (Bleecker, 1981)) and is only given a sketchy treatment.

4.2 *Strong Matching*

Definition 3. Two special Lagrangian systems $(\mathcal{Q}, L, \Lambda)$ and $(\mathcal{Q}, \tilde{L}, \tilde{\Lambda})$ with the same principal fiber bundle structure $\pi : \mathcal{Q} \to \mathcal{B}$ are said to be *strongly CL-equivalent* if $(\mathcal{Q}, L + \pi^* f, \Lambda)$ and $(\mathcal{Q}, \tilde{L} + \pi^* f, \tilde{\Lambda})$ are CL-equivalent as underactuated Lagrangian systems for every $f : \mathcal{B} \to \mathbb{R}$. (This is also expessed by saying that L and $\tilde{L}$ are *strongly matching.*)

In particular, strong CL-equivalence implies that $\mathfrak{L}_{\tilde{\Lambda}_A}\pi^* f = \mathfrak{L}_{\Lambda_A}\pi^* f$, identically in f, which in turn means that $\tilde{\Lambda}_A - \Lambda_A$ is *vertical*, *i.e.* tangential to the fibers. Since Λ is $\mathcal{G}$ -invariant, it is henceforth assumed, without loss of generality, that the basis fields Λ_A are $\mathcal{G}$-invariant. By (16_1) and the $\mathcal{G}$-invariance of L and $\tilde{L}$, this then holds for the $\tilde{\Lambda}_A$ as well.

For elements $\xi \in \mathfrak{g}$ of the Lie algebra of $\mathcal{G}$, $(\xi)_\mathcal{Q}$ denotes the corresponding vector field on $\mathcal{Q}$. The vertical vector field $\tilde{\Lambda}_A - \Lambda_A$ can be expressed as $(\Theta_A)_\mathcal{Q}$ for some unique $\Theta_A : \mathcal{Q} \to \mathfrak{g}$. The $\mathcal{G}$ -invariance of $\tilde{\Lambda}_A - \Lambda_A$ means that the Θ_A are *Ad*-equivariant.

By no loss of generality it can be assumed that $[\Lambda_A, \Lambda_B] = 0$, which is the condition that the Λ_A are the Λ-horizontal lifts of some coordinate vector fields $\frac{\partial}{\partial y^A}$ on $\mathcal{B}$. It follows from $\tilde{\Lambda}_A = \Lambda_A + (\Theta_A)_\mathcal{Q}$ that the $\tilde{\Lambda}_A$ then are the $\tilde{\Lambda}$-horizontal lifts of $\frac{\partial}{\partial y^A}$, and from this and the integrability of $\tilde{\Lambda}$ that $\left[\tilde{\Lambda}_A, \tilde{\Lambda}_B\right] = 0$. In terms of Θ_A and the bracket on $\mathfrak{g}$, the latter result reads

$$\mathfrak{L}_{\Lambda_A}\Theta_B - \mathfrak{L}_{\Lambda_B}\Theta_A + [\Theta_A, \Theta_B] = 0 \tag{20}$$

The *Ad*-equivariant 1-form $\Theta = \Theta_A \pi^* dy^A$ is independent of the particular coordinate system y^A. The equation (20) is a form of the Maurer-Cartan equations.

4.3 *Conditions for Strong Matching*

Consider the strong matching problem for Lagrangians of the form (8), without linear term in $\dot{q}$. As is easily checked, $L = \frac{1}{2}g - V$ and $\tilde{L} = \frac{1}{2}\tilde{g} - \tilde{V}$ are strongly matching precisely when $V = \tilde{V}$ and $\frac{1}{2}g$ and $\frac{1}{2}\tilde{g}$ are strongly matching. Hence, in the following only the case $V = 0$ is considered. From the discussion of the preceding paragraph, it follows that the necessary and sufficient conditions for strong matching is found by applying Theorem 9 with $\tilde{\Lambda}_A$ of the form $\Lambda_A + (\Theta_A)_{\mathcal{Q}}$.

Theorem 10. If g and $\tilde{g}$ restrict to the same metric on vertical vectors, Θ equals the Lie algebra valued one-form τ of (Bloch *et al.*, 1997*b*).

Proof. In a basis with the first $n - m$ vectors vertical and the last m vectors in Λ, the metrics take the block form $g = \begin{pmatrix} g_{11} & g_{12} \\ g_{21} & g_{22} \end{pmatrix}$ and similarly for $\tilde{g}$. Then the components of $\tau_{\mathcal{Q}}$ are given by $\tau_{\mathcal{Q}} = g_{11}^{-1}g_{12} - \tilde{g}_{11}^{-1}\tilde{g}_{12}$, while $\Theta_{\mathcal{Q}}$ is given by $g_{11}^{-1}g_{12} - \tilde{g}_{11}^{-1}\tilde{g}_{12} + (\tilde{g}_{11}^{-1} - g_{11}^{-1})\,\tilde{g}_{12}$. ■

Let ξ_α be a basis for $\mathfrak{g}$, $c^\gamma_{\alpha\beta}$ the structure constants, $[\xi_\alpha, \xi_\beta] = c^\gamma_{\alpha\beta}\xi_\gamma$, write $\Theta_A = \Theta^\alpha_A\ \xi_\alpha$and denote by $g_{\alpha\beta}$, $g_{\alpha B}$ and g_{AB} the components of g w.r.t. the anholonomic basis formed by the $(\xi_\alpha)_{\mathcal{Q}}$ and the Λ_A.

Theorem 11. In terms of the functions $\Theta^\alpha_A : \mathcal{Q} \to \mathbb{R}$, the conditions (9) read

$$\mathfrak{L}_{(\xi_\alpha)_{\mathcal{Q}}}\Theta^\gamma_A - c^\gamma_{\alpha\beta}\Theta^\beta_A = 0 \quad (21)$$

$$\mathfrak{L}_{\Lambda_A}\Theta^\gamma_B - \mathfrak{L}_{\Lambda_B}\Theta^\gamma_A + c^\gamma_{\alpha\beta}\Theta^\alpha_A\Theta^\beta_B = 0 \quad (22)$$

$$c^\gamma_{\alpha\beta}\Theta^\beta_A g_{\gamma B} - \Theta^\beta_B \mathfrak{L}_{\Lambda_A} g_{\alpha\beta} = 0 \quad (23)$$

$$(\mathfrak{L}_{\Lambda_C}\Theta^\alpha_A)\, g_{\alpha B} - \Theta^\beta_B\,(\mathfrak{L}_{\Lambda_A} g_{C\beta}) = 0 \quad (24)$$

$$\mathfrak{L}_{\Theta^\alpha_A(\xi_\alpha)_{\mathcal{Q}}}\mathfrak{L}_{\Lambda_B} g - \mathfrak{L}_{\Theta^\alpha_B(\xi_\alpha)_{\mathcal{Q}}}\mathfrak{L}_{\Lambda_A} g = 0 \quad (25)$$

Proof. This follows by substitution of $\tilde{\Lambda}_A = \Lambda_A + (\Theta_A)_{\mathcal{Q}}$ in (9). Condition (21) is the equivariance property, (22) and corresponds to (16_1), (23) and (24) are the condition (16_2) and (25) are the condition (16_3). ■

In future papers, these conditions will be discussed in more detail, in particular, they will be written as PDEs on the base space $\mathcal{B}$. Here only a degenerate example is given.

Corollary 12. If both the base space $\mathcal{B}$ and the symmetry group $\mathcal{G}$ are one-dimensional (q^1 is the fiber variable and q^2 is a base variable such that $\Lambda = \text{span}\left(\frac{\partial}{\partial q^2}\right)$), then the necessary and sufficient conditions on $\Theta = \vartheta\,(q^1, q^2)\, dq^2$ for strong matching is that either $\vartheta \equiv 0$ or that g_{11} is constant and $\vartheta = Cg_{12}$, for some constant C.

Proof. In this case, (22) and (25) are vacuous, and the remaining conditions read $\frac{\partial\vartheta}{\partial q^1} = 0$, $\vartheta\frac{\partial g_{11}}{\partial q^2} = 0$, and $\frac{\partial\vartheta}{\partial q^2}g_{12} - \vartheta\frac{\partial g_{12}}{\partial q^2} = 0$. ■

Applying this corollary to the inverted pendulum case, the solution in (Bloch *et al.*, 1997*b*) is recovered.

5. CONCLUDING REMARKS

The necessary and sufficient conditions for *strong matching* in this paper are of course sufficient but not necessary for *matching*. On one hand, they do not cover all cases covered by the master matching theorem of (Bloch *et al.*, 1999), on the other hand they work equally well for nonabelian structure groups.

6. REFERENCES

Auckly, D., L. Kapitanski and W. White (1998). Control of nonlinear underactuated systems. Preprint.

Bleecker, D. (1981). *Gauge Theory and Variational Principles.* Addison-Wesley.

Bloch, A., J. E. Marsden and G. Sánchez (1997*a*). Stabilization of relative equilibria of mechanical systems with symmetry. In: *Current and future directions in applied mathematics* (M. Alber, B. Hu and J. Rosenthal, Eds.). pp. 43–64. Birkhäuser.

Bloch, A., N. Leonard and J. E. Marsden (1997*b*). Stabilization of mechanical systems using controlled lagrangians. In: *Proc. of IEEE CDC.* Vol. 36. pp. 2356–2361.

Bloch, A., N. Leonard and J. E. Marsden (1998*a*). Controlled lagrangians and the stabilization of mechanical systems i. Preprint.

Bloch, A., N. Leonard and J. E. Marsden (1998*b*). Matching and stabilization by the method of controlled lagrangians. In: *Proc. of IEEE CDC.* Vol. 37. pp. 1446–1451.

Bloch, A., N. Leonard and J. E. Marsden (1999). Stabilization of the pendulum on a rotor arm by the method of controlled lagrangians. In: *Proc. of IEEE Int. Conference on Robotics and Automation.*

Hamberg, J. (1999). General matching conditions in the theory of controlled lagrangians. In: *Proc. of IEEE CDC.* Vol. 38. pp. 2519–2523.

Ortega, R., A. Loría, P. J. Nicklasson and H. Sira-Ramírez (1998). *Passivity-based Control of Euler-Lagrange Systems.* Springer.

TIME-VARYING STABILIZATION OF HAMILTONIAN SYSTEMS VIA GENERALIZED CANONICAL TRANSFORMATIONS

Kenji Fujimoto*, Toshiharu Sugie****

* *Faculty of Information Technology and Systems*
Department of Electrical Engineering
Delft University of Technology
P.O.Box 5031, 2600 GA Delft, The Netherlands
** *Department of Systems Science*
Graduate School of Informatics, Kyoto University
Uji, Kyoto 611-0011 Japan
fujimoto@i.kyoto-u.ac.jp
sugie@i.kyoto-u.ac.jp

Abstract: This paper focuses on the stabilization of port-controlled Hamiltonian systems employing possibly time-varying controllers. At first we refer to the generalized canonical transformation which preserves the structure of Hamiltonian systems and the passivity property that physical systems innately possess. Next we show a general stabilization strategy for port-controlled Hamiltonian systems based on it which is a natural generalization of well-known passivity based control. Finally we utilize this method to mechanical Hamiltonian systems with nonholonomic constraints by modifying the kinetic energy of the system. Furthermore some examples are given to show how this technique works for physical systems. *Copyright ©2000 IFAC*

Keywords: Physical models, nonlinear systems, time-varying systems, transformations, passive.

1. INTRODUCTION

Hamiltonian control systems (Nijmeijer and van der Schaft, 1990) are the systems described by the Hamilton's canonical equations which represent general physical systems. Recently port-controlled Hamiltonian systems are introduced as the generalization of Hamiltonian systems (Maschke and van der Schaft, 1992). They can represent Hamiltonian systems with a class of nonholonomic constraints (Maschke and van der Schaft, 1994; Khennouf *et al.*, 1995) as well as ordinary mechanical, electrical and electro-mechanical systems.

Generalized canonical transformations for port-controlled Hamiltonian systems are introduced as the generalization of conventional canonical transformations which are widely used for analysis in classical mechanics (Fujimoto and Sugie, 1998*a*; Fujimoto and Sugie, 1998*b*). This transformation preserves the structure of Hamiltonian systems and the passivity property which physical systems innately possess. This property gives us a general stabilization procedure for port-controlled Hamiltonian systems which is a natural generalization of well-known passivity based control (Takegaki and Arimoto, 1981; Ortega *et al.*, 1998). As an application, a stabilization method for a special class of nonholonomic port-controlled Hamiltonian systems are obtained which employ non-smooth Hamiltonian functions to obtain discontinuous stabilizing controllers (Fujimoto and Sugie, 1999*a*; Fujimoto *et al.*, 1999).

This paper will give a stabilization procedure based on possibly time-varying generalized canonical transformations. Firstly we refer to the generalized canonical transformation for port-controlled Hamiltonian systems which preserves the Hamiltonian structure and the passivity property. Secondly we introduce a general stabilization procedure for port-controlled Hamiltonian systems based on it which is a natural generalization of the passivity based control. Finally its application to the time-varying stabilization of nonholonomic Hamiltonian systems which modifies the kinetic energy of the original mechanical system is derived. In this method, we refer to (Pomet, 1992) for the time-varying stabilization of driftless nonholonomic systems and utilize this result to the stabilization of our nonholonomic port-controlled Hamiltonian systems.

2. CANONICAL TRANSFORMATIONS FOR HAMILTONIAN SYSTEMS

2.1 *Port-controlled Hamiltonian systems*

A possibly time-varying port-controlled Hamiltonian system with a Hamiltonian $H(x,t)$ is a system of the form

$$\begin{cases} \dot{x} = J(x,t)\dfrac{\partial H}{\partial x}(x,t)^T + g(x,t)u \\ y = g(x,t)^T\dfrac{\partial H}{\partial x}(x,t)^T \end{cases} \tag{1}$$

with $u, y \in \mathbb{R}^m$, $x \in \mathbb{R}^n$ and a skew symmetric matrix J, i.e. $J = -J^T$ holds (Maschke and van der Schaft, 1992; Fujimoto and Sugie, 1998*a*). The following properties of such systems are known.

Theorem 1 (Maschke and van der Schaft, 1992; Fujimoto and Sugie, 1998*a*) *Consider the system (1). Suppose Hamiltonian H satisfies $H(x,t) \geq H(0,t) = 0$ and $\partial H/\partial t \leq 0$. Then the input-output system is passive with respect to the storage function H, and the feedback*

$$u = -C\ y \tag{2}$$

with a matrix $C > 0$ renders $(u, y) \to 0$. Furthermore if H is a positive definite function and if the system is zero-state detectable, then the feedback (2) renders the system asymptotically stable.

The zero-state detectability and the positive definiteness of the Hamiltonian assumed in Theorem 1 do not always hold for general systems. In such a case, the generalized canonical transformation is useful. It provides a generalization of the stabilization method of exploiting virtual potential energy (Takegaki and Arimoto, 1981)

2.2 *Generalized canonical transformations and stabilization*

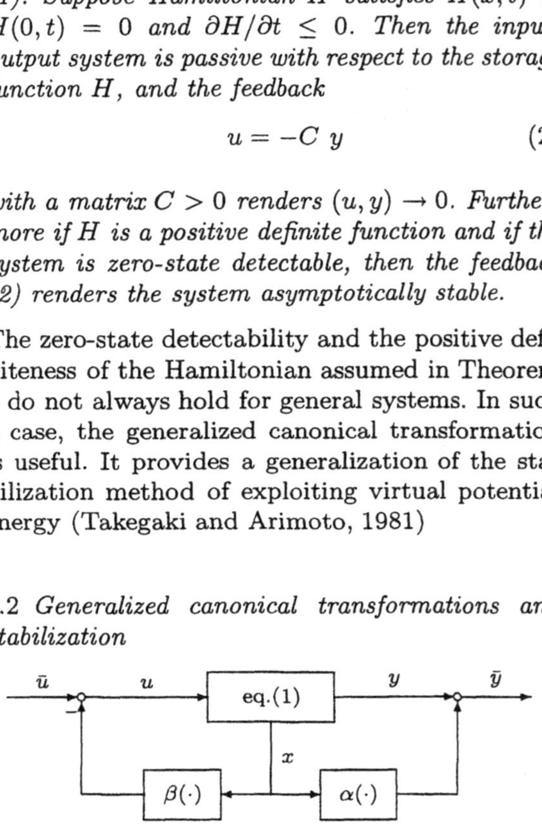

Fig. 1. Generalized canonical transformation

A generalized canonical transformation is a set of transformations

$$\begin{cases} \bar{x} = \Phi(x,t) \\ \bar{H} = H(x,t) + U(x,t) \\ \bar{y} = y + \alpha(x,t) \\ \bar{u} = u + \beta(x,t) \end{cases} \tag{3}$$

which preserves the structure of port-controlled Hamiltonian systems in (1). Here $\bar{x}$, $\bar{H}$, $\bar{y}$ and $\bar{u}$ denote the new state, the new Hamiltonian, the new output and the new input respectively and this transformation can be seen as in Figure 1 from the input-output point of view. The generalized canonical transformation is a natural generalization of a canonical transformation which is widely used for the analysis of conventional Hamiltonian systems in classical mechanics. In the figure, we employ a feedback β hence the conventional canonical transformation in the sense of classical mechanics can be obtained in the case $\beta = 0$. The properties of such transformations are summarized as follows.

Theorem 2 (Fujimoto and Sugie, 1998*a*; Fujimoto and Sugie, 1998*b*) *(i) Consider the system (1). For any functions U and β, there exists a pair of functions Φ and α such that the set (3) yields a generalized canonical transformation and the function α is given by $\alpha = g^T \frac{\partial U}{\partial x}^T$. Furthermore Φ yields a generalized canonical transformation if and only if*

$$\frac{\partial \Phi}{\partial (x,t)} \begin{pmatrix} J\frac{\partial U}{\partial x}^T + K\frac{\partial H+U}{\partial x}^T + g\beta \\ -1 \end{pmatrix} = 0 \tag{4}$$

holds with a skew-symmetric matrix $K(x,t)$.
(ii) Transform the system (1) by the generalized canonical transformation with U and β such that $H + U \geq 0$. Then the new input-output mapping $\bar{u} \mapsto \bar{y}$ is passive with the storage function $\bar{H}$ if and only if

$$\frac{\partial (H+U)}{\partial (x,t)} \begin{pmatrix} J\frac{\partial U}{\partial x}^T + g\beta \\ -1 \end{pmatrix} \geq 0. \tag{5}$$

Suppose that (5) holds and that $H + U$ is positive definite. Then the feedback $\bar{u} = -C\ \bar{y}$ with $C > 0$ renders the system stable. Suppose moreover that the transformed system is zero-state detectable with respect to x. Then the feedback renders the system asymptotically stable.

By using the generalized canonical transformation, we can change the property of systems without changing the intrinsic passive property of physical systems. The structure matrix $\bar{J}$ of the transformed system is given by

$$\bar{J} = \frac{\partial \Phi}{\partial x}(J + K)\frac{\partial \Phi}{\partial x}^T. \tag{6}$$

Therefore this transformation can also be used to change the structure matrix J. As mentioned in (Fujimoto and Sugie, 1998*a*), the (possibly) time-varying coordinate transformation $\bar{x} = \Phi(x,t)$ used in the generalized canonical transformation should satisfy the uniform boundedness condition.

If this property holds, then the (asymptotic) stabilization in the new coordinate coincides with that in the old coordinate. Now we discuss a stabilization procedure based on Theorem 2 using an example.

Example 1 Consider a conventional Hamiltonian system

$$\begin{cases} \begin{pmatrix} \dot{q} \\ \dot{p} \end{pmatrix} = \begin{pmatrix} 0 & I \\ -I & 0 \end{pmatrix} \begin{pmatrix} \frac{\partial H}{\partial q}^T \\ \frac{\partial H}{\partial p}^T \end{pmatrix} + \begin{pmatrix} 0 \\ I \end{pmatrix} u \\ y \quad = \begin{pmatrix} 0 & I \end{pmatrix} \begin{pmatrix} \frac{\partial H}{\partial q}^T \\ \frac{\partial H}{\partial p}^T \end{pmatrix} \end{cases} \tag{7}$$

with a time-varying Hamiltonian $H = \frac{1+a \sin t}{2} p^T p$, a constant a, $|a| < 1$ and the coordinate $x = (q, p)$. Consider the canonical transformation with $U = -\frac{a \sin t}{2} p^T p$. Then it follows from Theorem 2 that a coordinate transformation $\bar{x} = \Phi(x, t)$ yields a canonical transformation if and only if

$$\frac{\partial \Phi}{\partial (x,t)} \left(\begin{pmatrix} -(a \sin t)\, p \\ 0 \\ -1 \end{pmatrix} + K \begin{pmatrix} 0 \\ p \end{pmatrix} \right) = 0. \tag{8}$$

Now we choose $K = 0$. Then one solution of (8) gives the coordinate transformation of the generalized canonical transformation

$$\begin{pmatrix} \bar{q} \\ \bar{p} \end{pmatrix} = \Phi(q,p,t) = \begin{pmatrix} q + (a \cos t)\, p \\ p \end{pmatrix}. \tag{9}$$

In this coordinate $\bar{x} = (\bar{q}, \bar{p})$, the system reduces to another (time-varying) Hamiltonian system

$$\begin{cases} \begin{pmatrix} \dot{\bar{q}} \\ \dot{\bar{p}} \end{pmatrix} = \begin{pmatrix} 0 & I \\ -I & 0 \end{pmatrix} \begin{pmatrix} \frac{\partial \bar{H}}{\partial \bar{q}}^T \\ \frac{\partial \bar{H}}{\partial \bar{p}}^T \end{pmatrix} + \begin{pmatrix} a \cos t I \\ I \end{pmatrix} u \\ y \quad = \begin{pmatrix} a \cos t I & I \end{pmatrix} \begin{pmatrix} \frac{\partial \bar{H}}{\partial \bar{q}}^T \\ \frac{\partial \bar{H}}{\partial \bar{p}}^T \end{pmatrix} = \bar{p} = p \end{cases} \tag{10}$$

with the Hamiltonian $\bar{H} = \frac{1}{2} \bar{p}^T \bar{p}$. Thus the transformation with $U = -\frac{a \sin t}{2} p^T p$ and Φ in (9) is also the conventional canonical transformation in classical mechanics.

After all we obtain a Hamiltonian system (10) with the time invariant Hamiltonian $\bar{H}$, namely, Hamiltonian $\bar{H}$ does not depend explicitly on t in the new coordinate $(\bar{q}, \bar{p})$. This fact implies the port-controlled Hamiltonian system (10) is *passive* (*lossless*) with a storage function $\bar{H}$ (see (Fujimoto and Sugie, 1998*a*; Fujimoto and Sugie, 1998*b*) for more detail), whereas the original system (7) is not. Furthermore the *uniform boundedness* property

$$\frac{1}{\sqrt{1+a^2}} |x| \le |\bar{x}| \le \sqrt{1+a^2}\, |x| \tag{11}$$

holds under the coordinate transformation (9). Hence the (asymptotic) stabilization in the new coordinate $\bar{x} = (\bar{q}, \bar{p})$ is equivalent to that in the original coordinate $x = (q, p)$. Both analysis and synthesis (e.g. stabilization) of the new Hamiltonian system (10) will be easier than the original one (7) and this fact shows the advantage of utilizing canonical transformations.

In the above discussion, we did not use the feedback β. Now we try to use the feedback to obtain a positive definite Hamiltonian function for the asymptotic stabilization. Consider the port-controlled Hamiltonian system (10). This system is similar to the ordinary Hamiltonian system (7). Hence we assume U and β

$$U(\bar{q}, t) = \frac{\theta(t)}{2} \bar{q}^T \bar{q} \tag{12}$$

$$\beta(\bar{q}, t) = \theta(t)\, \bar{q} \tag{13}$$

with a scalar periodic positive function $\theta(t)$ will satisfy the passivity preserving condition (5)[1]. Then (5) reduces to

$$(2a \cos t)\, \theta(t)^2 - \dot{\theta}(t) \ge 0. \tag{14}$$

Solving this condition with the equality yields

$$\theta(t) = \frac{1}{c - 2a \sin t} \tag{15}$$

with an arbitrary positive constant $c > 2|a|$. Further, the relevant coordinate transformation satisfies the condition

$$\frac{\partial \bar{\Phi}}{\partial (\bar{q}, \bar{p}, t)} \begin{pmatrix} \frac{a \cos t}{c - 2a \sin t} \bar{q} \\ 0 \\ -1 \end{pmatrix} = 0 \tag{16}$$

by substituting $K = 0$ for (4) in Theorem 2. This yields the following solution

$$\begin{pmatrix} \hat{q} \\ \hat{p} \end{pmatrix} = \bar{\Phi}(\bar{q}, \bar{p}, t) = \begin{pmatrix} \frac{1}{\sqrt{c - 2a \sin t}} \bar{q} \\ \bar{p} \end{pmatrix}. \tag{17}$$

Similar to (11) the uniform boundedness property holds.

$$\min\left(1, \frac{1}{\sqrt{c+2|a|}}\right) |\bar{x}| \le |\hat{x}| \le \max\left(1, \frac{1}{\sqrt{c-2|a|}}\right) |\bar{x}| \tag{18}$$

At last we obtain another *lossless* port-controlled Hamiltonian system with Hamiltonian $\hat{H} = \frac{1}{2}(\hat{q}^T \hat{q} + \hat{p}^T \hat{p})$ which is positive definite and time invariant in the coordinate $\hat{x} = (\hat{q}, \hat{p})$. Since this system is zero-state detectable, the feedback $\hat{u} =$

[1] For conventional Hamiltonian system (7) with a kinetic energy Hamiltonian function $H = \frac{1}{2} p^T M(q)^{-1} p$, the choice of $U = \frac{c}{2} q^T q$ and $\beta = cq$ is well known as a method of exploiting potential energy (Takegaki and Arimoto, 1981).

$-C\ \hat{y}$ with $C > 0$ will render the system asymptotically stable. This feedback will also asymptotically stabilize the original system (7) from the uniform boundedness conditions (11) and (18). ◇

In this example, we focused on the stabilization of a time-varying system, but this stabilization method itself is of course useful for time invariant systems as well. If we want to design a time invariant controller for a time invariant system, then from the equations (4) and (5) we just should solve the equation

$$J\frac{\partial U}{\partial x}^T + K\frac{\partial (H+U)}{\partial x}^T + g\beta = 0. \qquad (19)$$

Then we can obtain a passivity preserving generalized canonical transformation without the coordinate transformation.

3. TIME-VARYING STABILIZATION OF NONHOLONOMIC HAMILTONIAN SYSTEMS

3.1 *Nonholonomic Hamiltonian systems*

This section discusses the asymptotic stabilization of nonholonomic port-controlled Hamiltonian systems via time-varying generalized canonical transformations. It is shown in (Maschke and van der Schaft, 1994) that a conventional mechanical Hamiltonian system with nonholonomic constraints is described by the following port-controlled Hamiltonian system

$$\begin{cases} \begin{pmatrix}\dot{q}\\ \dot{p}\end{pmatrix} = \begin{pmatrix} 0 & J_{12}(q)\\ -J_{12}(q)^T & J_{22}(q,p)\end{pmatrix}\begin{pmatrix}\frac{\partial H}{\partial q}^T\\ \frac{\partial H}{\partial p}^T\end{pmatrix} + \begin{pmatrix}0\\ G\end{pmatrix}u \\ y = G(q)^T\frac{\partial H}{\partial p}(q,p)^T \end{cases} \qquad (20)$$

with $q \in \mathbb{R}^n$, $p, u \in \mathbb{R}^m$ $(n > m)$ and

$$H(q,p) = \tfrac{1}{2}\, p^T M(q)^{-1} p. \qquad (21)$$

Here J_{12} has a full column rank, J_{22} is skew-symmetric, $M(q) > 0$ is symmetric and $G(q)$ is nonsingular. In the rest of the paper, we focus on this system.

3.2 *Modification of the kinetic energy and stabilization*

In the controller design given in Theorem 2 we have to choose an appropriate U such that the new Hamiltonian $\bar{H} = H + U$ is a positive definite function and that there exists a feedback β satisfying the passivity preserving condition (5). For conventional Hamiltonian systems as (7) it is easy to find an appropriate U as shown in Footnote 1 and for more complicated (but fully actuated) port-controlled Hamiltonian systems it becomes a bit more complicated but a method to solve the equation (19) as a PDE is proposed (Maschke *et al.*, 1998; Ortega *et al.*, 1999) [2] and we are often able to find an appropriate U. However for nonholonomic systems it is not so easy to find appropriate U (and β) because we have to employ time-varying or discontinuous feedback to achieve asymptotic stability (Brockett, 1983). This implies that we have to employ either non-smooth or time-varying U function. A non-smooth approach for the stabilization of a special class of nonholonomic Hamiltonian systems is shown in (Fujimoto and Sugie, 1999*a*; Fujimoto *et al.*, 1999) but time-varying approach was not obtained yet [3].

We now employ a time-varying Hamiltonian function U in order to achieve the asymptotic stability. To this end, we first employ a preliminary time-varying generalized canonical transformation to modify the kinetic energy, i.e. we change it into a time-varying one, and after that we will modify it again such that the new Hamiltonian becomes a positive definite function by employing a virtual potential energy in a similar way as the existing method (Takegaki and Arimoto, 1981).

Theorem 3 *(i) Consider the port-controlled Hamiltonian system (20) and a smooth vector valued function $\alpha(q,t)$. Suppose $\alpha(q,t)$ is an odd periodic function such that $\alpha(0,t) = 0$. Then there (locally) exists a coordinate transformation $\bar{q} = \Phi(q,t)$ and time periodic matrix valued functions $N_1(q,t)$ and $N_2(q,p,t)$ such that the following set of transformations*

$$\begin{cases} \bar{H} = H + p^T\alpha + \dfrac{1}{2}\alpha^T M\alpha \\ \bar{q} = \Psi(q,t) \\ \bar{p} = N_1(p + M\alpha) \\ \bar{y} = y + \alpha(q,t) \\ \bar{u} = u + G(q)^{-1}\times \\ \qquad \left(J_{12}^T\frac{\partial U}{\partial q}^T + M\frac{\partial \alpha}{\partial t} - (N_2 J_{12} + J_{22})\alpha\right) \end{cases} \qquad (22)$$

gives a generalized canonical transformation which transforms the port-controlled Hamiltonian system (20) into another lossless time-varying one

$$\begin{cases} \begin{pmatrix}\dot{\bar{q}}\\ \dot{\bar{p}}\end{pmatrix} = \begin{pmatrix} 0 & \bar{J}_{12}(\bar{q},t)\\ -\bar{J}_{12}(\bar{q},t)^T & \bar{J}_{22}(\bar{q},\bar{p},t)\end{pmatrix}\begin{pmatrix}\frac{\partial \bar{H}}{\partial \bar{q}}^T\\ \frac{\partial \bar{H}}{\partial \bar{p}}^T\end{pmatrix} + \begin{pmatrix}0\\ \bar{G}\end{pmatrix}\bar{u} \\ \bar{y} = \bar{G}(\bar{q},t)^T\frac{\partial \bar{H}}{\partial \bar{p}}^T \end{cases} \qquad (23)$$

with a time invariant Hamiltonian

$$\bar{H}(\bar{q},\bar{p}) = \tfrac{1}{2}\bar{p}^T\bar{M}(\bar{q})^{-1}\bar{p}. \qquad (24)$$

---

[2] It should be noted that the method in this paper can treat larger class of (time invariant) systems than (1).

[3] Preliminary results are found in (Khennouf *et al.*, 1995).

Here the coordinate transformation $\bar q = \Psi(q,t)$ *is the solution of a PDE*

$$\frac{\partial \Psi}{\partial(q,t)}\begin{pmatrix} J_{12}\,\alpha \\ -1 \end{pmatrix} = 0. \qquad (25)$$

N_1 *and* N_2 *can be directly obtained from (5) and the solution* $\Psi(q,t)$ *of (25), and they are uniformly bounded if so is* Ψ.

(ii) Consider another generalized canonical transformation in order to add a potential function $\bar U(\bar q)$ *such that the new Hamiltonian* $\hat H$ *is positive definite.*

$$\begin{cases} \hat H &= \bar H(\bar q,\bar p) + \bar U(\bar q) \\ \hat q &= \bar q \\ \hat p &= \bar p \\ \hat u &= \bar u + \bar G^{-1}(\bar q,t)\bar J_{12}^T \frac{\partial \bar U}{\partial \bar q}^T \\ \hat y &= \bar y \end{cases} \qquad (26)$$

Then the transformed system will be zero-state detectable, i.e. it can be asymptotically stabilized by the feedback $\hat u = -C\hat y$ *with* $C > 0$, *if* $\alpha(q,t)$ *and* $\bar U(\bar q)$ *satisfy*

$$\alpha(q,t) \equiv 0,\ \frac{\partial \bar U(\Psi(q,t))}{\partial q} J_{12} \equiv 0 \ \Rightarrow\ q = 0. \quad (27)$$

Theorem 3 can be easily obtained from Theorem 2. In Theorem 2, at first we choose U (and β) such that the passivity preserving condition (5) holds but now we choose the change of output $\alpha(q,t)$ at first then the rest variables are determined almost automatically from the passivity preserving condition (5). By the generalized canonical transformation (22) given in Theorem 3, we can transform the nonholonomic port-controlled Hamiltonian system (20) into another lossless *time-varying* one (23) with a *time invariant* Hamiltonian function which has a quadratic form with respect to the new momentum $\bar p$. Notice that the new Hamiltonian has the same form as the original kinetic energy and that the transformation preserves the passivity property although it is time-varying.

Remark This theorem contains the condition (27) which is not so easy to check in advance. However this stabilization procedure is very similar to the time-varying stabilization method for driftless systems by Pomet and the techniques in (Pomet, 1992) can be directly utilized for the stabilization of our nonholonomic Hamiltonian systems. Indeed it was shown that a sufficient condition for the existence of global uniformly bounded coordinate transformation Ψ is

$$|\alpha_1(q,t)|\,\|j_1(q,t)\| \le c(1 + \|q\|) \qquad (28)$$

with $\alpha = (\alpha_1, 0, \ldots, 0)$ where j_1 is the first column of J_{12}. Hence if we use α such that (28) holds then we can achieve the global asymptotic stability. We do not repeat such discussions here for the reason of space. See (Fujimoto and Sugie, 1999*b*) for detail and more practical sufficient conditions.

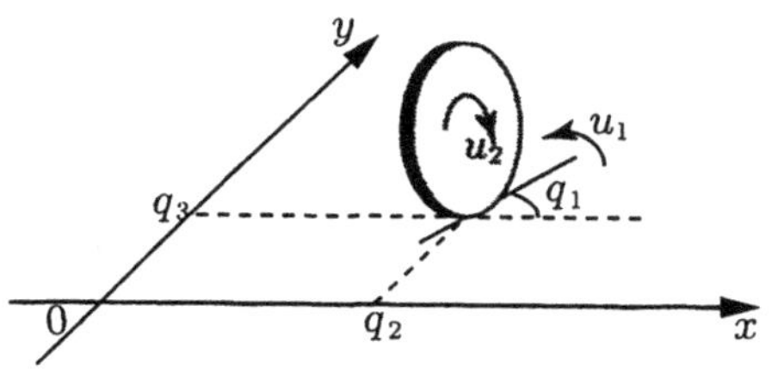

Fig. 2. A rolling coin.

Example 2 We consider a rolling coin on the x-y plane shown in Figure 2 which is an example of well-known knife edge systems. Let the configuration states q_1, q_2 and q_3 denote the heading angle, the position on x axis and the position on y axis respectively, let the inputs u_1 and u_2 denote the torque to increase the heading angle and that to increase the rolling angle and let the phase states p_1 and p_2 denote the momentum with respect to u_1 and u_2 respectively. By using this notations we can obtain the port-controlled Hamiltonian system of the form (20) with

$$J = \begin{pmatrix} 0 & 0 & 0 & 1 & 0 \\ 0 & 0 & 0 & 0 & \cos q_1 \\ 0 & 0 & 0 & 0 & \sin q_1 \\ -1 & 0 & 0 & 0 & 0 \\ 0 & -\cos q_1 & -\sin q_1 & 0 & 0 \end{pmatrix},\ g = \begin{pmatrix} 0 & 0 \\ 0 & 0 \\ 0 & 0 \\ 1 & 0 \\ 0 & 1 \end{pmatrix} \quad (29)$$

where $q = (q_1, q_2, q_3)$, $p = (p_1, p_2)$, $x = (q, p)$, $H = \frac{1}{2}p^T p$ (van der Schaft, 1996).

Now we choose the odd periodic function $\alpha(q,t)$ which defines the output function of the new port-controlled Hamiltonian system as

$$\alpha(q,t) = \begin{pmatrix} q_3 \sin t \\ 0 \end{pmatrix}. \qquad (30)$$

Then the generalized canonical transformation (22) reduces to

$$\bar q = \begin{pmatrix} q_1 - q_3 \cos t \\ q_2 \\ q_3 \end{pmatrix},\ \bar p = \begin{pmatrix} p_1 + q_3 \sin t \\ p_2 \end{pmatrix} \quad (31)$$

$$\bar u = u + \begin{pmatrix} 0 \\ (p_1 + q_3 \sin t)\sin q_1 \sin t \end{pmatrix}. \qquad (32)$$

This transformation transforms the system into another lossless (time-varying) port-controlled Hamiltonian system which has

$$\bar H = \frac{1}{2}\,\bar p^T \bar p, \quad \bar g = g$$

$$\bar J = \begin{pmatrix} 0 & 0 & 0 & 1 & -s_1 \cos t \\ 0 & 0 & 0 & 0 & c_1 \\ 0 & 0 & 0 & 0 & s_1 \\ -1 & 0 & 0 & 0 & s_1 \sin t \\ s_1 \cos t & -c_1 & -s_1 & -s_1 \sin t & 0 \end{pmatrix}$$

$$s_1 := \sin(\bar{q}_1 + \bar{q}_3 \cos t), \quad c_1 := \cos(\bar{q}_1 + \bar{q}_3 \cos t).$$

This system is zero-state detectable owing to the time-varying property of J_{12} matrix. (The fact that the null space of J_{12} matrix is time-varying is crucial.) Furthermore if we choose the adding potential function as $\bar{U} = \frac{1}{2}\bar{q}^T\bar{q}$ then we obtain the following stabilizing feedback.

$$u = -\begin{pmatrix} q_1 + q_3(\sin t - \cos t) + p_1 \\ (2q_3 - q_1 \cos t + p_1 \sin t)\sin q_1 + q_2 \cos q_1 + p_2 \end{pmatrix}$$

The responses from the initial condition $q(0) = (0,0,1)$ of the resulting feedback system in simulation are shown in Figure 3. The solid line, dashed line and dashed and solid line denote q_1, q_2 and q_3 respectively. Although the convergence is slow and oscillatory as usual, all states converges to the origin smoothly. ◇

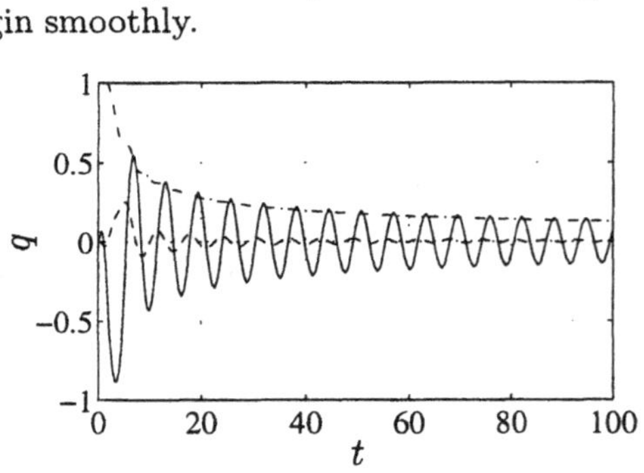

Fig. 3. Responses of q from $q(0) = (0,0,1)$.

It should be noted that the backstepping technique also can stabilize our nonholonomic Hamiltonian system by using the stabilizing controller for the driftless system (Fierro and Lewis, 1995; Jiang and Nijmeijer, 1997). However our result will be still useful because this stabilization procedure preserves the Hamiltonian structure and the resulting feedback is relatively simple. Also this method can be regarded as the *passivity based control of nonholonomic systems* hence it is expected that we can construct a robustly stable control systems. Of course we can utilize existing various tools for passivity based control, e.g. the output feedback stabilization of Hamiltonian systems (Stramigioli *et al.*, 1998) can be applied easily. Furthermore it is known that an interconnection of two Hamiltonian systems has a Hamiltonian structure as well (van der Schaft *et al.*, 1996). Hence it is also expected that our approach will be utilized to the stabilization of the combination of physical devices and nonholonomic systems.

ACKNOWLEDGMENT

The first author would like to thank Professor A. J. van der Schaft of the University of Twente and Professor B. M. J. Maschke of Conservatoire National des Arts et Métiers for the thoughtful discussions.

4. REFERENCES

Brockett, R. B. (1983). Asymptotic stability and feedback stabilization. In: *Differential Geometric Control Theory* (R. W. Brockett, R. S. Millmann and H. J. Sussmann, Eds.). pp. 181–191. Birkhäuser. Basel-Boston.

Fierro, R. and F. L. Lewis (1995). Control of a nonholonomic mobile robot: backstepping kinematics into dynamics. *Proc. 34th IEEE CDC* pp. 3805–3810.

Fujimoto, K. and T. Sugie (1998*a*). Canonical transformation and stabilization of generalized Hamiltonian systems. Submitted (1999); Preliminary version is in *Proc. 4th IFAC Symp. NOLCOS '98* pp. 544–549.

Fujimoto, K. and T. Sugie (1998*b*). Stabilization of generalized Hamiltonian systems –approach by canonical transformations–. *Trans. Institute of Systems, Control and Information Engineers* **11**(11), 623–629. (in Japanese).

Fujimoto, K. and T. Sugie (1999*a*). Stabilization of a class of Hamiltonian systems with nonholonomic constraints via canonical transformations. *Proc. ECC'99.*

Fujimoto, K. and T. Sugie (1999*b*). Time-varying stabilization of nonholonomic Hamiltonian systems via canonical transformations. Submitted.

Fujimoto, K., K. Ishikawa and T. Sugie (1999). Stabilization of a class of Hamiltonian systems with nonholonomic constraints and its experimental evaluation. To appear in *Proc. IEEE CDC'99.*

Jiang, Z. P. and H. Nijmeijer (1997). Tracking control of mobile robots: a case study in backstepping. *Automatica* **33**, 1393–1399.

Khennouf, H., C. Canudas de Wit, A. J. van der Schaft and J. Abraham (1995). Preliminary results on asymptotic stabilization of Hamiltonian systems with nonholonomic constraints. *Proc. 34th IEEE Conf. on Decision and Control* pp. 4305–4310.

Maschke, B. M. J. and A. J. van der Schaft (1992). Port-controlled Hamiltonian systems: modelling origins and system-theoretic properties. *IFAC Symp. NOLCOS* pp. 282–288.

Maschke, B. M. J. and A. J. van der Schaft (1994). A Hamiltonian approach to stabilization of nonholonomic mechanical systems. *Proc. 33rd IEEE Conf. on Decision and Control* pp. 2950–2954.

Maschke, B. M. J., R. Ortega and A. J. van der Schaft (1998). Energey-based lyapunov functions for forced Hamiltonian systems with dissipation. *Proc. 37th IEEE Conf. on Decision and Control.*

Nijmeijer, H. and A. J. van der Schaft (1990). *Nonlinear Dynamical Control Systems.* Springer-Verlag.

Ortega, R., A. J. van der Schaft, B. M. J. Maschke and G. Escobar (1999). Stabilization of port-controlled Hamiltonian systems: passivity and energy-balancing. Submitted (1999); To appear in *Proc. IEEE CDC '99.*

Ortega, R., A. Loría, P. J. Nicklasson and H. Sira-Rmírez (1998). *Passivity-based Control of Euler-Lagrange Systems.* Springer-Verlag. London.

Pomet, J.-B. (1992). Explicit design of time-varying stabilizing control laws for a class of controllable systems without drift. *Systems & Control Letters* **18**, 147–158.

Stramigioli, S., B. M. J. Maschke and A. J. van der Schaft (1998). Passive output feedback and port interconnection. *Proc. 4th IFAC Symp. NOLCOS* pp. 613–618.

Takegaki, M. and S. Arimoto (1981). A new feedback method for dynamic control of manipulators. *Trans. ASME, J. Dyn. Syst., Meas., Control* **103**, 119–125.

van der Schaft, A. J. (1996). *L_2-gain and Passivity Techniques in Nonlinear Control.* Vol. 218 of *LNCIS.* Springer-Verlag. Berlin.

van der Schaft, A. J., M. Dalsmo and B. M. J. Maschke (1996). Mathematical structures in the network representation of energy-conserving physical systems. *Proc. 35th IEEE CDC* pp. 201–206.

Stabilization of underactuated mechanical systems via interconnection and damping assignment

Romeo Ortega
Lab. des Signaux et Systèmes
CNRS-SUPELEC
Gif-sur-Yvette 91192
FRANCE
rortega@lss.supelec.fr

Mark W. Spong
Coordinated Science Lab
University of Illinois
1308 W. Main St.
Urbana, IL 61801
m-spong@uiuc.edu

Abstract: The present paper is concerned with the application of the newly developed interconnection and damping assignment (IDA) passivity-based control (PBC) to the problem of stabilization of *underactuated mechanical* systems. Our main contribution is the characterization of a class of systems for which IDA-PBC yields a smooth stabilizing controller. The class is given in terms of solvability of a partial differential equation. As an illustration we present a dynamic nonlinear output feedback IDA-PBC which "globally" stabilizes (except for a set of initial conditions that has zero measure) the upward position of a novel inverted pendulum. That is, we prove that it is possible to swing up this pendulum without switching nor measurement of velocities. *Copyright ©2000 IFAC*

Keywords: control of mechanical systems, Hamiltonian Systems, nonlinear control, energy shaping, passivity.

1 Introduction

The idea of energy–shaping control has its roots in the ground–breaking work of [13] in position regulation of robot manipulators, where robust control laws are derived with simple *potential* energy shaping.[1] Using the fundamental notion of *passivity*, the principle was later formalized in [7], where the term passivity-based control (PBC) was coined to define a controller design methodology whose aim is to render the closed–loop system passive with some *desired* storage function. The importance of linking passivity to energy–shaping can hardly be overestimated, since it shows that the approach does not rely on some particular structural properties of mechanical systems, but hinges upon the more fundamental (and universal) property of energy–balancing, hence can be extended to cover a wide range of engineering problems. In particular, we were interested in the application of PBC to electrical and electromechanical systems, which require shaping of the *total* energy. In carrying out this extension two approaches were pursued: the first one, closer to classical Lyapunov–based design, where we first select the storage–function to be assigned and then design the controller that ensures this objective. Extensive applications of this line of research may be found in [3]. A drawback of this approach is that the closed–loop storage functions (typically taken as quadratic in errors) *are not* energy functions in any meaningful physical sense. Actually, it has been shown in [3] that the stabilization mechanism is akin to systems inversion instead of energy–balancing, hence an unnatural stable invertibility assumption is required.

To overcome this problem we have recently introduced a new PBC, called interconnection and damping assignment (IDA) PBC, where the closed–loop storage function is now a *bona fide* energy–function which is obtained as a result of our choice of desired subsystems interconnection and damping. IDA–PBC was first reported in [4] and [5].[2] Since then many successful applications, including mass–balance systems [6], electrical motors [8], magnetic levitation systems [10], and power converters [9] have been reported. IDA–PBC is briefly reviewed in Section 2.

The present paper is concerned with the application of IDA–PBC to the problem of stabilization of *underactuated mechanical* systems. Our main contribution, which is given in Section 3, is the characterization of a class of systems for which IDA–PBC yields a smooth stabilizing controller. The class is given in terms of solvability of a partial differential equation (PDE). As an illustration we present in Section 4 a dynamic nonlinear output feedback IDA–PBC which "globally" stabilizes (except for a set of initial conditions that has zero measure) the upward position of a

[1] Simultaneously and independently of [13] the utilization of these ideas for Euler–Lagrange systems was suggested in [2]. See also the innovative paper [11].

[2] See also [14] for a tutorial account.

novel inverted pendulum. That is, we prove that it is possible to swing up this pendulum without switching nor measurement of velocities.

2 Interconnection and damping assignment PBC

The theory for IDA–PBC has been derived for port–controlled Hamiltonian (PCH) models, which include Euler–Lagrange models, and encompass a very large class of physical nonlinear systems, see [14] for a list of references. PCH systems are of the form

$$\Sigma : \left\{ \begin{array}{rcl} \dot{x} & = & [J(x) - \mathcal{R}(x)]\frac{\partial H}{\partial x}(x) + g(x)u \\ y & = & g^\top(x)\frac{\partial H}{\partial x}(x) \end{array} \right. \quad (2.1)$$

where $x \in \mathcal{R}^n$ is the state vector, $u \in \mathcal{R}^m$ and $y \in \mathcal{R}^m$ are the power variables, which are conjugate in the sense that their product has units of power, for instance currents and voltages in electrical circuits, or forces and velocities in mechanical systems. $H(x)$ is the energy function (which by definition is bounded from below); the skew–symmetric matrix $J(x) = -J^\top(x)$ captures the interconnection structure, and $\mathcal{R}(x) = \mathcal{R}^\top(x) \geq 0$ is the dissipation matrix –due to resistances and frictions.

It is clear that PCH systems satisfy the energy–balancing equation

$$\underbrace{H[x(t)] - H[x(0)]}_{stored\ energy} = \underbrace{\int_0^t u^\top(s)y(s)ds}_{supplied} - \underbrace{d(x(t))}_{dissipated} \quad (2.2)$$

with $d(x(t)) \triangleq \int_0^t \left(\frac{\partial H}{\partial x}[x(s)]\right)^\top \mathcal{R}[x(s)]\frac{\partial H}{\partial x}[x(s)]ds$. From which we easily see that the energy of the uncontrolled system (i.e., with $u(t) \equiv 0$) is non–increasing, that is, $H[x(t)] \leq H[x(0)]$, and it will actually decrease in the presence of dissipation. Since the energy function is bounded from below, the system will eventually stop in a point of minimum energy. This point is usually not the one of practical interest, and control is introduced to operate the system around some non–zero equilibrium point, say x_*. Motivated by (2.2) PBC poses this stabilization problem in terms of the following:

Passivation objective Select a control action[3] $u = \beta(x) + v$ so that the closed–loop dynamics satisfies the new energy–balancing equation

$$H_d[x(t)] - H_d[x(0)] = \int_0^t v^\top(s)y'(s)ds - d_d(x(t)) \quad (2.3)$$

[3] We consider here only static state feedback control laws and refer the reader to the references for the dynamic and output feedback cases. See also Section 4.

where $H_d(x)$ is the desired total energy function, which has a *minimum* at x_*, y' (which may be equal to y) is the new passive output, and we have replaced the natural dissipation term by some desired function $d_d(x)$ to increase the convergence rate.

In IDA–PBC for PCH systems the passivation objective is achieved aiming at the closed–loop dynamics

$$\Sigma_d : \left\{ \begin{array}{rcl} \dot{x} & = & [J_d(x) - \mathcal{R}_d(x)]\frac{\partial H_d}{\partial x}(x) + g(x)v \\ y' & = & g^\top(x)\frac{\partial H_d}{\partial x}(x) \end{array} \right. \quad (2.4)$$

where $J_d(x) = -J_d^\top(x)$ and $\mathcal{R}_d(x) = \mathcal{R}_d^\top(x) \geq 0$ are some *desired* interconnection and damping matrices, respectively. It is clear that the solutions of (2.4) satisfy (2.3) with a suitably modified $d_d(x(t))$.

The proposition below shows that the control problem reduces to the solution of a PDE.

Proposition 1 [5] Given $J(x), \mathcal{R}(x), H(x), g(x)$ and the desired equilibrium to be stabilized x_*. Assume we can find functions $\beta(x), \mathcal{R}_a(x), J_a(x)$ such that $J_d(x) \triangleq J(x) + J_a(x) = -J_d^\top(x)$, $\mathcal{R}_d(x) \triangleq \mathcal{R}(x) + \mathcal{R}_a(x) = \mathcal{R}_d^\top(x) \geq 0$, and a vector function $K(x) : \mathcal{R}^n \to \mathcal{R}^n$ satisfying

$$[J_d(x) - \mathcal{R}_d(x)]K(x) = - [J_a(x) - \mathcal{R}_a(x)]\frac{\partial H}{\partial x}(x) + g(x)\beta(x) \quad (2.5)$$

and such that

(i) *(Integrability)* $K(x)$ is the gradient of a scalar function. That is, $\frac{\partial K}{\partial x}(x) = \left[\frac{\partial K}{\partial x}(x)\right]^\top$.

(ii) *(Equilibrium assignment)* $K(x)$, at x_*, verifies $K(x_*) = -\frac{\partial H}{\partial x}(x_*)$.

(iii) *(Lyapunov stability)* The Jacobian of $K(x)$, at x_*, satisfies the bound $\frac{\partial K}{\partial x}(x_*) > -\frac{\partial^2 H}{\partial x^2}(x_*)$.

Under these conditions, the system (2.1) in closed-loop with $u = \beta(x) + v$ will be a PCH system with dissipation of the form (2.4) with new energy function $H_d(x) \triangleq H(x) + H_a(x)$, and

$$\frac{\partial H_a}{\partial x}(x) = K(x) \quad (2.6)$$

Furthermore, if $v(t) \equiv 0$, x_* will be a (locally) *stable* equilibrium. It will be *asymptotically* stable if, in addition, the largest invariant set under the closed-loop dynamics contained in

$$\left\{ x \in \mathcal{R}^n \cap \mathcal{B} \mid \left[\frac{\partial H_d}{\partial x}(x)\right]^\top \mathcal{R}_d(x)\frac{\partial H_d}{\partial x}(x) = 0 \right\}$$

equals $\{x_*\}$.

Remark In [5] it is shown that the IDA–PBC methodology generates *all* asymptotically stabilizing state feedback laws for the PCH system (2.1). More precisely, it is shown that if there exists a (continuously differentiable) $\beta(x)$ such that $\dot{x} = [J(x) - \mathcal{R}(x)]\frac{\partial H}{\partial x}(x) + g(x)\beta(x)$ is asymptotically stable, then there exist (continuous) $J_a(x), \mathcal{R}_a(x), H_a(x)$ which satisfy the conditions of Proposition 1.

3 IDA–PBC of mechanical systems

In this section we will apply IDA–PBC to mechanical systems with total energy

$$H(x) = \frac{1}{2}x_p^\top M^{-1}(x_q)x_p + U(x_q) \qquad (3.1)$$

where we have partitioned the state into its position and momenta components as $x \triangleq [x_q^\top, x_p^\top]^\top$, $M(x_q) = M^\top(x_q) > 0$ is the inertia matrix, and $U(x_q)$ is the potential energy, which is bounded from below. The PCH description (2.1) corresponds to constant interconnection and input matrices of the form

$$J = \begin{bmatrix} 0 & I \\ -I & 0 \end{bmatrix}, \quad g(x_q) = \begin{bmatrix} 0 \\ g_p(x_q) \end{bmatrix} \qquad (3.2)$$

and we assume the system has no natural damping, i.e., $\mathcal{R}(x) = 0$.

We want to stabilize an equilibrium of the form $x_* = [x_{q*}^\top, 0]^\top$. Towards this end we propose to take

$$J_a(x) = \begin{bmatrix} 0 & J_1(x_q) \\ -J_1^\top(x_q) & J_2(x) \end{bmatrix} = -J_a^\top(x)$$
$$\mathcal{R}_a(x) = \begin{bmatrix} 0 & 0 \\ 0 & \mathcal{R}_2(x) \end{bmatrix} = \mathcal{R}_a^\top(x) \geq 0$$

with $J_1(x_q), J_2(x), R_2(x)$ free parameters to be defined later. Under these conditions, the PDE (2.5), (2.6) reduces to[4]

$$\begin{aligned} K_p &= -J_1[\frac{\partial H}{\partial x_p} + K_p] && (3.3) \\ K_q &= -J_1^\top[\frac{\partial H}{\partial x_q} + K_q] - \\ &- (J_2 - \mathcal{R}_2)[\frac{\partial H}{\partial x_p} + K_p] - g_p\beta && (3.4) \end{aligned}$$

where we have also partitioned the gradient of $H_a(x)$ (2.6) as $K(x) \triangleq [K_q^\top(x), K_p^\top(x)]^\top$. Let us first observe that if we do not modify the interconnection matrix then we can only shape the potential energy. Indeed, if $J_a(x) = 0$ we have that $K_p(x) = 0$ and consequently $H_a = H_a(x_q)$, and we cannot modify the kinetic energy. The control then takes the well-known Proportional+Derivative form of "standard" PBC as the solution of

$$g_p(x_q)\beta(x) = -K_q(x_q) - \mathcal{R}_2(x)M^{-1}(x_q)x_p$$

In the fully actuated case we have $g_p(x_q) = I$ and we can assign an arbitrary potential energy and add damping in all channels. Underactuation puts, however, a severe constraint on the method as we have shown in Section 3.2 of [3], see also next section.

Having realized that we have to modify the kinetic energy, the key idea is to chose the free parameters in such a way that the PDE (3.3), (3.4) admits a solution of the form

$$H_a(x) = \frac{1}{2}x_p^\top N(x_q)x_p + H_a'(x_q) \qquad (3.5)$$

for some symmetric matrix $N(x_q)$. To shape the x_p components of $H_d(x)$ we require $M^{-1}(x_q) + N(x_q)$ to be positive definite. On the other hand, $H_a'(x_q)$ will need to dominate the natural potential energy term $U(x_q)$. From (3.3) we obtain an expression of the matrix $N(x_q)$ in terms of $M(x_q)$ and $J_1(x_q)$, similar to the *matching conditions* of [1]. To solve (3.4) we have to eliminate from them the dependence with respect to x_p –this is the task of $J_2(x), R_2(x)$– which is complicated, of course, by the lack of full actuation.

We are in position to present our main result.

Proposition 2 Consider the mechanical system (2.1), (3.1), (3.2) and the desired equilibrium to be stabilized $x_* = [x_{q*}^\top, 0]^\top$. Assume we can find functions $\beta(x)$, $\mathcal{R}_2(x) = \mathcal{R}_2^\top(x) \geq 0$, $J_1(x_q), J_2(x) = -J_2^\top(x)$ and a vector function $K_q'(x_q) : \mathcal{R}^{n/2} \to \mathcal{R}^{n/2}$ satisfying

$$\begin{aligned} &[I + J_1^\top]K_q'(x_q) = -J_1^\top\frac{\partial H}{\partial x_q}(x) + \\ &+\frac{1}{2}[I + J_1^\top]\frac{\partial}{\partial x_q}\{x_p^\top[I + J_1^\top]^{-1}J_1M^{-1}x_p\} - \\ &-[J_2 - \mathcal{R}_2][I + J_1^\top]^{-1}M^{-1}x_p - g_p(x_q)\beta(x) \quad (3.6) \end{aligned}$$

and such that

(i) *(Integrability)* $K_q'(x_q)$ is the gradient of a scalar function.

(ii) *(Equilibrium assignment)* $K_q'(x_q)$, at x_{q*}, verifies $K_q'(x_{q*}) = -\frac{\partial U}{\partial x_q}(x_{q*})$.

(iii) *(Lyapunov stability)* The Jacobian of $K_q'(x_q)$, at x_{q*}, satisfies the bound $\frac{\partial K_q'}{\partial x_q}(x_{q*}) > -\frac{\partial^2 U}{\partial x_q^2}(x_{q*})$, and $[I + J_1^\top(x_q)]^{-1}M^{-1}(x_q)$ is a symmetric positive definite matrix.

[4] When clear from the context the arguments will be ommited.

Under these conditions, the system (2.1), (3.1), (3.2) in closed-loop with $u = \beta(x)$ is a PCH system with total energy

$$H_d(x) = \frac{1}{2}x_p^\top [I + J_1]^{-1} M^{-1} x_p + U(x_q) + H'_a(x_q) \tag{3.7}$$

where

$$\frac{\partial H'_a}{\partial x_q}(x_q) = K'_q(x_q) \tag{3.8}$$

Furthermore, x_* will be a (locally) *stable* equilibrium. It will be *asymptotically* stable if, in addition, the largest invariant set under the closed-loop dynamics contained in

$$\left\{ x \in \mathcal{R}^n \cap \mathcal{B} \mid \left[\frac{\partial H_d}{\partial x}(x)\right]^\top \mathcal{R}_d(x) \frac{\partial H_d}{\partial x}(x) = 0 \right\}$$

equals $\{x_*\}$.

Proof To carry-out the proof we will show that, for the particular case of the mechanical system (2.1), (3.1), (3.2), the conditions of the general Proposition 1 reduce to those of Proposition 2.

As pointed out above the key PDE (2.5), (2.6) takes in this case the form (3.3), (3.4). Postulating a solution $H_a(x)$ of the form (3.5) we have that $K_p(x) = N(x_q)x_p$, and consequently (3.3) holds iff $N(x_q) = -J_1(x_q)[M^{-1}(x_q) + N(x_q)]$. After some simple calculation, and assuming invertibility of $I + J_1(x_q)$, we see that this is equivalent to

$$M^{-1}(x_q) + N(x_q) = [I + J_1(x_q)]^{-1} M^{-1}(x_q) \tag{3.9}$$

which, under the conditions of the proposition, is symmetric and positive definite. Hence, in its x_p coordinates, $H_d(x)$ (3.7) has a minimum at the desired point.

For (3.4) notice first that, with the postulated solution (3.5), we have

$$K_q(x) = \frac{1}{2}\frac{\partial}{\partial x_q}\left\{x_p^\top N(x_q) x_p\right\} + K'_q(x_q)$$

where $K'_q(x_q)$ is defined by (3.8). Now, from (3.9) it is possible to show that

$$N(x_q) = -[I + J_1(x_q)]^{-1} J_1(x_q) M^{-1}(x_q)$$

Replacing the last two expressions in (3.4) and after some lengthy, but straightforward, calculations we obtain (3.6).

The proof is completed noting that under the remaining conditions of the proposition, the function $H_d(x)$ (3.7) has a (local isolated) minimum in x_*.

4 Inverted pendulum system

In this section we apply Proposition 2 to stabilize the upward position of a pendulum device which consists of a physical pendulum with a rotating mass at the end. The motor torque produces an angular acceleration of the end-mass which generates a coupling torque at the pendulum axis.

Model The dynamic equations of the device can be written in standard form as

$$\begin{bmatrix} m_1 + m_2 & -m_1 \\ m_1 & m_1 \end{bmatrix} \ddot{q} = \begin{bmatrix} mgl\sin(q_1) \\ u \end{bmatrix}$$

The change of coordinates $x_q \triangleq [x_{q1}, x_{q2}]^\top = [q_1, q_1 + q_2]^\top$ leads to the simplified description

$$\begin{bmatrix} m_2 & 0 \\ 0 & m_1 \end{bmatrix} \ddot{x}_q + \begin{bmatrix} -m_3\sin(x_{q1}) \\ 0 \end{bmatrix} = \begin{bmatrix} -1 \\ 1 \end{bmatrix} u$$

where $m_3 \triangleq mgl$. Defining the state $x = [x_q^\top, x_p^\top] = [x_q^\top, M\dot{x}_q^\top]$, with $M = \text{diag}\{m_2, m_1\}$, the total energy is

$$H(x) = \frac{1}{2}x_p^\top M^{-1} x_p + m_3 \cos x_{q1}$$

and the PCH model takes the form (2.1), (3.2) with $g_p = [-1, 1]^\top$. The equilibrium to be stabilized is the upward position with the inertia disk aligned, which corresponds to $x_* = 0$.

Controller design We will design our IDA–PBC in three steps. First, we will obtain a state–feedback that shapes the energy to globally stabilize the upward position, then we add the damping for asymptotic stability. Finally, we show that it is possible to replace the velocity feedback by its dirty derivative preserving the global stability property.

1. Energy shaping First, notice that the inertia matrix is independent of x_q. Hence, recalling the discussion above Proposition 2, we can take $J_2(x) = \mathcal{R}_2(x) = 0$. As will become clear below, we can also take J_1 to be a constant matrix. With these considerations the PDE to be solved takes the form

$$\begin{bmatrix} 0 & I + J_1 \\ -(I + J_1)^\top & 0 \end{bmatrix} K = \begin{bmatrix} -J_1 M^{-1} x_p \\ J_1^\top \begin{bmatrix} -m_3 \sin x_1 \\ 0 \end{bmatrix} \end{bmatrix} \tag{4.1}$$

Spelling out the equations we obtain, for the first two rows $J_1(M^{-1}x_p + K_p) = -K_p$. Let us postulate now a solution of the form (3.5) with *constantN* $= N^\top$. In this case we get $J_1(M^{-1} + N)x_p = -Nx_p$ which, under the assumption of invertibility of $I + J_1$, is satisfied iff $M^{-1} + N = (I + J_1)^{-1}M^{-1}$. Let us fix

$$J_1 = \begin{bmatrix} a & d \\ b & 0 \end{bmatrix}$$

with a, b, d constants to be defined. Some simple calculations show that the symmetry condition $N = N^\top$ imposes $b = \frac{m_1}{m_2}d$, while the inequality

$$1 + a - \frac{m_1}{m_2}d^2 > 0 \tag{4.2}$$

ensures $M^{-1} + N > 0$, and consequently the shaping of the x_p component of $H_d(x)$.

Let us now look at the third and fourth equations of (4.1), which take the form

$$-(I + J_1)^\top K_q = J_1^\top \begin{bmatrix} -m_3 \sin(x_{q1}) \\ 0 \end{bmatrix} + \begin{bmatrix} -u \\ u \end{bmatrix} \tag{4.3}$$

Replacing J_1, b and adding up both rows we get the simple PDE

$$(1 + a + d)K_{q1} + (1 + \frac{m_1}{m_2}d)K_{q2} = (a + d)m_3 \sin(x_{q1})$$

If $1 + a + d \neq 0$ and $1 + \frac{m_1}{m_2}d \neq 0$, then this PDE admits a family of solutions

$$H_a'(x_q) = -\frac{a + d}{1 + a + d}m_3 \cos(x_{q1}) + \Phi(z)$$

where $\Phi(\cdot)$ is an arbitrary (continuously differentiable) function with argument

$$z \triangleq x_{q1} + \frac{1 + a + d}{1 + \frac{m_1}{m_2}d}x_{q2}$$

It is easy to check that the new potential energy

$$m_3 \cos(x_{q1}) + H_a'(x_q) = \frac{m_3}{1 + a + d}\cos(x_{q1}) + \Phi(z)$$

will have an extremum at $x_q = 0$ iff

$$\frac{\partial \Phi}{\partial z}(0) = 0 \tag{4.4}$$

It will, furthermore, be a minimum if the following inequalities hold

$$\begin{aligned} \frac{\partial^2 \Phi}{\partial z^2}(0) &> \frac{m_3}{1 + a + d} \\ -(1 + a + d)\frac{\partial^2 \Phi}{\partial z^2}(0) &> 0 \end{aligned}$$

Clearly, a necessary condition for the latter is

$$1 + a + d < 0 \tag{4.5}$$

in which case it is sufficient to take

$$\frac{\partial^2 \Phi}{\partial z^2}(0) > 0 \tag{4.6}$$

To verify (4.4) and (4.6) we propose the quadratic function $\Phi(z) = \frac{K_p'}{2}z^2$, with $K_p' > 0$. Finally, to verify the remaining conditions of Proposition 2, the parameters a and d should satisfy (4.2) and (4.5), whose intersection is clearly a non–empty set.

The energy–shaping stage of the design is completed evaluating the control law from (either one of the equations of) (4.3) and replacing $H_a'(x_q), \Phi(z)$ and z to get

$$u = \gamma_1 \sin(x_{q1}) + K_p\,(x_{q1} + \gamma_2 x_{q2})$$

where $K_p \triangleq \frac{1+a-\frac{m_1}{m_2}d^2}{1+\frac{m_1}{m_2}d}K_p' > 0$ and

$$\gamma_1 \triangleq \frac{-dm_3}{1 + a + d}, \quad \gamma_2 \triangleq -\frac{1 + a + d}{1 + \frac{m_1}{m_2}d}, \tag{4.7}$$

2. Damping injection The controller above yields the *conservative* closed–loop system $\dot{x} = J_d \frac{\partial H_d}{\partial x}(x)$, $J_d = -J_d^\top$, with the origin a stable equilibrium, because $\{0\} = \arg\min H_d(x)$. To make this equilibrium *asymptotically* stable we propose to inject damping feeding back the passive output $y' = g^\top(x)\frac{\partial H_d}{\partial x}(x)$, which satisfies

$$y' = \kappa\left(x_{p1} + \frac{m_1}{m_2}\gamma_2 x_{p2}\right)$$

for some $\kappa > 0$. Now, it is possible to show that, under the action of the control law

$$u = \gamma_1 \sin(x_{q1}) + K_p\,(x_{q1} + \gamma_2 x_{q2}) + K_v y' \tag{4.8}$$

where $K_p, K_v > 0$, the trajectories of the closed–loop system satisfy the implication $\dot{H}_d[x(t)] \equiv 0 \Rightarrow x(t) \equiv \bar{x}$, where $\bar{x}$ are the equilibria of the closed–loop system, which are given by $\bar{x} = [k\pi, -\frac{k\pi}{\gamma_2}, 0, 0]^\top$, for $k \in \mathcal{Z}$. We have already shown that the equilibria for k even (which includes the upward position) are stable. Therefore invoking LaSalle's invariance principle we conclude that the upward position is asymptotically stable. To prove that the result is *"almost" global*[5] we notice that the equilibria with k odd are unstable. The latter stems from the fact that the Hessian of $H_d(x)$ at those points is sign–indefinite with one negative and three positive eigenvalues. Thus the system linearized around the unstable equilibrium points will have a one–dimensional stable subspace and a three–dimensional unstable subspace. The trajectories starting on the stable manifold will converge to the downward position, however, this set of initial conditions has zero measure, and consequently the basin of attraction of our controller is an open dense set in the state space.

3. Output feedback It is well–known that in PBC designs it is possible to obviate velocity measurement feeding back instead the dirty derivative of

[5] By "almost" we mean that the set of initial conditions which do not converge to $\{0\}$ is of zero measure.

positions [3]. This feature stems from that fact that for passive maps we can replace a *constant* feedback by a feedback through any positive real transfer function preserving stability. In particular we can use the feedback $u_{di} = \frac{K_f p}{\tau p+1} z$, with $p \triangleq \frac{d}{dt}$ and $K_f, \tau > 0$, to implement the damping injection. Notice that z depends only on positions, hence u_{di} is implementable without velocity feedback. This consideration leads us to our final result contained in the proposition below, whose proof is available upon request to the authors.

Proposition 3 Consider the gyroscopic pendulum model (2.1) in closed-loop with the dynamic output feedback controller

$$\begin{aligned} u &= -\gamma_1 \sin(x_{q1}) + K_p(x_{q1} + \gamma_2 x_{q2}) + \zeta + \frac{K_f}{\tau} z \\ \dot{\zeta} &= -\frac{1}{\tau}\zeta - \frac{K_f}{\tau^2} z \\ z &= x_{q1} + \gamma_2 x_{q2} \end{aligned}$$

where $K_p > 0$, γ_1, γ_2 are given by (4.7), $\tau, K_f > 0$, and a, d are constants such that (4.2) and (4.5) hold. Then, the origin is an "almost" globally asymptotically stable equilibrium of the closed-loop.

Simulation results We simulated the response of the pendulum with the IDA-PBC (4.8) with $m_1 = 2, m_2 = 1, m_3 = 10$ and $\gamma_1 = 4, \gamma_2 = 1/8, K_p = 1, K_v = 10$. Figure 1 shows the swingup (almost dead-beat) response of the pendulum starting at nearly the vertically downward position, with the remaining initial conditions zero.

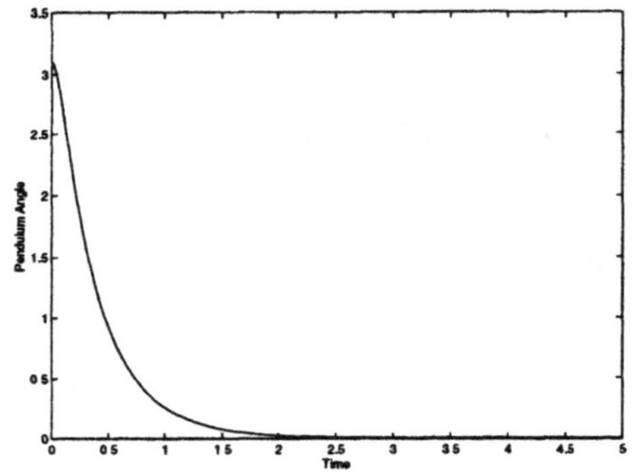

Figure 1: Swingup response of the Pendulum.

References

[1] A. Bloch, N. Leonhard and J. Marsden, Controlled Lagrangians and the stabilization of mechanical systems, *Proc. IEEE Conf. Decision and Control*, Tampa, FL, USA, Dec. 1998.

[2] E. Jonckheere, Lagrangian theory of large scale systems, *Proc. European Conf. Circuit Th. and Design*, The Hague, The Netherlands, 1981, pp. 626–629.

[3] R. Ortega, A. Loria, P. J. Nicklasson and H. Sira–Ramirez, **Passivity–based Control of Euler–Lagrange Systems**, Springer-Verlag, Berlin, Communications and Control Engineering, Sept. 1998.

[4] R. Ortega, A. van der Schaft, B. Maschke and G. Escobar, Energy–shaping of port–controlled Hamiltonian systems by interconnection, *IEEE Conf. Dec. and Control*, Phoenix, AZ, USA, Dec. 1999, pp. 1646–1650.

[5] R. Ortega, A. van der Schaft, B. Maschke and G. Escobar, Stabilization of port–controlled Hamiltonian systems: Energy–balancing and passivation, *Automatica*, (to appear).

[6] R. Ortega, A. Astolfi, G. Bastin and H. Rodriguez, Output–feedback regulation of mass–balance systems, in **New Directions in Nonlinear Observer Design**, eds. H. Nijmeijer and T. Fossen, Springer–Verlag, Berlin, 1999.

[7] R. Ortega and M. Spong, Adaptive motion control of rigid robots: A tutorial, *Automatica*, Vol. 25, No.6, pp. 877-888, 1989.

[8] V. Petrovic, R. Ortega and A. Stankovic, A globally convergent energy–based controller for PM synchronous motors, *CDC'99*, Phoenix, AZ, USA, Dec. 7–10, 1999, pp. 334–339.

[9] H. Rodriguez, R. Ortega and G. Escobar, A Robustly Stable Output Feedback Saturated Controller for the Boost DC–to–DC Converter, *Systems and Control Letters*, (to appear). See also: *CDC'99*, Phoenix, AZ, USA, Dec. 7–10, 1999, pp. 2100-2105.

[10] H. Rodriguez, R. Ortega, I. Mareels and G. Espinosa, Nonlinear control of magnetic levitation systems via interconnection and damping assignment, *ACC 2000*, Chicago, June 2000, (submitted).

[11] J. J. Slotine, Putting physics in control –The example of robotics, *IEEE Control Syst. Magazine*, Vol. 8, No. 6, 1988, pp. 12–17.

[12] M.W. Spong and L. Praly, Control of underactuated mechanical systems using switching and saturation, in **Control Using Logic–based Switching**, Ed. A. S. Morse, Springer, Lecture Notes No. 222, pp. 162–172.

[13] M. Takegaki and S. Arimoto, A new feedback method for dynamic control of manipulators, *ASME J. Dyn. Syst. Meas. Cont.*, Vol. 102, pp. 119-125, 1981.

[14] A. J., van der Schaft, **L_2–Gain and Passivity Techniques in Nonlinear Control**, Springer-Verlag, Berlin, 1999.

CONTROL OF ELASTIC SYSTEMS, A HAMILTONIAN APPROACH

Kurt Schlacher * Andreas Kugi *

* *Institute of Automatic Control and Electrical Drives, Johannes Kepler University of Linz, Austria*

Abstract: This contribution deals with control techniques for mechanical structures with piezoelectric sensor and actuator layers, which are spatially distributed. Hamiltonian systems are used to describe these structures. It turns out that collocated control using a suitable actuator and sensor pairing simplifies the controller design a lot. This method, well known for finite dimensional systems, seems to be even more important for infinite dimensional ones, because one can find simple solutions for the PD- and special H_2-, H_∞-design problems in the collocated case. Since this approach is based on the so called Poisson bracket, one can unify the controller design for finite and for infinite dimensional systems. Of course, the stability investigations are much more complicated in the latter case. *Copyright ©2000 IFAC*

Keywords: Hamiltonian systems, piezoelasticity, PD-, H_2- and H_∞-design

1. INTRODUCTION

Smart structures based on piezoelectricity represent an important new group of actuators and sensors for active vibration control of mechanical systems. In contrast to conventional techniques this technology allows to construct spatially distributed devices (see, e.g., (Tzou, 1992)). This fact requires special control techniques to improve the dynamical behavior of this kind of smart structures (see, e.g., (Kugi *et al.*, 1999), (Schlacher *et al.*, 1996)), since the design of the spatial distribution of the actuators and sensors adds an additional freedom to the design of the control law. Therefore, the design of the controller has to be considered together with the design of the actuators and sensors.

It is well known that the dynamical equations of elasticity take often the form

$$\rho_{\mathrm{Ref}}\frac{\partial^2}{\partial t^2}u^i + \sum_{j=1}^{3}\frac{\mathrm{d}}{\mathrm{d}x^j}\frac{\partial}{\partial u^i_j}W = 0\ , \quad i = 1,2,3$$

with the mass density ρ_{Ref} in the reference configuration, the displacement u and the stored energy function W. The dynamics of smart structures is described by equations of the same type, where the stored energy function W is replaced by a more complicated one, provided that hysteretic and polarization effects are ignored. Furthermore, these equations can be put into Hamiltonian form with the total energy

$$H = \int_{\mathcal{B}}\left(\frac{1}{2\rho_{\mathrm{Ref}}}\|p\|^2 + W\right)\mathrm{d}x^1\mathrm{d}x^2\mathrm{d}x^3$$

and canonical coordinates u and $p = \rho_{\mathrm{Ref}}\partial_t u$. The integral is taken over the reference configuration $\mathcal{B}$ of the elastic body. Therefore, the second part of this contribution is concerned with an introduction to Hamiltonian systems. Since we propose an approach based on the Poisson bracket, we can unify the controller design for finite dimensional and infinite dimensional systems in the third part. This unification concerns the structure of the control law only, of course, the stability investigations are much more complicated in the infinite dimensional case (Marsden and Hughes, 1994). We restrict ourselves to the time invariant case and use the fact that the change of the total energy of the system with respect to the time equals the flow of power into the system minus the dissipated energy. All stability considerations rely on the simple assumption that the decrease of the total energy implies stability of the controlled system. Based on this assumption we present solutions for a special class of H_2- and H_∞-design problems because we are able to convert the partial differential equations of the Hamilton-Jacobi type into simple algebraic ones (Schlacher, 1997). The collocation of sensors and actuators is the price which one has to pay. The fourth part presents a smart beam, a structure with attached piezoelectric layers, which act as sensors and actuators. A summary finishes this contribution.

2. HAMILTONIAN SYSTEMS

Before we start with infinite dimensional Hamiltonian systems, we repeat some facts of finite dimensional ones. We consider the $(1+q)$-dimensional, smooth manifold $\mathcal{E} = R \times \mathcal{M}$, called the total manifold, with local coordinates t, u^j, $j = 1, \ldots, q$. t denotes the independent variable and u the dependent ones, which form a local chart of the manifold $\mathcal{M}$. Let $h : \mathcal{M} \rightarrow R$ denote a function on $\mathcal{M}$ and ϕ_τ be a one parameter group $(\tilde{t}, \tilde{u})(\tau) = \phi_\tau(t, u)$, which acts on $\mathcal{E}$ such that $\tilde{t}(\tau) = t + \tau$ is met. The infinitesimal generator $X \in \mathcal{T}(\mathcal{E})$ of ϕ_τ takes the form $X = \partial_t + \hat{X}$, $\hat{X} = \sum_{i=1}^q \mu^i \partial_{u^i}$. Iff h is invariant with respect to ϕ_τ, then the relation

$$\hat{X}(h) = \sum\nolimits_{i=1}^q \mu^i \partial_{u^i} h = 0 \qquad (1)$$

is met. The choice $\hat{X}_h = J(\mathrm{d}h, \cdot)$, $\mathrm{d}h \in \mathcal{T}^*(\mathcal{E})$ with an alternating bi-vector J ensures that (1) holds because of $\hat{X}_h(h) = J(\mathrm{d}h, \mathrm{d}h) = 0$. Furthermore, J induces a bracket operation by

$$\{f, g\} = J(\mathrm{d}f, \mathrm{d}g) , \qquad (2)$$

which assigns to each pair of smooth, real valued functions f, g the function $\{f, g\}$. The bracket (2) is bilinear, skew symmetric and meets the Leibniz rule. If the Jacobi identity is fulfilled in addition, then (2) is called the Poisson bracket. The Jacobi identity imposes the additional conditions

$$\{J_{ij}, u^k\} + \{J_{ki}, u^j\} + \{J_{jk}, u^i\} = 0$$

on J. In this case, $\hat{X}_h = J(\mathrm{d}h, \cdot)$ is called the *Hamiltonian vector field* for the Hamiltonian function h, the unique vector field X on $\mathcal{M}$ that fulfills the equations

$$X_h(f) = \{f, h\} = -\{h, f\} . \qquad (3)$$

Roughly speaking, the Hamiltonian function above must be replaced by a Hamiltonian functional in the infinite dimensional case. Therefore, we consider an $(p+1+q)$-dimensional smooth manifold $\mathcal{E} = R \times \mathcal{D} \times \mathcal{M}$ with local coordinates t, x^i, u^j, $i = 1, \ldots, p$, $j = 1, \ldots, q$. Here t, x denote the independent coordinates and u denotes the dependent coordinates, $\mathcal{E}^{(n)}$ is the n-th order jet space of $\mathcal{E}$ with local coordinates $t, x^i, u^j, u^j_{i_1}, \ldots, u^j_{i_1,\ldots,i_n} = t, x, u^{(n)}$. To shorten the notation, we use the symmetric multi-indices notation with $J = (j_1, \ldots, j_k)$, $0 \leq j_i \leq p$ and $k = \#J$ to describe the kth order partial derivatives

$$f_J = f_{j_1,\ldots,j_k} = \frac{\partial^k}{\partial x^{j_1} \cdots \partial x^{j_k}} f , \quad f_0 = \frac{\partial}{\partial t} f$$

of a smooth function $f : R^{p+1} \rightarrow R$ (Olver, 1993). The case $\#J = 0$, or the zero order partial derivative of f, is defined by $\partial_J f = f$.

Let us consider the functional

$$H = \int_{\mathcal{D}} h\left(x, u^{(n)}\right) \omega , \qquad (4)$$

$\omega = \mathrm{d}x^1 \wedge \ldots \wedge \mathrm{d}x^p$, where the density h is well defined on $\mathcal{D}$ for $t \geq 0$. To complete the problem, we have to add suitable conditions for $u^{(n)}$ on the boundary $\partial\mathcal{D}$. For the sake of simplicity, we will specify the boundary conditions in the applications only, because they are not relevant for the rest. Again let ϕ_τ be a one parameter group $\left(\tilde{t}, \tilde{u}^{(n)}\right)(\tau) = \phi_\tau\left(t, u^{(n)}\right)$, which acts on $\mathcal{E}^{(n)}$ such that $\tilde{t}(\tau) = t + \tau$ is met. The infinitesimal generator $X \in \mathcal{T}(\mathcal{E})$ of ϕ_τ takes the form $X = \partial_t + \mathrm{pr}^{(n)}\hat{X}$ with the evolutionary field $\hat{X} = \sum_{i=1}^q \mu^i\left(t, x, u^{(n)}\right) \partial_{u^i}$. Thereby, $\mathrm{pr}^{(n)}\hat{X}$ denotes the prolongation of $\hat{X}$ given by

$$\mathrm{pr}^{(n)}\hat{X} = \sum\nolimits_{j=1}^q \sum\nolimits_J D_J \mu^j \frac{\partial}{\partial u^j_J} \qquad (5)$$

and

$$D_J = \frac{\mathrm{d}}{\mathrm{d}x^{j_1}} \cdots \frac{\mathrm{d}}{\mathrm{d}x^{j_k}}$$

with $J = (j_1, \ldots, j_k)$ and the total derivatives

$$\begin{aligned} \frac{\mathrm{d}}{\mathrm{d}x^i} &= \frac{\partial}{\partial x^i} + \sum_{\alpha=1}^q \sum_J u^\alpha_{J,i} \frac{\partial}{\partial u^\alpha_J} \\ \frac{\mathrm{d}}{\mathrm{d}t} &= \frac{\partial}{\partial t} + \sum_{\alpha=1}^q \sum_J u^\alpha_{J,0} \frac{\partial}{\partial u^\alpha_J} \end{aligned} \qquad (6)$$

with $i = 1, \ldots, p$. The sum in (5) and (6) is over all symmetric multi-indices J up to order n. The functional (4) is invariant with respect to ϕ_τ, iff the relation

$$\begin{aligned} \mathrm{pr}^{(n)}\hat{X}(H) &= \int_{\mathcal{D}} \mathrm{pr}^{(n)}\hat{X}(h)\, \omega \\ &= \int_{\mathcal{D}} \sum_{i=1}^q \mu^i \mathsf{E}_i(h)\, \omega = 0 \end{aligned} \qquad (7)$$

with the Euler-Lagrange operators

$$\mathsf{E}_i = \sum\nolimits_J (-1)^{\#J} D_J \frac{\partial}{\partial u^i_J} , \quad i = 1, \ldots, q \qquad (8)$$

is met.

If D is a matrix differential operator, then its adjoint D^* is the differential operator, which fulfills the relation

$$\int_{\mathcal{D}} \sum_{i=1}^{n} \sum_{j=1}^{m} b^i D_{ij}\left(a^j\right)\omega = \int_{\mathcal{D}} \sum_{i=1}^{n} \sum_{j=1}^{m} a^i D_{ij}^*\left(b^j\right)\omega .$$

D is called self-adjoint if $D^* = D$ is met, and it is called skew-adjoint, if $D^* = -D$ is fulfilled, respectively. Let A denote a skew-adjoint differential operator, then the choice

$$\mu^i = \sum_{j=1}^{q} A_{ij}\left(E_j(h)\right)$$

guarantees that (7) is met. Now, we are ready to define the bracket

$$\{G,H\} = \int_{\mathcal{D}} \sum_{i,j=1}^{q} E_i(g)\, A_{ij}\left(E_j(h)\right)\omega , \quad (9)$$

which is bilinear and skew-symmetric. The Leibniz rule has no counterpart, because there exists no well defined multiplication of functionals. Iff the Jacobi identity is fulfilled, then (9) is called the Poisson bracket. Again, the Jacobi identity imposes additional conditions on A, e.g., see (Olver, 1993).

Following the considerations above, we derive the Hamiltonian vector field $\hat{X}_H$,

$$\hat{X}_H = \sum_{j=1}^{q} A_{ij}\left(E_j(h)\right)\frac{\partial}{\partial u^i} \quad (10)$$

the unique evolutionary field, which satisfies

$$\mathrm{pr}^{(n)}\hat{X}_H(F) = \{F,H\} = -\{H,F\} \quad (11)$$

and induces the Hamiltonian system of evolution equations by

$$u_0^i = \sum_{j=1}^{q} A_{ij}\left(E_j(h)\right) , \quad i = 1,\dots,q . \quad (12)$$

Obviously the extensions of Hamiltonian systems from the finite to the infinite dimensional case requires the replacements of Hamiltonian function, the partial derivatives and the bi-vector by a Hamiltonian functional, the Euler-Lagrange operators and a skew-adjoint differential operator, respectively.

3. HAI-SYSTEMS

Let us consider a dynamical system $\mathcal{H}$ with the special Hamiltonian functional H,

$$\begin{aligned} H &= H^0 - \sum_{i=1}^{m} U^i H^i \\ H^i &= \int_{\mathcal{D}} h^i\left(x, u^{(n)}\right)\omega , \quad i = 0,\dots,m \end{aligned} \quad (13)$$

and the m-dimensional control input acting on the system. Since the inputs U^i enter H in an affine way, this type of control system is called a Hamiltonian affine input system or HAI-system for short.

If H^0 denotes the total energy of the system, then its change according to the evolution of $\mathcal{H}$ is given by $\{H^0, H\}$. From

$$\begin{aligned} \{H^0,H\} &= \left\{H + \sum_{i=1}^{m} U^i H^i, H\right\} \\ &= \sum_{i=1}^{m} U^i \{H^i, H\} \end{aligned} \quad (14)$$

follows that it is equal to the flow of power caused by the input U. Therefore, a natural choice for the output Y of the system $\mathcal{H}$ is given by

$$Y^i = H^i , \quad i = 1,\dots,m . \quad (15)$$

Y is called the collocated output, too (see, e.g., (Nijmeijer and van der Schaft, 1990)).

Let S denote a self-adjoint non negative differential operator, then a possible generalization of (12) is given by

$$\begin{aligned} u_0^i &= \sum_{j=1}^{q} \left(A_{ij} - S_{ij}\right)\left(E_j\left(h^0\right)\right) \\ &+ \sum_{j=1}^{m} B_{ij} U^j \end{aligned} \quad (16)$$

for $i = 1,\dots,q$. The change of the total energy H^0 according to the evolution of (16) is given by

$$\begin{aligned} \mathrm{pr}^n \hat{X}\left(H^0\right) &= \sum_{j=1}^{m} Y^j U^j \\ &- \int_{\mathcal{D}} \sum_{i,j=1}^{q} S_{ij}\left(E_j\left(h^0\right)\right) E_i\left(h^0\right)\omega \end{aligned}$$

with the an alternative choice for the output Y^j,

$$Y^j = \int_{\mathcal{D}} \sum_{i=1}^{m} E_i\left(h^0\right) B_{ij}\omega ,$$

$j = 1,\dots,m$. The analogous situation in the finite dimensional case can be found e.g. in (Ortega *et al.*, 1998) or (van der Schaft, 2000).

3.1 *Control of HAI-Systems*

We consider the Hamiltonian AI-system (13) with collocated output (15). Let H^0 be a positive definite functional of the state of $\mathcal{H}$. From now on, we assume that the requirement

$$\{H^0, H\} = \sum_{i=1}^{m} U^i \dot{Y}^i = \frac{d}{dt} H^0 \leq 0 , \quad (17)$$

$\{Y^i, H\} = \dot{Y}^i$, $i = 1,\dots,m$ implies the stability of the system $\mathcal{H}$. In contrast to the finite dimensional case, more investigations are necessary for infinite dimensional systems $\mathcal{H}$ (see, e.g., (Marsden and Hughes, 1994)). But this blanket hypothesis applies for all examples which will be presented later.

According to the hypothesis above, we are able to summarize some nice control strategies for HAI-systems. The simple *PD*- law (see, e.g., (Nijmeijer and van der Schaft, 1990))

$$u^i = -\sum_{j=1}^{m} \left(K_{ij} Y^j + D_{ij} \dot{Y}^j\right) \quad (18)$$

with positive (semi-)definite matrices K and D preserves stability. The P-part adds nothing else than an additional potential to H^0, because it meets the relations

$$\mathsf{E}_j\left(\hat{H}^k\right) = \mathsf{E}_j\left(H^k\right) U^k\ ,$$

$j = 1, \dots, q$ for a suitable functional $\hat{H}^k$.

We are able to determine a solution of (13), such that the objective function

$$\begin{aligned} J_2 &= \sup_{T\in[0,\infty)} \inf_{u\in L_2^m[0,T]} \int_0^T l dt \\ 2l &= \left\|\dot{Y}\right\|^2 + \|U\|^2 \end{aligned} \qquad (19)$$

with euclidean norm $\| \ \|$ is minimized (Schlacher and Kugi, 1999). The Hamilton-Jacobi-Bellman inequality of this H_2-problem is given by

$$\inf_u \left(\{V, H\} + l\right) \leq 0\ . \qquad (20)$$

A short calculation shows that the ansatz $V = \rho H^0$, $\rho > 0$ leads to the simple control law

$$U^i = -\rho \dot{Y}^i\ ,\ i = 1, \dots, m$$

and converts (20) to

$$\frac{1-\rho^2}{2}\left\|\dot{Y}\right\|^2 \leq 0\ .$$

The choice $\rho = 1$ solves the problem exactly and the objective function J_2,

$$J_2 = -\int_0^\infty \sum_{j=1}^m \dot{Y}^j U^j \mathrm{d}t$$

is equal to the energy dissipated by the controller.

Let us assume that we can split the input of (13) into two parts such that U^i, $i = 1, \dots, m$ acts as the control input and D^i, $i = 1, \dots, m$ acts as the disturbance input on the structure. We consider the H_∞-design problem (see, e.g., (van der Schaft, 2000))

$$\begin{aligned} J_\infty &= \sup_{T\in[0,\infty)} \inf_{u\in L_2^m[0,T]} \sup_{d\in L_2^m[0,T]} \int_0^T l dt \\ 2l &= \left\|\dot{Y}\right\|^2 + \|U\|^2 - \gamma\|D\|^2 \end{aligned} \qquad (21)$$

$\gamma > 0$, with the Hamilton-Jacobi-Bellman-Isaacs inequality

$$\inf_u \sup_d \left(\{V, H\} + l\right) \leq 0\ . \qquad (22)$$

Again, the ansatz $V = \rho H^0$, $\rho > 0$ (Schlacher and Kugi, 1999) leads to the simple equations

$$U^i = -\rho\dot{Y}^i\ ,\quad \gamma D^i = \rho\dot{Y}^i\ ,\quad i = 1, \dots, m$$

and simplifies (22) to

$$\sum_{i=1}^m \frac{1}{2}\left(1 - \rho^2\frac{\gamma-1}{\gamma}\left\|\dot{Y}\right\|^2\right) \leq 0\ .$$

Of course, one can reach even equality, iff the relation $\gamma > 1$ is met. Again, the objective function J_∞,

$$J_\infty = -\sqrt{\frac{\gamma-1}{\gamma}}\int_0^\infty \sum_{j=1}^m \dot{Y}^j U^j \mathrm{d}t$$

is proportional to the energy dissipated in the controller.

It is worth mentioning that all proposed feedback strategies depend on the Poisson bracket only. Therefore, one can use them for finite dimensional as well as for infinite dimensional systems. Of course, the stability test is much more complicated in the latter case.

4. THE SMART BEAM

Let us consider the beam of fig. 1, which moves in a 2-dimensional Euclidean space with standard orthonormal basis $\mathcal{B} = \{e_1, e_2\}$, $(e_i, e_j) = \delta^{ij}$ and coordinates x^i, $i = 1, 2$ with $x = \sum_{i=1}^2 x^i e_i$. The independent coordinates are $t = x^0$, $S = x^1$ and the dependent coordinates are the displacements u^j, $j = 1, 2$. L denotes the length of the beam in the reference configuration with $u^1 = u^2 = 0$ for $S \in [0, L]$. Furthermore, we assume that the line mass density ρ_{Ref} is constant in this configuration. The beam is equipped with piezoelectric layers

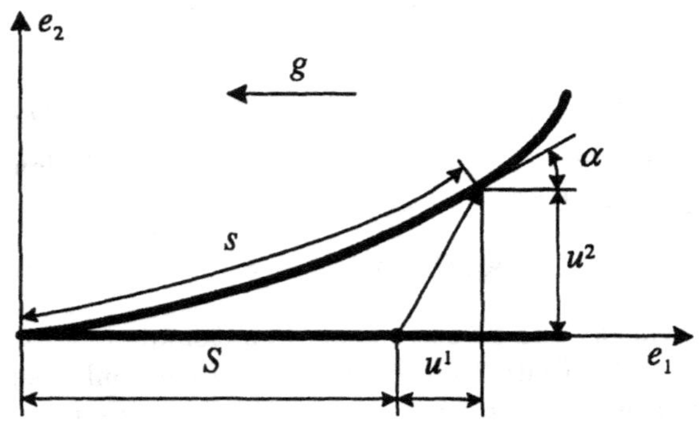

Fig. 1. The considered smart beam.

to counteract the acceleration of gravity g. The design of the spatial distributions of the actuator and sensor layers is part of the controller design itself.

Let φ denote the map from the reference configuration to the current one, then we have

$$\varphi(S, t) = (i + u)(S, t)$$

with the identity map i. It is well known that the length s and the curvature $\kappa(s)$ of $\varphi(\bullet, t)$ determine the position of the curve unambiguously up to a rotation. Let S denote the length in the reference position, then we have

$$\frac{1}{\varepsilon+1}\frac{\partial}{\partial S}(i+u) = \begin{bmatrix} 1+u_1^1 \\ u_1^2 \end{bmatrix} = \begin{bmatrix} \cos\alpha \\ \sin\alpha \end{bmatrix}\ ,$$

because of $\varphi^*(\mathrm{d}s) = (\varepsilon + 1)\,\mathrm{d}S$ with

$$\left\| \frac{\partial}{\partial S}(i + u)\, e_1 \right\| = \sqrt{(1 + u_1^1)^2 + (u_1^2)^2} = \varepsilon + 1 .$$

Here, φ^* denotes the pullback of $\mathrm{d}s$ to the reference configuration. In addition, the relation

$$\frac{\partial}{\partial S}\alpha = \frac{\partial}{\partial s}\alpha \frac{\partial s}{\partial S} = \kappa(\varepsilon + 1)$$

is fulfilled.

The derivation of the evolutionary equations of the smart beam is based on the conservation of mass and the balance laws of momentum as well as of moment of momentum (Marsden and Hughes, 1994). Furthermore, the rotational inertia is neglected. According to this assumptions, we assume that the kinetic energy W_k, the elastic potential W_p and the gravitational potential W_g are given by

$$\begin{aligned} W_k &= \int_L \frac{\rho_{\mathrm{Ref}}}{2}\left((u_0^1)^2 + (u_0^2)^2\right)\mathrm{d}S \\ W_p &= \int_L \frac{\rho_{\mathrm{Ref}}}{2}\left(k_\varepsilon \varepsilon^2 + k_\kappa \kappa^2\right)\mathrm{d}S \\ W_g &= \int_L g\rho_{\mathrm{Ref}}\left(S + u^1\right)\mathrm{d}S \end{aligned}$$

with positive real numbers k_ε, k_κ. The conservation of mass takes the infinitesimal form $\rho_{\mathrm{Ref}}\mathrm{d}S = \varphi^*(\rho \mathrm{d}s)$ with the density ρ in the current configuration. This law was used to transform the different energies and potentials to the reference position. It is worth mentioning that the elastic potential W_p is derived from nonlinear constitutive laws. The total energy H^0 of the free beam is given by $H^0 = W_k + W_p + W_g$. The kinematic boundary conditions are

$$u(0,t) = 0 \qquad \text{and} \qquad \alpha(0,t) = 0 .$$

One can show (see, e.g., (Kugi *et al.*, 1999)) that the piezoelectric actuator layers add terms with the structure

$$U^i H^i = U^i \int_L \mu^1(S)\, \rho_{\mathrm{Ref}} f(\varepsilon, \kappa)\,\mathrm{d}S$$

to the Hamiltonian functional (13). A straightforward design (Kugi *et al.*, 1999) of these actuators leads to the following two types of actuators

$$\begin{aligned} H_A &= \int_L \mu_A(S)\, \varepsilon \rho_{\mathrm{Ref}} \mathrm{d}S \\ H_B &= \int_L \mu_B(S)\, \kappa \rho_{\mathrm{Ref}} \mathrm{d}S , \end{aligned} \tag{23}$$

with voltages as physical control inputs. Using piezoelectric sensors, we are able to measure the outputs

$$\begin{aligned} Y_A &= \int_L \lambda_A(S)\, \varepsilon \rho_{\mathrm{Ref}} \mathrm{d}S \\ Y_B &= \int_L \lambda_B(S)\, \kappa \rho_{\mathrm{Ref}} \mathrm{d}S , \end{aligned} \tag{24}$$

which are electrical charges (see, e.g., (Kugi *et al.*, 1999)). It is worth mentioning that the total time derivatives $\dot{Y}_A$, $\dot{Y}_B$ of Y_A, Y_B are electrical currents, which are easy to measure. Furthermore, we have the possibility to choose the functions μ_A, μ_B, λ_A, λ_B according to the design problem. It is easy to see that this technology allows actuator-sensor collocation in a simple way.

Now we are able to pose the design problem. Find a control law with suitable actuators and sensors such that the reference position is stabilized and such that the influence of the gravitation is eliminated at least in this position. We have to distinguish two cases. If the beam is stiff enough that there exists only one equilibrium, of course, with $\alpha = 0$, then we skip the following step, otherwise we choose the control laws

$$U_B^i = -k_B H_B^i , \quad i = 1, \dots, m$$

and actuators with

$$\mu_B^i = \begin{cases} 1 \text{ if } S \in [S^i, S^i + \Delta S] \\ 0 \text{ otherwise,} \end{cases}$$

and $0 \leq S^1$, $S^i + \Delta S < S^{i+1}$, $i = 1, \dots, m-1$, $S^m + \Delta S \leq L$, $\Delta S > 0$ of the type

$$H_B^i = \int_L \mu_B^i \rho_{\mathrm{Ref}} \kappa \mathrm{d}S = \rho_{\mathrm{Ref}} \Delta\alpha^i , \tag{25}$$

$\Delta\alpha^i = \alpha(S^i + \Delta S) - \alpha(S^i)$. These actuators can be realized with patches of piezoelectric layers in a straightforward manner. In addition, this type of control law is derivable from the potential

$$W_{U_B^i} = \frac{k_B}{2}\left(H_B^i\right)^2$$

because of

$$\begin{aligned} \mathsf{E}_j\left(\frac{k_B}{2}\left(H_B^i\right)^2\right) &= k_B H_B^i \int_L \mathsf{E}_j\left(\mu_B^i \rho_{\mathrm{Ref}} \kappa\right)\mathrm{d}S \\ &= -U_B^i \int_L \mathsf{E}_j\left(\mu_B^i \rho_{\mathrm{Ref}} \kappa\right)\mathrm{d}S . \end{aligned}$$

For sufficiently large $k_B > 0$ and sufficiently enough patches we can always achieve that there exists only one equilibrium with $\alpha = 0$.

To get a closer insight into the influence of the gravitation on the beam, we rewrite $W_g / g\rho_{\mathrm{Ref}}$ as

$$\begin{aligned} \int_L \left(S + u^1\right)\mathrm{d}S &= \int_L (L - S)\, u_1^1 \mathrm{d}S + \frac{L^2}{2} \\ &= \int_L (L - S)\left((\varepsilon + 1)\cos\alpha - 1\right)\mathrm{d}S + \frac{L^2}{2} \end{aligned}$$

because $(L - S)\,u^1$ vanishes for $S \in \{0, L\}$. Now, it is straightforward to see that the control law $U_A = -g$ with the actuator (23)

$$H_A = \int_L (L - S)\, \varepsilon \rho_{\mathrm{Ref}} \mathrm{d}S \tag{26}$$

cancels the influence of the gravitation on the beam at the equilibrium $\alpha = 0$. The collocated sensors for the actuators (25) and (26) follow directly from (24) as $Y_B^i = H_B^i$ and $Y_A = H_A$.

According to this construction, the global minimum $W_{\min}$ of the function

$$W = W_p + W_g + \sum_{i=1}^{m} W_{U_B^i} + H_A g$$

occurs at $\alpha = 0$, $\varepsilon = 0$. To finish the controller design, we have to add damping to the system. According to the results of the third section we choose the control laws

$$\begin{aligned} U_B^i &= -k_B Y_B^i - d_B \dot{Y}_B^i \ , \quad i = 1, \ldots, m \\ U_A &= -g - d_A \dot{Y}_A \end{aligned}$$

with $d_A > 0$, $d_B > 0$.

Finally, the set of the evolutionary equations of the closed loop in Hamiltonian form can be derived from

$$\begin{aligned} H^0 &= \int_L \frac{1}{2\rho_{\mathrm{Ref}}} \left(\left(p^1\right)^2 + \left(p^2\right)^2 \right) \mathrm{d}S \\ &\quad + W - W_{\min} \\ H &= H^0 + H_A d_A \dot{Y}_A + \sum_{i=1}^{m} H_B^i d_B \dot{Y}_B^i \end{aligned}$$

with $p^i = \rho_{\mathrm{Ref}} u_0^i$, $i = 1, 2$ in canonical coordinates u^1, u^2, p^1, p^2. The skew-adjoint differential operator degenerates to a skew-symmetric matrix A,

$$\mathsf{A} = \begin{bmatrix} 0 & I \\ -I & 0 \end{bmatrix}$$

with the identity matrix I of dimension 2. The matrix A fulfills the Jacobi identity in a trivial way. The Euler-Lagrange operators are given by

$$\begin{aligned} \mathsf{E}_i &= \frac{\partial}{\partial u^i} \\ &\quad - \sum_{j=1}^{2} \left(\frac{\mathrm{d}}{\mathrm{d}x^j} \frac{\partial}{\partial u_j^i} - \sum_{k=1}^{2} \frac{\mathrm{d}}{\mathrm{d}x^k} \frac{\mathrm{d}}{\mathrm{d}x^j} \frac{\partial}{\partial u_{jk}^i} \right) \\ \mathsf{E}_{i+2} &= \frac{\partial}{\partial p^i} \ , \end{aligned}$$

$i = 1, 2$ in this case. It is worth deriving these equations by the help of a computer algebra program to see for oneself the enormous complexity of this set of equations. Of course, we can show only that $\{H^0, H\} \leq 0$ is met for the proposed controllers. Asymptotic stability cannot be proven here, but follows from the insight into the physics of the smart beam.

5. SUMMARY

This contribution presents control techniques for mechanical structures with piezoelectric sensor and actuator layers. Smart structures, e.g., based on piezoelectricity, allow to improve the dynamical behavior of mechanical devices significantly, because spatially distributed sensors and actuators can be constructed. Hamiltonian systems were used to describe the mechanical model. It turns out that collocated control using a suitable actuator and sensor pairing simplifies the controller design a lot. This method is well established for finite dimensional systems, but seems to be even more important for infinite dimensional ones. The presented mathematical models of the smart beam takes into account a nonlinear formulation of the kinematics, which is based on the natural rotational invariants of a plain curve, as well as simple nonlinear constitutive laws. It turns out that for this example passivity of the control law guarantees stability of the closed loop, if the Hamiltonian system itself is stable. Last but not least, one should mention that there are many open problems in the control of smart structures.

6. REFERENCES

Kugi, A., K. Schlacher and H. Irschik (1999). Infinite dimensional control of nonlinear beam vibrations by piezoelectric actuator and sensor layers. *Nonlinear Dynamics* **66**, 267–269.

Marsden, J.E. and T.J. Hughes (1994). *Mathematical Foundations of Elasticity.* Dover Publications. New York.

Nijmeijer, H. and A.J. van der Schaft (1990). *Nonlinear Dynamical Control Systems.* Springer Verlag. Berlin.

Olver, P.J. (1993). *Applications of Lie Groups to Differential Equations.* Springer Verlag. Wien.

Ortega, R., A. Loria, P.J. Nicklasson and H. Sira-Ramirez (1998). *Passivity-based Control of Euler-Lagrange Systems.* Springer Verlag. Berlin.

Schlacher, K. (1997). Mathematical strategies common to mechanics and control. *Zeitschrift für angewandte Mathematik und Mechanik, ZAMM* **78**, 723–730.

Schlacher, K. and A. Kugi (1999). Control of mechanical structures by piezoelectric actuators and sensors. In: *Lectures Notes in Control and Information Sciences 246, Stability and Stabilization of Nonlinear Systems* (D. Aeyels, F. Lamnabhi-Lagarrigue and A. van der Schaft, Eds.). pp. 275–292. Springer Verlag.

Schlacher, K., H. Irschik and A. Kugi (1996). Control of nonlinear beam vibrations by multiple piezoelectric layers. In: *IUTAM Symposium on Interaction between Dynamics and Control in Advanced Mechanical Systems* (D.H. van Campen, Ed.). pp. 355–363. Kluwer Academic Publishers.

Tzou, H.S. (1992). Active piezoelectric shell continua. In: *Intelligent Structural Systems* (H.S. Tzou and G.L. Anderson, Eds.). pp. 9–74. Kluwer AcademicPublishers.

van der Schaft, A.J. (2000). *L_2−Gain and Passivity Techniques in Nonlinear Control.* Springer Verlag. London.

AN ENERGY-BASED LYAPUNOV FUNCTION FOR PHYSICAL SYSTEMS

Marwan U. Bikdash and Richard A. Layton

North Carolina A&T State University
Greensboro, NC

Abstract: A new expression of the first law of thermodynamics is obtained for nonholonomic lumped-parameter systems. The input power due to the constraints is expressed in the framework of Lagrangian differential algebraic equations using Lagrange multipliers. General procedures to prove the stability of autonomous systems and to design control laws for them using the method of Lyapunov are then outlined and discussed. The approach in based on (a) postulating a Lyapunov function candidate consisting of the total energy stored in the system (expressed in terms of momenta not flows) augmented by a simple positive definite function of the displacements, (b) computing its time-derivative using the first law of thermodynamics, and (c) designing control laws as to make the time-rate of the candidate function negative definite. The procedure is applicable to systems where the control actuation appears through effort sources (such as voltage sources and applied forces), or flow sources (such as current or velocity sources). *Copyright © 2000 IFAC*

Keywords: Lyapunov function, stabilization methods, nonlinear systems theory, nonlinear control systems.

1. INTRODUCTION

Lyapunov functions form a cornerstone of dynamic stability theory and control design. Unfortunately, there is no general and reliable procedure to find Lyapunov functions even when they are known to exist. For physical systems, the total energy stored in the system is often suggested as a Lyapunov function candidate. Indeed, Lyapunov's first inspiration for his functions was precisely the energy stored in the system.

Unfortunately, using the stored energy is often problematic. Mainly there are two difficulties. One is to "compute" the time-derivative of the Lyapunov function $L(x)$ of the system state x along the system trajectories. We intend to show that this difficulty stems from our choices of the representational framework; that is, from representing and modelling the system as a set of first-order differential equations $\dot{x} = f(x, u, t)$ where $x \in \Re^n$, and $u \in \Re^m$. Indeed we will show that if the Lagrangian Differential Algebraic Equation (LDAE) formulation (Layton, 1998) is used, then this difficulty will be circumvented.

The second difficulty is to ascertain the negative definiteness of $\dot{L}(x)$. Usually, the first difficulty is the more important and the more immediate. Clearly, if $\dot{L}(x)$ cannot be evaluated or expressed symbolically then its sign definiteness cannot be ascertained.

Recently, a systematic modelling procedure for physical multidisciplinary system was introduced using an energy-work-constraint formulation. The procedure yields a system of differential algebraic equations (DAEs). The analyst's choice of coordinates remains faithful to the physics and avoids the introduction of independent coordinates. Independent coordinates are needed for the classical derivation of Lagrange's equations, or in general for all ordinary differential equation (ODE) descriptions of a system. This is achieved at the price of introducing artificial combinations of the physical coordinates and necessitating the explicit elimination of constraints. This elimination is impossible in general, and, when possible, it makes the physical interpretation of the equations and variables very difficult.

As a result, the Lyapunov approach to stability and stabilization becomes difficult because the expression of the time-rate of the stored energy in terms of the power and heat rate that are input to the system is not readily available in the ODE formulation.

A notable exception is in the field of robotics where the special nature of some robotic systems is very effectively used (Slotine and Li, 1991). Here the stored energy is augmented by a virtual potential energy and the time-rate of energy is obtained through a power balance equation, much like what is done in this paper. Indeed this paper generalizes this approach to all systems that can be formulated using the LDAE approach, including multidisciplinary lumped parameters subject to nonholonomic constraints.

The history of Lyapunov's stability and stabilization is long and venerable. Careful discussions of Lyapunov's theory can be found in (Kahlil, 1989) and references therein. We simply mention here those new developments that are immediately relevant or have a claim to generality. Johansson and Rantzer (1998) discussed a class of Lyapunov functions which are quadratic around the origin and piecewise-linear in the other regions of interest. The solution procedure is in terms of linear matrix inequalities and is related to the circle and Popov criteria. Blanchini and Miani (1999) introduced a new class of Lyapunov functions for linear uncertain systems in a method that extends linear matrix inequalities approach.

Maschke *et al.* (1998) discussed using the total stored energy of nonlinear Hamiltonian systems as a Lyapunov function in the presence of dissipation and nonzero forcing. They argue that the constant forcing induces a forced equilibrium which may be not a minimum of the total energy. To resolve this complication, they proposed as a candidate Lyapunov function the difference between the stored energy and the energy externally supplied to the system. It is not clear how positive definiteness of the candidate Lyapunov function can be shown for general systems, but Maschke *et al.* (1998) used Casimir's functions to complete the proof for systems satisfying some favorable integrability conditions.

There are computer-aided approaches to finding Lyapunov functions such as that proposed in (Ohta *et al.*, 1993), where the construction of a Lyapunov function is reduced to the construction of a polytope satisfying some conditions. Similarly, McConley (1997) discussed a polytope-based construction which is computable in polynomial time.

In this paper, we show that the LDAE formulation simplifies considerably the application of the Lyapunov approach to physical systems through the judicious use of an energy-based Lyapunov function candidate and the first law of thermodynamics.

2. SYSTEM DYNAMICS

In this paper, we concentrate on the stability and stabilization of autonomous systems. A system of nonlinear ODE's is said to be autonomous if the dynamics have no explicit dependence on time, as shown in the notation $\dot{x} = f(x)$, where $x \in \Re^N$. A nonautonomous system in ODE form is written as $\dot{x} = f(x,t)$, where $x \in \Re^N$. Clearly this accommodates control inputs and disturbance and noise inputs as well. In the context of control systems, the system $\dot{x} = F(x,u)$ is autonomous when u is a state-feedback control law given by $u = g(x)$. Then the closed-loop dynamics $\dot{x} = F(x, g(x))$ is equivalent to the form $\dot{x} = f(x)$.

A more physically meaningful description of the system is provided by the Lagrangian DAE formulation (Layton, 1998), where the modelling is performed in the framework of energy, virtual work, and constraints. The Lagrangian equations of motion can be written as

$$\frac{d}{dt}\left(\frac{\partial T^*}{\partial \dot{q}_i}\right) - \frac{\partial T^*}{\partial q_i} + \frac{\partial V}{\partial q_i} + \frac{\partial D}{\partial \dot{q}_i} + \sum_l \frac{\partial \phi_l}{\partial q_i}\kappa_l + \sum_l \frac{\partial \psi_l}{\partial \dot{q}_i}\mu_l = Q_i(q,\dot{q},u), \quad (1)$$

for $i = 1, \cdots, n$. These equations are subject to the holonomic constraints $\phi \in \Re^{l_1}$

$$\phi_l(q,u,t) = 0, \quad (2)$$

and the nonholonomic constraints $\psi \in \Re^{l_2}$ of the Pfaffian form

$$\psi_l(\dot{q},q,u,t) = 0. \quad (3)$$

Here $T^*(\dot{q},q,t)$ is the kinetic coenergy of the system, $V(q,t)$ is any acceptable definition of potential energy, $D(\dot{q}, q, t)$ is the content function representing the time-rate of heat dissipation, $q(t) \in \Re^n$ represents displacements, $\dot{q}(t) \in \Re^n$ represents the corresponding flows, $Q(t)$ is a vector of generalized forces and often represents effort sources, and $u(t) \in \Re^m$ is a known vector possibly time-varying parameters.

To the constraints correspond Lagrange multipliers $\kappa \in \Re^{l_1}$ and $\mu \in \Re^{l_2}$. The system dynamics is then described in terms of what we call a descriptor vector $X^T = [\dot{q}^T, q^T, \kappa^T, \mu^T]$. This is a generalization of the state vector of the ODE formulation where the assumption of "minimal realization" (a state vector of minimum dimension) is enforced. Clearly, the descriptor vector is not of minimum dimension, since eliminating one constraint would eliminate at least one coordinate q_1, the corresponding flow $\dot{q}_1$, and the multiplier corresponding to the constraint.

Another DAE can be obtained using the Hamiltonian formulation where the momenta p of kinetic stores are used instead of the flows. In this case, the descriptor vector becomes $[p^T, q^T, \kappa^T, \mu^T]^T$. See (Layton, 1998) for a detailed discussion. In this formulation, the kinetic energy $T(p, q, t)$ is used instead of the coenergy $T^*(\dot{q}, q, t)$. The two are known to satisfy the Legendre transformation

$$T(p, q, t) + T^*(\dot{q}, q, t) = \sum p_i \dot{q}_i, \qquad (4)$$

and they are numerically equal when all kinetic stores have linear constitutive laws.

To obtain an ODE description from a DAE description, the constraints must be completely eliminated. This is effectively done when a minimal set of coordinates ξ is expressed as a function of the old displacements $\xi = h(q)$, where $h(q)$ satisfies reasonable smoothness conditions. The complete discussion of this reduction is beyond the scope of this paper. In this paper, we simply assume that such reduction exists, and that the state x is equivalent to $[p^T, q^T]$ or alternatively to $[\dot{q}^T, q^T]$. Note that although the multipliers appear in the descriptor vector, they are totally dependent on $[p^T, q^T]$ (or $[\dot{q}^T, q^T]$) and the system parameters. *The multipliers are not needed to specify the state; only the displacements and momenta are.*

3. LYAPUNOV STABILITY AND CONTROL DESIGN

3.1 *Lyapunov Functions*

We assume without loss of generality that the origin ($x = 0$, or equivalently, $q = 0$ and $\dot{q} = 0$) is an equilibrium point of the system whose stability is to be studied. For autonomous systems, Lyapunov stability theory can be summarized in the following theorem (Slotine and Li, 1991).

Theorem 3.1. The origin of the system $\dot{x} = f(x)$ is stable in the sense of Lyapunov if there exists a positive-definite function $L(x)$ whose time derivative along the system trajectories $\dot{L}(x)$ is a negative semi-definite function of x. If $\dot{L}$ is a negative definite then the origin is asymptotically stable. If $\dot{L}(x)$ is negative semi-definite but it is zero only at points x not belonging to the invariant sets of the system, then the stability is asymptotic. If in addition $L(x)$ is radially unbounded ($L(x) \to \infty$ as $||x|| \to \infty$) then the stability is global.

In the DAE formulation, the definition of positive definiteness must be consistent with that of the ODE formulation:

Definition 3.1. A function $L(x)$ is said is be a positive-definite (positive semi-definite) function of the state $x \in \Re^n$ if $L(x) \geq 0$ and $L(x) = 0$ if (and only if) $x = 0$.

In the LDAE framework, the Lyapunov function must be a positive definite function of the state, not necessarily of the descriptor. Hence the multipliers do no contribute to the positive definiteness of the Lyapunov function candidate. Equally importantly, the sign definiteness must be shown for admissible values of the descriptor vector; in particular for admissible displacements q and momenta p.

Constraints will generally impact sign definiteness, but not always. For example, the quantity $q_1 q_2$ is not sign-definite in general. When the constraint $q_1 + q_2 = 0$ is enforced, however, that quantity becomes negative definite. The constraint $q_1 - q_2 = 0$ will make it positive definite. On the other hand, a positive-definite function retains its sign definiteness when a constraint is added. For example, the term q_1^2 is positive definite in q_1 regardless of any constraint.

Hence we will seek in our designs to work with functions which are sign-definite in the state of the system regardless of the constraints. This is most naturally done using the DAE formulations. If no such functions can be found, the effect of constraints on sign definiteness must be ascertained.

3.2 *Energy-Based Lyapunov Function*

Hence we propose the Lyapunov function candidate

$$L(p, q) = T(p, q) + V(q) + S(q) \qquad (5)$$

where $S(q)$, referred to as a virtual potential function in (Slotine and Li, 1991), and $V(q)$ are positive semidefinite functions of q, but $S(q)$ is chosen such that $V(q) + S(q)$ is positive definite in q (if possible, or in q subject to constraints). A typical choice is $S(q) = \frac{1}{2}q^T S_0 q$ where S_0 is a positive semi-definite matrix.

The assumption that $V(q)$ is a positive semi-definite function of q must be carefully pursued. Often a popular choice of coordinates makes it negative. In other cases, it may be advantageous to remove efforts which contribute terms with nondefinite sign from $V(q)$ and model them among the generalized forces Q. An example is the potential energy of a pendulum due to gravity when the reference datum is chosen at the hub of the pendulum. Possible fixes include redefining the datum by shifting it downward to the level of the pendulum mass at rest, or by modelling the force of gravity as an applied force instead of as a potential field.

The kinetic energy is usually a positive-definite function of the momenta p, but is generally a positive semi-definite function of q. It is often independent of q (and written $T(p)$).

The time derivative of the proposed Lyapunov function is

$$\dot{L} = \dot{T} + \dot{V} + \dot{S} \tag{6}$$

and it must be evaluated along the system trajectories prescribed by the system dynamics. Fortunately, the time derivative of $S(q)$ along the system trajectories is easy to express because $\dot{S} = \frac{\partial S}{\partial q}\dot{q} = \frac{\partial S}{\partial q} f$, with f being the flows. We emphasize that expressing $\dot{S}$ symbolically *is independent of the constraints and of Lagrange's equations of motion* in Eq. (1). In the next section we will show how to compute $\dot{T} + \dot{V}$ along the system trajectories from the first law of thermodynamics.

4. FIRST LAW OF THERMODYNAMICS

The first law of thermodynamics can be written in the rate form

$$\dot{E} + \dot{\mathbf{Q}} = P, \tag{7}$$

where E is the total energy stored in the system, including the kinetic energy $T(p, q, t)$ expressed in terms of coordinates q, momenta p, and time t, the potential energy $V(q, t)$, and the thermodynamic internal energy here assumed constant or subsumed in the potential energy V, $\dot{\mathbf{Q}}$ is the time-rate of heat dissipated by the system to the environment, and P is the total power input to the system. In other words,

$$\frac{dE}{dt} + \left\{ \begin{array}{c} \text{power dissipated as heat} \\ \text{and crossing system} \\ \text{boundaries quasi-isothermally} \end{array} \right\} = \left\{ \begin{array}{c} \text{input power to the} \\ \text{system including that} \\ \text{due to the constraints} \end{array} \right\}. \tag{8}$$

We assume $E = T + V$ where T is the kinetic energy and V is any acceptable expression of potential energy. We recall that V can be made to represent the energy stored in all potential stores; i.e., those elements whose efforts are conservative or path-independent. But the definition of the potential need not be exhaustive: Conservative efforts can be treated as applied efforts and accounted for outside the potential energy definitions.

Following (Layton, 1998), we consider the class of systems where the input power and the rate of heat dissipation can be written in the form

$$P - \dot{\mathbf{Q}} = \sum_j e_j^n \dot{q}_j \tag{9}$$

where e_j^n is the resultant of all (conservative or nonconservative) efforts acting in the direction of the coordinate q_j and not already accounted for in the kinetic and potential energies. Hence such an effort e_j^n can be classified as one of the following or a combination thereof:

- An applied effort Q_j, also called a generalized force. The virtual work done by all external sources can be expressed as $\delta W = \sum_j Q_j \delta q_j$ where δq_j is a virtual displacement. This term may be used to account for a conservative effort such as gravitational force that the analyst chooses to exclude from the expression of the potential energy.
- A dissipative effort e_j^d contributing to the heat dissipation term $\dot{\mathbf{Q}}$. This is usually represented in the so-called content function $D(\dot{q}, q, t)$ such that
$$e_j^d = -\frac{\partial D}{\partial \dot{q}_j}. \tag{10}$$
- A holonomic constraint effort $e_j^\phi = -\sum_l \frac{\partial \phi_l}{\partial q_j}\kappa_l$, where $\phi_l(q, t) = 0$ is the lth holonomic constraint acting on the system and κ_l is the corresponding multiplier. Clearly $\frac{d\phi_l}{dt} = 0$ and hence
$$\frac{\partial \phi_l}{\partial t} + \sum_j \frac{\partial \phi_l}{\partial q_j}\dot{q}_j = 0. \tag{11}$$
- A nonholonomic Pfaffian constraint effort $e_j^\psi = -\sum_l \frac{\partial \psi_l}{\partial \dot{q}_j}\mu_l$ where
$$\psi_l(\dot{q}, q, t) = \sum_j b_j^l(q, t)\,\dot{q}_j + b_0^l(q, t) = 0 \tag{12}$$

is the lth Pfaffian constraint acting on the system, and μ_l is the corresponding multiplier. Note that since $b_j^l(q,t) = \frac{\partial \psi_l}{\partial \dot{q}_j}$ we can write

$$\sum_j \frac{\partial \psi_l}{\partial \dot{q}_j}\dot{q}_j + b_0^l(q,t) = 0. \qquad (13)$$

4.1 *Power Supplied by Constraints*

The power supplied through the holonomic constraints can then be expressed as

$$\begin{aligned} P_{\text{hol.}} &= \sum_j e_j^{\phi}\dot{q}_j = -\sum_j\sum_l \frac{\partial \phi_l}{\partial q_j}\kappa_l \dot{q}_j \\ &= -\sum_l \kappa_l \sum_j \frac{\partial \phi_l}{\partial q_j}\dot{q}_j \\ &= \sum_l \kappa_l \frac{\partial \phi_l}{\partial t}. \end{aligned} \qquad (14)$$

where Eq. (11) has been used. Similarly, the power supplied to the system through nonholonomic Pfaffian constraints is

$$\begin{aligned} P_{\text{nonhol.}} &= \sum_j e_j^{\psi}\dot{q}_j = -\sum_j\sum_l \frac{\partial \psi_l}{\partial q_j}\mu_l \dot{q}_j \\ &= -\sum_l \mu_l \sum_j \frac{\partial \psi_l}{\partial q_j}\dot{q}_j \\ &= \sum_l \mu_l b_0^l(q,t) \end{aligned} \qquad (15)$$

where Eq. (13) has been used. In short:

- The power input to the system by the lth holonomic constraint is $\kappa_l \frac{\partial \phi_l}{\partial t}$.
- The power input to the system by the lth nonholonomic constraint is $\mu_l b_0^l(q,t)$.

Example 4.1. The slider-crank mechanism (Example 2.22 in (Layton, 1998)) has a holonomic constraint of the form

$$\phi = (x - r\cos\theta)^2 + (r\sin\theta)^2 - l^2 = 0 \qquad (16)$$

Since $\frac{\partial \phi}{\partial t} = 0$, this constraint produces no power.

Example 4.2. The nonholonomic constraint of the boat motion in the direction of its heading (See Example 2.26 in (Layton, 1998)) is of the form

$$\dot{x}\tan\theta - \dot{y} = 0 \qquad (17)$$

where x and y are planar position coordinates and θ is the boat heading; all are components of the displacement vector q. Hence the corresponding $b(q,t)$ is zero. This constraint produces no input power to the boat.

Example 4.3. Consider the bead sliding on rotating rod (Example 2.32 in (Layton, 1998)). The constraint can be written as (with $c = 0$),

$$\phi(x,y,t) = x\tan\omega_0 t - y = 0 \qquad (18)$$

and it represents the fact that the bead cannot move perpendicularly off of the rod, and the rod is made to rotate at a constant angular speed ω_0. This constraint must produce input power equal to

$$\kappa\frac{\partial \phi}{\partial t} = \kappa\omega_0 x\left(1 + \tan^2\omega_0 t\right) = \frac{\kappa\omega_0 x}{\cos^2\theta} \qquad (19)$$

where κ is the corresponding multiplier. It is not obvious what κ or $\frac{\partial \phi}{\partial t}$ represent. If the rod is massless, the input power at the shaft of rotation is $\omega_o rN = \omega_o xN/\cos\theta$. This must be the same as the power supplied by the constraint because all other powers are already accounted for. Comparing we obtain $\kappa = N\cos\theta$ which is the vertical component of the normal force! Alternatively, the constraint can be written as $\phi(x,y,t) = x\sin\omega_0 t - y\cos\omega_0 t = 0$, in which case $\kappa\frac{\partial \phi}{\partial t} = \omega_0\kappa\left(x\cos\omega_0 t + y\sin\omega_0 t\right)$. Comparing with $\omega_o rN = \omega_o N\sqrt{x^2+y^2}$, we obtain that $\frac{N}{\kappa} = \frac{x}{\sqrt{x^2+y^2}}\cos\omega_0 t + \frac{y}{\sqrt{x^2+y^2}}\sin\omega_0 t = 1$. Hence for this expression of the constraint the multiplier is simply the normal force.

4.2 *Dissipation Term*

To determine the heat dissipation terms, we make use of the Legendre transformation (Layton, 1998)

$$D(\dot{q},q,t) + G(e,q,t) = -\sum_j e_j^d\dot{q}_j = \dot{\mathbf{Q}} \qquad (20)$$

relating the content function to the co-content $G(e,q,t)$. But $-\sum e_j^d\dot{q}_j$ is nothing but the total heat dissipation rate. Hence $\dot{\mathbf{Q}} = D + G$.

5. STABILITY AND STABILIZATION

Substituting Eqs. (20), (14), and (15) into Eq. (8) and rearranging we obtain the power balance equation

$$\dot{T} + \dot{V} = P - (D + G) \qquad (21)$$

where

$$\begin{aligned} P &= \sum_j Q_j\dot{q}_j + P_{\text{hol.}} + P_{\text{nonhol.}} \\ &= \sum_j Q_j\dot{q}_j + \sum_l \frac{\partial \phi_l}{\partial t}\kappa_l + \sum_l b_0^l(q,t)\mu_l \end{aligned} \qquad (22)$$

is the power input to the system, and is directly expressible in terms of the descriptor vector and the systems inputs and parameters.

At this stage we examine the structure of the time-derivative $\dot{L}$ of the proposed L. Substituting Eq. (21) into Eq. (6) yields

$$\dot{L} = P - (D + G) + \dot{S} \qquad (23)$$

and discuss its use for stability determination and for stabilization using control. Clearly we have bypassed the first difficulty of the Lyapunov method, namely computing $\dot{L}$ effectively and in a concise form. The resulting expression of $\dot{L}$ is simple in two important respects: (a) The energy storing elements of the system (masses, capacitors, etc...) do not contribute explicit terms to $\dot{L}$; e.g., *masses, inductances, and capacitances do not appear in the expression of* $\dot{L}$. (b) constraints do not contribute explicit terms to $\dot{L}$ unless they contain sources (usually flow sources).

We are now faced with the second difficulty of the Lyapunov approach; namely, to ascertain the negative definiteness of $\dot{L}$ as a function of the state x, or alternatively, as a function of $\dot{q}$ and q subject to the constraints. It is crucial to realize that this problem (a) is purely algebraic; and (b) is application dependent.

We will however make some general remarks.

- The term $(D + G)$ is usually a positive definite (or semi-definite) function of $\dot{q}$ and hence contributes a negative definite (or semi-definite) term to $\dot{L}$.
- If the potential function is positive definite in q, there is no need to introduce the function S and hence stability is determined by the negative definiteness of the input power term P.
- The power input by an effort source (such as an applied forces or a a voltage source) appears as a $Q_i \dot{q}_i$ term in Eq. (22).
- The power input by a flow sources (such as an applied velocity or a current source) appears (usually) in the holonomic constraints. For example, a current source f_s is connected between ground and node $\mathbf{N}$ appears in a kinematic constraint $\phi_{\mathbf{N}}(q,t) = \sum_{i \in C} q_i + q_s(t) = 0$ where $\dot{q}_s(t) = f_s(t)$ and $q_i, i \in C_{\mathbf{N}}$, are charges flowing in the branches connected to $\mathbf{N}$. Note that the power input due to this source is $\kappa_{\mathbf{N}} \frac{\partial \phi_{\mathbf{N}}}{\partial t}$. Clearly, $\frac{\partial \phi_{\mathbf{N}}}{\partial t} = f_s(t)$ is the injected current, and hence $\kappa_{\mathbf{N}}$ is the node voltage.
- If the flow and/or effort sources can be controlled through an information feedback mechanism, then the sign definiteness of P or even $\dot{L}$ can be specified. In general, it is sufficient that the input power never exceeds the dissipation in absolute value to guarantee an asymptotically stable feedback system.
- When the constraints contribute input power, the corresponding multipliers must be known to determine stability. This may prove to be a difficulty for stability determination.
- For Lyapunov control design, however, the multipliers can be assumed to be measurable or observable, since they usually have physical interpretation. In the previous example, the multiplier $\kappa_{\mathbf{N}}$ is the voltage at node $\mathbf{N}$ and can be measured and fed back directly to control the magnitude of the current source.

6. CONCLUSIONS

In this paper, a simple expression for the power supplied to a system by kinematic constraints is derived. The first law of thermodynamics is then used to express the time rate of the stored energy of the system in terms of the descriptor vector of the Lagrangian Differential Algebraic Equations (LDAEs). The stored energy is then used as the basic block of a Lyapunov function candidate to determine system stability or to design stabilizing control laws. The simplifications introduced by this approach are significant and very promising.

Acknowledgement. This work is partly supported by the National Renewable Energy Laboratories through the contract NREL RCX716469 and by the US Office of Naval Research through the contract N00014-96-1123. We gratefully acknowledge this support.

7. REFERENCES

Blanchini, F. and S. Miani (1999). A new class of universal Lyapunov functions for the control of uncertain linear systems. *IEEE Trans. Automatic Control* **44**(3), 641–647.

Johansson, M. and A. Rantzer (1998). Computation of piecewise quadratic Lyapunov functions and hybrid systems. *IEEE Trans. Automatic Control* **43**(4), 555–559.

Kahlil, H. (1989). *Nonlinear Systems.* Prentice Hall. Englewood Cliffs.

Layton, R.A. (1998). *Principles of Analytical System Dynamics.* Springer-Verlag. New York.

Maschke, B., R. Ortega and A. van der Schaft (1998). Energy-based Lyapunov functions of forced Hamiltonian with dissipation. In: *37th Conference on Decision and Control.* Tampa. pp. 3599–3604.

McConley, M. (1997). Polytopic control Lyapunov functions for robust stabilization of a class of nonlinear systems. In: *American Control Conference.* Albuquerque. pp. 416–419.

Ohta, Y., H. Imanishi, L. Gong and H. Haneda (1993). Computer generated Lyapunov functions for a class for nonlinear systems. *IEEE Trans. Circuits and Systems-I: Fundamental Theory and Applications* **40**(5), 343–354.

Slotine, J. and W. Li (1991). *Applied Nonlinear Control.* Prentice Hall. Englewood Cliffs.

AN ALMOST POISSON STRUCTURE FOR THE GENERALIZED RIGID BODY EQUATIONS

Anthony M. Bloch [*,1] **Peter E. Crouch** [**,2]
Jerrold E. Marsden [***,3] **Tudor S. Ratiu** [****,4]

* *Department of Mathematics, University of Michigan, Ann Arbor, MI 48109*
** *Center for Systems Science and Engineering, Arizona State University, Tempe, AZ 85287*
*** *Control and Dynamical Systems, California Institute of Technology 107-81, Pasadena, CA 91125*
**** *Départément de Mathématiques, École polytechnique fédérale de Lausanne, CH - 1015 Lausanne, Switzerland*

Abstract: In this paper we introduce almost Poisson structures on Lie groups which generalize Poisson structures based on the use of the classical Yang-Baxter identity. Almost Poisson structures fail to be Poisson structures in the sense that they do not satisfy the Jacobi identity. In the case of cross products of Lie groups, we show that an almost Poisson structure can be used to derive a system which is intimately related to a fundamental Hamiltonian integrable system — the generalized rigid body equations. *Copyright© 2000 IFAC*

Keywords: Almost Poisson structure, Poisson Lie groups, Sklyanin bracket, reduction, the rigid body, integrable system.

1. INTRODUCTION

The theory of Poisson-Lie groups is of great interest in the theory of both classical and quantum systems. The basic theory of Poisson manifolds is discussed, for example, in Weinstein [1983], while the theory of Poisson-Lie groups (groups endowed with a Poisson structure) was developed in the work of Drinfeld [1983], Semenov-Tian-Shansky [1985], and Lu and Weinstein [1990], among other works.

A key aspect of much of this work is the so-called modified Yang-Baxter equation. Operators satisfying this equation are classical r-matrices and may be used to define Poisson structures on Lie groups. This yields, for example, the Sklyanin bracket which is used in Deift and Li [1991] to analyze the Toda lattice flow and its relationship to the SVD (singular value decomposition) flow.

To our knowledge no such Poisson-Lie structure has been found for other key classical integrable systems such as the generalized rigid body equations. In this paper we define a structure on a group which is not Poisson-Lie, but is very close to being Poisson in a way which we will make precise and which yields the generalized rigid body equations in a new form. This structure is based on a form of the equations originally derived in the work of Bloch and Crouch [1996] and Bloch, Brockett, and Crouch [1997] and inspired by op-

[1] Research partially supported by NSF grant DMS-9803181, AFOSR grant F49620-96-1-0100, and an NSF group infrastructure grant at the University of Michigan
[2] Work supported in part by NSF grant DMS 91011964 and NATO grant CRG 910926
[3] Research partially supported by the California Institute of Technology and NSF grant DMS-9802106
[4] Work supported in part by NSF grant DMS-98-02378 and the Swiss NSF.

timal control problems (see also Brockett [1994]). A connection of this work to discrete rigid body equations may be found in Bloch, Crouch, Marsden, and Ratiu [1998], [2000].

2. PRELIMINARY DEFINITIONS AND BACKGROUND THEORY

In this section we introduce some notation and review some principal results on Poisson structures. Let G be a Lie group with Lie algebra $\mathfrak{g}$, and let $\langle\, , \,\rangle$ be an Ad-invariant inner product on $\mathfrak{g}$. A key object in defining Poisson structures on groups is the notion of left and right derivatives of C^∞ functions on G (see Semenov-Tian-Shansky [1985] and Deift and Li [1991]).

Definition 1. Let $\phi \in C^\infty(G)$. Then the ***right derivative*** $D_R\phi_g^\#$ of ϕ at g is given by

$$D_R\phi_g(X) = \frac{d}{dt}\bigg|_{t=0} \phi(ge^{tX}) = \langle D_R\phi_g^\#, X\rangle \quad (2.1)$$

where $X \in \mathfrak{g}$, $D_R\phi_g \in \mathfrak{g}^*$ and $D_R\phi_g^\# \in \mathfrak{g}$. Similarly, the ***left derivative*** of $\phi \in C^\infty(G)$ is given by

$$D_L\phi_g(X) = \frac{d}{dt}\bigg|_{t=0} \phi(e^{tX}g) = \langle D_L\phi_g^\#, X\rangle \quad (2.2)$$

where $X \in \mathfrak{g}$, $D_L\phi_g \in \mathfrak{g}^*$ and $D_L\phi_g^\# \in \mathfrak{g}$.

Let $Ad_g : \mathfrak{g} \to \mathfrak{g}$, $Ad_g(X) = \frac{d}{dt} ge^{tX}g^{-1}|_{t=0}$ denote the adjoint action of G on its Lie algebra (see, e.g., Marsden and Ratiu [1994]). Thus $Ad_g = L_{g*}R_{g^{-1}*}$ where L_{g*} and R_{g*} denote the left and right action of G on its tangent bundle respectively; L_g and R_g are the left and right actions of the group on itself. Then we have:

Lemma 2.

$$D_R\phi_g \circ Ad_{g^{-1}} = D_L\phi_g \quad (2.3)$$

$$Ad_g(D_R\phi_g^\#) = (D_L\phi_g^\#)\,. \quad (2.4)$$

Proof. We have

$$D_R\phi_g(X) = \frac{d}{dt}\bigg|_{t=0} \phi(L_g e^{tx}) = \phi_* L_{g*}X$$

$$D_L\phi_g(X) = \frac{d}{dt}\bigg|_{t=0} \phi(R_g e^{tx}) = \phi_* R_{g*}X\,.$$

Hence $D_R\phi_g \circ Ad_{g^{-1}} = D_L\phi_g$ proving 2.3. Thus

$$\langle D_R\phi_g^\#, Ad_{g^{-1}}X\rangle = \langle D_L\phi_g^\#, X\rangle$$

or

$$\langle Ad_g D_R\phi_g^\#, X\rangle = \langle D_L\phi_g^\#, X\rangle$$

by Ad-invariance of the inner product. This gives (2.4). ■

Now in order to check the Jacobi identity in our Poisson bracket we also need the second derivatives of functions $\phi \in C^\infty(G)$. We make the following (natural) definitions:

Definition 3. Let $X, Y \in \mathfrak{g}$. We define

$$D_R^2\phi_g(X,Y) = \frac{d}{dt}\bigg|_{t=0} \frac{d}{ds}\bigg|_{s=0} \phi(ge^{sY}e^{tX}) \quad (2.5)$$

$$D_L^2\phi_g(X,Y) = \frac{d}{dt}\bigg|_{t=0} \frac{d}{ds}\bigg|_{s=0} \phi(e^{tX}e^{sY}g) \quad (2.6)$$

$$D_RD_L\phi_g(X,Y) = \frac{d}{dt}\bigg|_{t=0} \frac{d}{ds}\bigg|_{s=0} \phi(e^{tX}ge^{sY}) \quad (2.7)$$

$$D_LD_R\phi_g(X,Y) = \frac{d}{dt}\bigg|_{t=0} \frac{d}{ds}\bigg|_{s=0} \phi(e^{sY}ge^{tX})\,. \quad (2.8)$$

Then we have

Lemma 4.

(i) $D_RD_L\phi_g(X,Y) = D_LD_R\phi_g(Y,X)$ (2.9)

(ii) $D_R^2\phi_g(X,Y) - D_R^2\phi_g(Y,X) = D_R\phi_g([X,Y])$ (2.10)

(iii) $D_L^2\phi_g(X,Y) - D_L^2\phi_g(Y,X) = D_L\phi_g([Y,X])\,.$ (2.11)

Proof. (i) follows directly from the definitions. To prove (ii) and (iii) note that $(D_R\phi_g)(X) = X_g^L(\phi)$ and $(D_L\phi_g)(X) = X_g^R(\phi)$ where $X_g^L = L_{g*}X$ and $X_g^R = R_{g*}X$ are the left- and right-invariant vector fields determined by $X \in \mathfrak{g}$, respectively. Thus, $D_R^2\phi_g(X,Y) = X_g^L(Y^L(\phi))$ and, by a similar argument, $D_R^2\phi_g(Y,X) = Y_g^L(X^L(\phi))$. This gives (ii) and (iii) follows in the same way. ■

2.1 *The Poisson Bracket Structure*

We now introduce (see Semenev-Tian-Shansky [1985], Deift and Li [1991]) an important class of Poisson structures on Lie groups via classical r-matrices, or, more specifically, via operators satisfying the Yang-Baxter identity (sometimes called the modified Yang-Baxter equation). Operators of this type yield a Poisson bracket usually known as the Sklyanin bracket.

Definition 5. A linear operator A is said to satisfy the ***Yang-Baxter identity*** if $A : \mathfrak{g} \to \mathfrak{g}$ satisfies

(i) $[AX, AY] - A([AX,Y] + [X,AY]) = [X,Y]$

(ii) $\langle AX, Y\rangle + \langle X, AY\rangle = 0\,.$ (2.12)

An operator A is called a ***classical r-matrix*** if

$$[X,Y]_A = \frac{1}{2}([AX,Y] + [X,AY]) \qquad (2.13)$$

defines a Lie bracket on $\mathfrak{g}$. $(\mathfrak{g}, [\cdot,\cdot]_A)$ is called a ***Baxter Lie algebra***.

Semenov-Tian-Shansky showed the following:

Lemma 6. An operator satisfying the Yang-Baxter identity is a classical r-matrix.

Lemma 7. If A satisfies the Yang-Baxter identity, then

$$\{\phi,\psi\}_S = \frac{1}{2}(AD_R\phi^{\#}, D_R\phi^{\#}) - \frac{1}{2}(AD_L\phi^{\#}, D_L\phi^{\#}) \qquad (2.14)$$

defines a Poisson structure on G called the ***Sklyanin bracket***.

Deift and Li [1991] used an extension of this bracket to analyze the Toda lattice equations. This extension employed operators satisfying the Yang-Baxter identity. Here we describe a generalization of the Sklyanin bracket making explicit the exact role of the Yang-Baxter identity in the proof of the Jacobi identity for the corresponding Poisson structure.

Definition 8. A linear mapping $J : \mathfrak{g} \to \mathfrak{g}$ is called an ***equivariant Yang-Baxter operator*** if, for some smooth function F_J,

$$[JX, JY] - J[JX,Y] - J[X,JY] = F_J(X,Y),$$

where

$$Ad_g F_J(X,Y) = F_J(Ad_g X, Ad_g Y)$$

and if

$$\langle JX, Y\rangle + \langle X, JY\rangle \equiv 0\,. \qquad (2.15)$$

We have the following result.

Theorem 9. Let J_1 and J_2 be equivariant Yang-Baxter operators with $F_{J_1} = F_{J_2} \equiv F_J$. Then

$$\{\phi,\psi\}^{\pm} = \frac{1}{2}(\langle D_L\phi^{\#}, J_1 D_L\psi^{\#}\rangle \pm \langle D_R\phi^{\#}, J_2, D_R\psi^{\#}\rangle) \qquad (2.16)$$

is a Poisson bracket on G.

Proof. The bracket is clearly skew symmetric and bilinear. To prove the Jacobi identity, proceed as follows:

$$\begin{aligned}
&4\{\phi_1,\{\phi_2,\phi_3\}^{\pm}\}^{\pm}\\
&= 2\langle D_L\phi_1^{\#}, J_1(D_L\{\phi_2,\phi_3\}^{\pm})\rangle\\
&\quad \pm 2\langle D_R\phi_1^{\#}, J_2(D_R\{\phi_2,\phi_3\}^{\pm})^{\#}\rangle\\
&= \langle D_L\phi_1^{\#}, J_1(D_L\langle D_L\phi_2^{\#}, J_1 D_L\phi_3^{\#}\rangle)^{\#}\rangle\\
&\quad \pm \langle D_L\phi_1^{\#}, J_1(D_L\langle D_R\phi_2^{\#}, J_2 D_R\phi_3^{\#}\rangle)^{\#}\rangle\\
&\quad \pm \langle D_R\phi_1^{\#}, J_2(D_R\langle D_L\phi_2^{\#}, J_1, D_L\phi_3^{\#}\rangle)^{\#}\rangle\\
&\quad + \langle D_R\phi_1^{\#}, J_2(D_R\langle D_R\phi_2^{\#}, J_2 D_R\phi_3^{\#}\rangle)^{\#}\rangle\,.
\end{aligned}$$

Since J_1 and J_2 are skew and $\langle D_{L(R)}\phi^{\#}, X\rangle = D_{L(R)}\phi(X)$ by definition, we may rewrite this expression as

$$\begin{aligned}
4\{\phi_1,\{\phi_2,\phi_3\}^{\pm}\}^{\pm} &= -D_L^2\phi_2(J_1D_L\phi_3^{\#}, J_1D_L\phi_1^{\#})\\
&\quad + D_L^2\phi_3(J_1D_2\phi_2^{\#}, J_1D_L\phi_1^{\#})\\
&\quad - D_R^2\phi_2(J_2D_R\phi_3^{\#}, J_2D_R\phi_1^{\#})\\
&\quad + D_R^2\phi_3(J_2D_R\phi_2^{\#}, J_2D_R\phi_1^{\#})\\
&\quad \mp D_LD_R\phi_2(J_2D_R\phi_3^{\#}, J_1D_L\phi_1^{\#})\\
&\quad \pm D_LD_R\phi_3(J_2D_R\phi_2^{\#}, A_1D_2\phi_1^{\#})\\
&\quad \mp D_RD_L\phi_2(J_1D_L\phi_3^{\#}, J_2D_R\phi_1^{\#})\\
&\quad \pm D_RD_L\phi_3(J_1D_L\phi_2^{\#}, J_2D_R\phi_1^{\#})\,.
\end{aligned} \qquad (2.17)$$

We need to show that this expression plus its cyclic permutations is identically zero. To do this, represent the first two terms in (2.17) in obvious fashion as $-(2,3,1) + (3,2,1)$. This plus its permutations is the expression

$$\begin{aligned}
&-(2,3,1) + (3,2,1)\\
&-(3,1,2) + (1,3,2)\\
&-(1,2,3) + (2,1,3)\,.
\end{aligned}$$

Using Lemma 4 (iii) this equals

$$\begin{aligned}
&\langle D_L\phi_2^{\#}, [J_1D_L\phi_3^{\#}, J_1D_L\phi_1^{\#}]\rangle\\
&\quad + \langle D_L\phi_1^{\#}, [J_1D_L\phi_2^{\#}, J_1D_L\phi_3^{\#}]\rangle\\
&\quad + \langle D_L\phi_3^{\#}, [J_1D_L\phi_1^{\#}, J_1D_L\phi_2^{\#}]\rangle\,.
\end{aligned}$$

Now use the facts that J_1 is skew and $\langle [X,Y],Z\rangle + \langle Y,[X,Z]\rangle = 0$ to rewrite this as

$$\begin{aligned}
&\langle D_L\phi_1^{\#}, [J_1D_L\phi_2^{\#}, J_1D_L\phi_3^{\#}]\rangle\\
&-\langle D_L\phi_1^{\#}, J_1[D_L\phi_2^{\#}, J_1D_L\phi_3^{\#}]\rangle\\
&-\langle D_L\phi^{\#}, J_1[J_1D_L\phi_2^{\#}, D_L\phi_3^{\#}]\rangle\,.
\end{aligned} \qquad (2.18)$$

Since J_1 is an equivariant Yang-Baxter operator this equals

$$-\langle D_L\phi_1^{\#}, F_J(D_L\phi_2^{\#}, D_L\phi_3^{\#})\rangle\,. \qquad (2.19)$$

Similarly, now using Lemma 4 (ii), we obtain for the second two terms of (2.17),

$$+\langle D_R\phi_1^{\#}, F_J(D_R\phi_2^{\#}, D_R\phi_3^{\#})\rangle\,. \qquad (2.20)$$

By Lemma 2, we have $(D_L\phi_g)^{\#} = Ad_g(D_R\phi_g^{\#})$. Thus, since $\langle\ ,\ \rangle$ is Ad-invariant and F_J is Ad-equivariant, the cyclic sum of terms 1 and 2 in (2.17) is identically zero.

Now consider the last four terms of (2.17). By (i) of Lemma 4 we may rewrite these terms as

$$\begin{array}{l}\mp D_L D_R \phi_2 (J_2 D_R \phi_3^{\#}, J_1 D_L \phi_1^{\#}) \\ \pm D_L D_R \phi_3 (J_2 D_R \phi_2^{\#}, J_1 D_L \phi_1^{\#}) \\ \mp D_L D_R \phi_2 (J_2 D_R \phi_1^{\#}, J_1 D_L \phi_3^{\#}) \\ \pm D_L D_R \phi_3 (J_2 D_R \phi_1^{\#}, J_1 D_L \phi_2^{\#}) .\end{array}$$

Again, represent this as

$$\mp(2,3,1) \pm (3,2,1) \mp (2,1,3) \pm (3,1,2).$$

Then the cyclic sum is

$$\begin{array}{l}\mp(2,3,1) \pm (3,2,1) \mp (2,1,3) \pm (3,1,2) \\ \mp(3,1,2) \pm (1,3,2) \mp (3,2,1) \pm (1,2,3) \\ \mp(1,2,3) \pm (2,1,3) \mp (1,3,2) \pm (2,3,1)\end{array}$$

which is identically zero. Hence the Jacobi identity is satisfied. ■

The proof of lemma 7 follows directly from this result by noting that $F_J(X,Y) = [X,Y]$, is clearly *Ad*-invariant.

3. DYNAMICS WITH RESPECT TO AN ALMOST POISSON STRUCTURE

In this section we consider dynamics on groups under the bracket (2.16), but in the case where F_J is generally *not* Ad_g equivariant – such structures will be called ***almost Poisson structures*** on Lie groups. We begin by considering the general case and then specialize to the case of particular interest to us – cross products of Lie groups.

Such a bracket is an instance of an *almost* Poisson structure (see da Silva and Weinstein [1999]):

Definition 10. An ***almost Poisson manifold*** is a pair $(M, \{,\})$ where M is a smooth manifold and (i) $\{,\}$ defines an ***almost Lie algebra structure*** on the C^∞ functions on M, i.e. the bracket satisfies all conditions for a Lie algebra except that the Jacobi identity is not satisfied and (ii) $\{,\}$ is a derivation in each factor.

Given an almost Poisson structure with local coordinate expression $\pi^{ij}(z)$ (the generalization of the Poisson tensor) and a function H on M, one defines an ***almost Poisson*** vector field on M by

$$\dot{z}^i = \pi^{ij}(z)\frac{\partial H}{\partial z^j}. \tag{3.1}$$

Now in our setting we have:

Theorem 11. The equation

$$\dot{\phi} = \{\phi, H\}^{\pm} \tag{3.2}$$

where $\{\ ,\ \}^{\pm}$ is the bracket (2.16) on a Lie group G yields the equation

$$\dot{g} = \frac{1}{2}R_{g*}J_1 D_L H^{\#} \pm \frac{1}{2}Lg_* J_2 D_R H^{\#}. \tag{3.3}$$

If $H \in I(G)$, the functions on G invariant under conjugation, then $D_L H = D_R H = DH$ and (3.2) becomes

$$\dot{g} = \frac{1}{2}R_{g*}J_1 DH^{\#} \pm \frac{1}{2}L_{g*}J_2 DH^{\#} \tag{3.4}$$

Proof. Equation (3.2) may be written as

$$\frac{d}{dt}\phi(g) = \frac{1}{2}D_L\phi(J_1 D_L H^{\#}) \pm \frac{1}{2}D_R\phi(J_2 D_R H^{\#}).$$

But

$$D_L\phi = \phi_* R_{g^*}, D_R\phi = \phi_* L_{g^*} \text{ and } \frac{d}{dt}\phi(g) = \phi_*\dot{g}.$$

Hence we obtain (3.3).

Now, if $\phi \in I(G)$, i.e., $\phi(g) = \phi(hgh^{-1})$, for all $h \in G$, then $\phi_* X = \phi_* Ad_h X$, for all $h \in G$, $X \in \mathfrak{g}$. Hence $\phi_* = \phi_* Ad_h$ or $\phi_* R_{h_*} = \phi_* L_{h_*}$ and thus

$$D_L\phi_g = \phi_* R_{g_*} = \phi_* L_{g_*} = D_R\phi_g . \quad ■$$

We now consider the case of interest to us, where the Lie group is

$$D = G \times G \tag{3.5}$$

with Lie algebra $\mathfrak{g} \times \mathfrak{g} \equiv \mathfrak{g}^D$. We define the map $J : \mathfrak{g}^D \to \mathfrak{g}^D$ given by

$$J(X_1, X_2) = (X_2, -X_1) \tag{3.6}$$

for $(X_1, X_2) \in \mathfrak{g}^D$.

We shall shortly define a bracket using J but note that J does *not* define a Poisson structure, for while J is clearly skew, if $X, Y \in \mathfrak{g}^D$

$$\begin{aligned}F_J(X,Y) &= [JX, JY] - J[JX, Y] - J[X, JY] \\ &= \begin{pmatrix}[X_1 + X_2, Y_1 + Y_2] \\ [X_1 + X_2, Y_1 + Y_2]\end{pmatrix} \\ &\quad - \begin{pmatrix}[X_1, Y_1] \\ [X_2, Y_2]\end{pmatrix}\end{aligned}$$

which is clearly not Ad_g invariant.

We now make the following definitions in this setting in order to set up the equations of motion:

Definition 12. Let e be the identity in G, $g \in D$, $(X_1, X_2) \in \mathfrak{g}^D$, then we define

$$\begin{aligned}(D_R^1\phi)_g(X_1) &= \left.\frac{d}{dt}\right|_{t=0} \phi(g(e^{tX_1}, e)) \\ (D_R^2\phi)_g(X_2) &= \left.\frac{d}{dt}\right|_{t=0} \phi(g(e, e^{tX_2})) \\ (D_L^1\phi)_g(X_1) &= \left.\frac{d}{dt}\right|_{t=0} \phi((e^{tX_1}, e)g) \\ (D_L^2\phi)_g(X_2) &= \left.\frac{d}{dt}\right|_{t=0} \phi((e, e^{tX_2})g) \\ (D_R\phi)_g(X_1, X_2) &= ((D_R^1\phi)_g, (D_R^2\phi)_g)(X_1, X_2) \\ (D_L\phi)_g(X_1, X_2) &= ((D_L^1\phi)_g, (D_L^2\phi)_g)(X_1, X_2) .\end{aligned}$$

Using the inner product on $\mathfrak{g}$ and $\mathfrak{g}^D$ we get obvious definitions of $(D_L^1\phi_g)^{\#} \in \mathfrak{g}$, etc. Then the + bracket (2.16) becomes

$$\{\phi,\psi\} = \frac{1}{2}\langle D_L^1\phi^\#, D_L^2\psi^\#\rangle - \frac{1}{2}\langle D_L^2\phi^\#, D_L^1\psi^\#\rangle$$
$$+\frac{1}{2}\langle D_R^1\phi^\#, D_R^2\psi^\#\rangle$$
$$-\frac{1}{2}\langle D_R^2\phi^\#, D_R^1\psi^\#\rangle\,. \qquad (3.7)$$

Using (3.3), we have

Corollary 13. The equations $\dot{\phi} = \{\phi, H\}$ induced by the bracket (3.7) on D are given by

$$\dot{g}_1 = \frac{1}{2}R_{g_1^*}(D_L^2 H_{(g_1,g_2)})^\# + \frac{1}{2}L_{g_1^*}(D_R^2 H_{(g_1,g_2)})^\#$$
$$\dot{g}_2 = -\frac{1}{2}R_{g_2^*}(D_L^1 H_{(g_1,g_2)})^\# - \frac{1}{2}L_{g_2^*}(D_R^1 H_{(g_1,g_2)})^\#\,. \qquad (3.8)$$

We now prove the main result of the paper, which provides a special case of these equations and turns out to be intimately related to the generalized rigid body equations on $so(n)$.

Theorem 14. Consider the flow (3.8) in the case

$$H = 4\,\mathrm{Trace}(I_1 g_2^{-1} g_1 + I_2 g_1 g_2^{-1}) \qquad (3.9)$$

where I_1 and I_2 are symmetric, positive definite matrices. Let J_1 and J_2 be defined by

$$J_1^{-1}(X) = I_1 X + X I_1$$
$$J_2^{-1}(X) = I_2 X + X I_2.$$

Then the flow (3.8) is given by

$$\dot{g}_1 = J_2^{-1}(g_2 g_1^{-1} - g_1 g_2^{-1})g_1 - g_1 J_1^{-1}(g_2^{-1}g_1 - g_1^{-1}g_2)$$
$$\dot{g}_2 = J_2^{-1}(g_2 g_1^{-1} - g_1 g_2^{-1})g_2$$
$$- g_2 J_1^{-1}(g_2^{-1}g_1 - g_1^{-1}g_2)\,. \qquad (3.10)$$

Proof. Let $g = (g_1, g_2)$ and compute as follows:

$$D_R^2 H_g(X) = 4\,\mathrm{Trace}(-I_1 X g_2^{-1} g_1 - I_2 g_1 X g_2^{-1})$$
$$= 4\,\mathrm{Trace}(-J_1^{-1}(X) g_2^{-1} g_1 + X I_1, g_2^{-1}g_1 - g_2^{-1} I_2 g_1 X)$$
$$= 4\,\mathrm{Trace}(-X J_1^{-1}(g_2^{-1}g_1) + X I_1, g_2^{-1}g_1 - g_2^{-1} I_2 g_1 X)$$
$$= 2\,\mathrm{Trace}(-X J_1^{-1}(g_2^{-1}g_1 - g_1^{-1}g_2)$$
$$+ X(I_1 g_2^{-1} g_1 - g_1^{-1} g_2 I_1) + (g_1^{-1} I_2 g_2 - g_2^{-1} I_2 g_1)X)$$

since $X \in so(n)$ and $g_1 g_2 \in SO(n)$. Therefore,

$$(D_R^2 H_g)^\# = -2J_1^{-1}(g_2^{-1}g_1 - g_1^{-1}g_2)$$
$$+ 2(I_1 g_2^{-1} g_1 - g_1^{-1} g_2 I_1) + 2(g_1^{-1} I_2 g_2 - g_2^{-1} I_2 g_1).$$

Computing similarly we find:

$$(D_L^2 H_g)^\# = -2J_2^{-1}(g_1 g_2^{-1} - g_2 g_1^{-1})$$
$$+ 2(g_1 g_2^{-1} I_2 - I_2 g_2 g_1^{-1}) + 2(g_2 I_1 g_1^{-1} - g_1 I_1 g_2^{-1})$$
$$(D_L^1 H_g)^\# = 2J_2^{-1}(g_1 g_2^{-1} - g_2 g_1^{-1})$$
$$+ 2(g_2 g_1^{-1} I_2 - I_2 g_1 g_2^{-1}) + 2(g_1 I g_2^{-1} - g_2 I_1 g_1^{-1})$$
$$(D_R^1 H_g)^\# = 2J_1^{-1}(g_2^{-1}g_1 - g_1^{-1}g_2)$$
$$+ 2(I_1 g_1^{-1} g_2 - g_2^{-1} g_1 I_1) + 2(g_2^{-1} I_2 g_1 - g_1^{-1} I_2 g_2)\,.$$

Substitution into (3.8) gives the result. ■

4. THE RIGID BODY EQUATIONS

In this last section we relate the system (3.10) to the rigid body system on $SO(n)$. We show

Theorem 15. The equations (3.10) in the case $J_1^{-1} = 0$, $J_2 = J$ are locally equivalent to the generalized rigid body equations.

To do this, we recall some background information from Bloch, Brockett, and Crouch [1997] (see also Bloch, Crouch, Marsden, and Ratiu [1998, 2000]). We recall that the rigid body equations on $SO(n)$ (or generally on any compact Lie group – see e.g. Marsden and Ratiu [1994], Ratiu [1980]) may be written as

$$\dot{Q} = \Omega Q$$
$$\dot{M} = [\Omega, M] \qquad (4.1)$$

where $Q \in SO(n)$ denotes the configuration space variables, $\Omega \in so(n)$ is the angular velocity, and $M = J\Omega = \Lambda\Omega + \Omega\Lambda$ is the angular momentum. Here J is a symmetric positive definite operator defined by the diagonal positive definite matrix Λ. We remark that the rigid body equations here are written in right-invariant as opposed to the commonly used left-invariant form in order to be consistent with the conventions used in the remainder of the paper. This results in a sign change in the second of equations (4.1). The classical rigid body equations (4.1) are of course Hamiltonian on $T^*SO(n)$ with respect to the canonical symplectic structure. We now consider the following equations:

$$\dot{Q} = \Omega Q$$
$$\dot{P} = \Omega P, \qquad (4.2)$$

where $\Omega = J^{-1}M$ and $M = PQ^T - QP^T$ for $Q, P \in SO(n)$. We then can easily check that:

Proposition 16. The mapping $(Q, P) \mapsto (Q, M)$, from $SO(n) \times SO(n)$ to $T^*SO(n)$ takes all solutions of equation (4.2) onto solutions of the generalized rigid body equations (4.1).

Proof. Differentiating $M = PQ^T - QP^T$ and using the equations (4.2) gives the second of equations (4.1). ■.

Conversely, given the rigid body equations (4.1) we may solve for the variable P in the expression

$$M = PQ^T - QP^T$$

in a neighborhood of $M = 0$. Locally, in a neighborhood of $M = 0$, where $\sinh^{-1}$ is well defined

$$P = \left(e^{\sinh^{-1} M/2}\right) Q\,. \qquad (4.3)$$

This follows from the observation that

$$M = e^{\sinh^{-1} M/2} - e^{-\sinh^{-1} M/2}.$$

For $so(n)$, however, sinh is many to one, so the two representations are not entirely equivalent. (For more details giving the precise region for M for which one can solve for P, see Bloch, Crouch, Marsden, and Ratiu [2000]. This reference also contains a discussion of how this system is related to certain optimal control problems.)

Observe, however, that this proves Theorem 15, as this new form of the rigid body equation is exactly of the type given to us by our almost Poisson structure, and the system of equations (3.10), under the conditions of theorem 15

We would like to say some more, however, about the Hamiltonian structure of these equations, again following Bloch, Brockett, and Crouch [1997].

Note firstly that the generalized rigid body flow naturally reduces to a flow in the variable M on an adjoint orbit of $so(n)$ and we can view the map which takes $PQ^T - QP^T$ to M as reduction. In fact, the map $(Q, P) \mapsto (Q, M)$ given above is a canonical transformation from the symplectic structure on $T^*gl(n)$ to that on $T^*SO(n)$ which intertwines the Hamiltonian equations (4.2) on $T^*gl(n)$ with the Hamiltonian equations (4.1) on $T^*SO(n)$.

While the classical rigid body equations (4.1) are Hamiltonian on $T^*SO(n)$ with respect to the canonical symplectic structure, on the group we have

Proposition 17. The generalized rigid body equations in the form (4.2) are Hamiltonian on $T^*gl(n)$ with respect to the canonical symplectic structure and the Hamiltonian

$$H = \frac{1}{4}\left\langle J^{-1}(PQ^T - QP^T), PQ^T - QP^T\right\rangle, \tag{4.4}$$

where $\langle \xi, \eta \rangle = \text{Trace}(\xi^T \eta)$.

This is a straightforward computation.

We remark that here P and Q are natural coordinates for $T^*gl(n)$ and, for $P(0), Q(0) \in SO(n)$, $P(t)$ and $Q(t)$ evolve in $SO(n)$ under the flow of H. Hence $SO(n) \times SO(n)$ is an invariant manifold for the flow of H. Note also that this Hamiltonian is equivalent to $H = (1/4)\langle J^{-1}M, M\rangle$, as in Ratiu [1980].

REFERENCES

Bloch, A.M., R.W. Brockett, and T.S. Ratiu [1992] Completely integrable gradient flows. *Comm. Math. Phys.* **147**, 57–74.

Bloch, A.M., R. W. Brockett, and P.E. Crouch [1997] Double bracket equations and geodesic flows on symmetric spaces. *Comm. Math. Phys.* **187**, 357–373.

Bloch, A.M. and P.E. Crouch [1995] On the geometry of optimal control and geodesic flows *Proc. CDC* **34**, IEEE, 3283-3288.

Bloch, A. M. and P. E. Crouch [1996] Optimal control and geodesic flows *Systems and Control Letters* **28**, 65-72.

Bloch, A.M., P.E. Crouch, J.E. Marsden, and T.S. Ratiu [1998] Discrete rigid body dynamics and optimal control *Proc. CDC* **37**, IEEE, 2249–2254.

Bloch, A.M., P.E. Crouch, J.E. Marsden and T.S. Ratiu [2000] Optimal control and discrete rigid body equations, to appear.

Bloch, A.M., H. Flaschka, and T.S. Ratiu [1990] A convexity theorem for isospectral sets of Jacobi matrices in a compact Lie algebra. *Duke Math. J.* **61**, 41–66.

Brockett, R.W. [1994] The double bracket equation as a solution of a variational problem. *Fields Institute Comm.* **3**, 69-76.

Cannas da Silva, A. and A. Weinstein [1999] *Geometric Models for Noncommutative Algebras*, American Mathematical Society.

P. Deift and L.-C, Li [1991] Poisson geometry of the Analog of the Miura maps and Backlund-Darboux Transformations for equations of Toda type and periodic Toda flows, *Communications in Mathematical Physics* **143**, 201–214.

Drinfeld, V. [1983], Hamiltonian structures on Lie groups, Lie bialgebras and the geometric meaning of the classical Yang-Baxter equations, *Sov. Math. Dokl.* **27**, 68-71.

Lu, J. and A. Weinstein [1990] Poisson Lie groups, dressing transformations, and the Bruhat decomposition *J. Diff. Geom* **31**, 501-526.

Marsden, J.E. and T.S. Ratiu [1994] *Introduction to Mechanics and Symmetry*. Texts in Applied Mathematics, **17**, Springer-Verlag, 1994. Second Edition, 1999.

Ratiu, T. [1980] The motion of the free n-dimensional rigid body. *Indiana U. Math. J.*, **29**, 609-627.

Semenov-Tain Shansky [1985], Dressing transformations and Poisson group actions *Publ. RIMS, Kyoto University*, **21**, 1237-1260.

Weinstein, A [1983], Local structure of Poisson manifolds, *J. Diff. Geometry*, **18**, 523-558.

SYMMETRIES AND CONSERVATION LAWS FOR IMPLICIT PORT-CONTROLLED HAMILTONIAN SYSTEMS

G. Blankenstein * A.J. van der Schaft *

* *Faculty of Mathematical Sciences, Department of Systems, Signals and Control, University of Twente, P.O.Box 217, 7500 AE Enschede, The Netherlands, e-mail: {g.blankenstein,a.j.vanderschaft}@math.utwente.nl*

Abstract: In this paper we describe the correspondence between symmetries and conservation laws of implicit port-controlled generalized Hamiltonian systems. Furthermore, symmetries of interconnected implicit Hamiltonian systems are studied.

Keywords: conserved quantities, conservation laws, implicit systems, interconnected systems, mechanical systems, nonlinear control systems, symmetry

1. INTRODUCTION

A well-known result in mechanics is Noether's theorem, stating that to any symmetry of a mechanical system there corresponds a conserved quantity, that is, a function of the state which is constant along solutions of the system. This result has been extended to general explicit Hamiltonian (or "Poisson") systems, and has been very important in the development of the theory of reduction of these systems, see (Abraham and Marsden, 1978; Marsden and Ratiu, 1986). In previous work (Blankenstein and van der Schaft, 1999*a*; Blankenstein and van der Schaft, 1999*b*) the classical Noether result was extended to *implicit* Hamiltonian systems, e.g. mechanical systems with (nonholonomic) constraints. This led to a theory extending the reduction theory for explicit Hamiltonian systems as well as constrained mechanical systems (e.g. (Bloch *et al.*, 1996; Cantrijn *et al.*, 1999)) to a general unified reduction theory for implicit generalized Hamiltonian systems (Blankenstein and van der Schaft, 1999*a*; Blankenstein and van der Schaft, 1999*b*).

In this paper we consider symmetries of implicit *port-controlled* generalized Hamiltonian systems, i.e. implicit generalized Hamiltonian systems with external (port) variables. Examples are constrained mechanical systems with inputs being forces (torques) and outputs being generalized velocities. In general, the presence of inputs breaks the symmetry of the system. However, allowing for what is called *external symmetries* one can in some cases recover the original symmetries of the system without inputs. The particle in $\mathbb{R}^3$, with rotational symmetry, subject to external forces will serve as an example.

The main result of the paper is a generalization of Noether's theorem to this class of systems. It will be shown that to a symmetry of an implicit port-controlled generalized Hamiltonian system there corresponds a *conservation law*. This is defined as a function of the state of the system such that its derivative along solutions of the system is a function of the (integrated) external variables only. This means that the evolution of this function is completely determined by the external variables, that is, the measurements, of the system. The result generalizes the previous Noether type of results where, in the absence of external variables, the function is constant along solutions of the system, i.e. a conserved quantity. The idea is based on the results given in (van der Schaft, 1981) where the correspondence between symmetries and conservation laws, in (explicit, symplectic) Hamiltonian systems with inputs and outputs, was first developed.

Finally the paper addresses the interconnection properties of symmetries. In (van der Schaft and Maschke, 1997; Maschke and van der Schaft, 1997; Dalsmo and van der Schaft, 1999) it has been shown that implicit port-controlled generalized Hamiltonian systems are well suited for describing complex energy-conserving physical systems by using a "modular approach", that is, considering the system as a power-conserving interconnection of smaller subsystems. Again, the symmetry properties of these interconnected systems are very important for their analysis and possible reduction. In this paper we give a basic result, giving necessary and sufficient conditions under which symmetries of the subsystems give rise to a symmetry of the interconnected system.

2. IMPLICIT PORT-CONTROLLED HAMILTONIAN SYSTEMS

The underlying geometric structure of an implicit port-controlled generalized Hamiltonian system is a Dirac structure. Consider a smooth n-dimensional manifold M, and define $TM \oplus T^*M$ as the smooth vector bundle over M with fiber $T_zM \times T_z^*M$ at $z \in M$. Let D be a smooth vector subbundle of $TM \oplus T^*M$. Let Δ be a smooth vector field on M and δ a smooth one-form on M. We say the pair (Δ, δ) is in D, denoted by $(\Delta, \delta) \in D$, if $(\Delta(z), \delta(z)) \in D(z)$ for every $z \in M$. A generalized Dirac structure D is defined as follows.

Definition 1. (Courant, 1990; Dorfman, 1993) A *generalized Dirac structure* D on M is defined as a smooth vector subbundle $D \subset TM \oplus T^*M$ satisfying the property $D = D^\perp$, where $D^\perp$ is defined as

$$D^\perp := \{(\Delta_2, \delta_2) \in TM \oplus T^*M \mid \langle \delta_1, \Delta_2 \rangle + \langle \delta_2, \Delta_1 \rangle = 0, \ \forall (\Delta_1, \delta_1) \in D\}.$$

Here $\langle \cdot, \cdot \rangle$ denotes the natural pairing between a one-form and a vector field. Actually, from the property $D = D^\perp$ it follows that $\dim D(z) = n$ for every $z \in M$, and so D is a vector subbundle of $TM \oplus T^*M$ (with n-dimensional fibers $D(z)$, $z \in M$). By taking $(\Delta_1, \delta_1) = (\Delta_2, \delta_2) = (\Delta, \delta)$ it follows directly from $D = D^\perp$ that for all $(\Delta, \delta) \in D$

$$\langle \delta, \Delta \rangle = 0. \tag{1}$$

A generalized Dirac structure is called *closed*, or just Dirac structure, if the following additional property is satisfied ((Courant, 1990; Dorfman, 1993))

$$\langle L_{\Delta_1} \delta_2, \Delta_3 \rangle + \langle L_{\Delta_2} \delta_3, \Delta_1 \rangle + \langle L_{\Delta_3} \delta_1, \Delta_2 \rangle = 0,$$

for all $(\Delta_i, \delta_i) \in D$, $i = 1, 2, 3$. Here L denotes the Lie derivative. This property generalizes the Jacobi identity for Poisson structures.

Now we can define implicit port-controlled generalized Hamiltonian systems. Consider a smooth manifold $\mathcal{X}$, being the state space of the system. Furthermore, consider the linear space $\mathcal{F}$ of external flows and dually the space $\mathcal{F}^*$ of external efforts (i.e. $\mathcal{F} \times \mathcal{F}^*$ is the space of external variables of the system). Let there be a smooth function $H : \mathcal{X} \rightarrow \mathbb{R}$, called the *Hamiltonian* (or energy) function. Finally, consider a generalized Dirac structure D on the product manifold $M = \mathcal{X} \times \mathcal{F}$, only depending on $x \in \mathcal{X}$ (i.e. the fibers to the base M only depend on the point $x \in \mathcal{X}$). Notice that $D(x,y) = D(x) \subset T_x\mathcal{X} \times \mathcal{F} \times T_x^*\mathcal{X} \times \mathcal{F}^*$ for every point $(x, y) \in \mathcal{X} \times \mathcal{F}$, with x local coordinates for $\mathcal{X}$ and y (local) coordinates for $\mathcal{F}$, where we used the identification $T_y\mathcal{F} \simeq \mathcal{F}$, $\forall y \in \mathcal{F}$ (because $\mathcal{F}$ is a linear space).

Definition 2. (Dalsmo and van der Schaft, 1999; van der Schaft and Maschke, 1997) An *implicit port-controlled (generalized) Hamiltonian system*, defined by a 4-tuple $(\mathcal{X}, \mathcal{F}, D, H)$, where $\mathcal{X}$ is the state space, $\mathcal{F}$ is the linear space of external flows, D is the (generalized) Dirac structure on $\mathcal{X} \times \mathcal{F}$ (only depending on $x \in \mathcal{X}$) and H is the Hamiltonian, is defined as the set of time functions $(x(t), f(t), e(t)) \in \mathcal{X}^{\mathbb{R}} \times \mathcal{F}^{\mathbb{R}} \times (\mathcal{F}^*)^{\mathbb{R}}$ (the behavior) satisfying

$$(\dot{x}(t), f(t), dH(x(t)), -e(t)) \in D(x(t)), \ \forall t \in \mathbb{R}. \tag{2}$$

From (1) it follows that the system satisfies the following energy balance

$$\frac{dH}{dt}(x(t)) = \langle dH(x(t)), \dot{x}(t) \rangle = \langle e(t), f(t) \rangle. \tag{3}$$

Remark 1. The minus sign in front of $e(t)$ is to assure that the incoming power $\langle e, f \rangle$ is counted positively.

In this paper we will specifically look at implicit port-controlled generalized Hamiltonian systems of the form

$$\dot{x} = J(x)\frac{\partial H}{\partial x}(x) + g(x)f + b(x)\lambda \tag{4}$$

$$e = g^T(x)\frac{\partial H}{\partial x}(x), \tag{5}$$

$$0 = b^T(x)\frac{\partial H}{\partial x}(x), \tag{6}$$

where for clarity of notation we left out the argument t. Here, $J : T^*\mathcal{X} \rightarrow T\mathcal{X}$ is a skew-symmetric vector bundle map, $g(x) : \mathcal{F} \rightarrow T_x\mathcal{X}$ represents the input vector fields, $b(x) : \mathbb{R}^k \rightarrow T_x\mathcal{X}$ represents the constraint vector fields and $\lambda \in \mathbb{R}^k$ are the Lagrange multipliers required to keep the algebraic constraints (6) to be satisfied for all time. In (Dalsmo and van der Schaft, 1999) this class of systems is thoroughly investigated, especially their closedness properties (of the corresponding Dirac structure D). Notice that the condition that D only depends on $x \in \mathcal{X}$ is exactly matched by the fact that the maps J, g and b only depend on $x \in \mathcal{X}$ and not on $y \in \mathcal{F}$.

Example 1. An important example of the previous class of systems is formed by mechanical systems with kinematic constraints $A^T(q)\dot{q} = 0$ and inputs (forces, torques) and outputs (generalized velocities). They can be written as

$$\begin{bmatrix} \dot{q} \\ \dot{p} \end{bmatrix} = \begin{bmatrix} 0 & 1 \\ -1 & 0 \end{bmatrix} \begin{bmatrix} \dfrac{\partial H}{\partial q} \\ \dfrac{\partial H}{\partial p} \end{bmatrix} + \begin{bmatrix} 0 \\ F(q) \end{bmatrix} f + \begin{bmatrix} 0 \\ A(q) \end{bmatrix} \lambda,$$

$$0 = [0 \; A^T(q)] \begin{bmatrix} \dfrac{\partial H}{\partial q}(q,p) \\ \dfrac{\partial H}{\partial p}(q,p) \end{bmatrix},$$

$$e = [0 \; F^T(q)] \begin{bmatrix} \dfrac{\partial H}{\partial q}(q,p) \\ \dfrac{\partial H}{\partial p}(q,p) \end{bmatrix}.$$

In (Dalsmo and van der Schaft, 1999) it is shown that the corresponding generalized Dirac structure is closed only if the constraints are holonomic.

3. SYMMETRIES AND CONSERVATION LAWS

Symmetries of Dirac structures are defined in (Dorfman, 1993; Courant, 1990) as follows.

Definition 3. Consider a generalized Dirac structure D on M. A *symmetry* of D is defined as a smooth vector field Δ on M having the property that if $(\Delta_1, \delta_1) \in D$ then also $(L_\Delta \Delta_1, L_\Delta \delta_1) \in D$.

The following result is given in (Courant, 1990; Dorfman, 1993).

Proposition 1. Consider a Dirac structure D on M (i.e. which is closed). Let Δ be a smooth vector field on M, and assume that there exists a smooth function $P : M \to \mathbb{R}$ such that $(\Delta, dP) \in D$. Then Δ is a symmetry of D.

Symmetries of generalized Dirac structures are studied in (van der Schaft, 1998; Blankenstein and van der Schaft, 1999*a*; Blankenstein and van der Schaft, 1999*b*). In those papers we consider symmetries of implicit generalized Hamiltonian systems, that is, systems without external variables. A symmetry of an implicit generalized Hamiltonian system $(\mathcal{X}, D, H)$, where D is a generalized Dirac structure on $\mathcal{X}$, is defined as a vector field X on $\mathcal{X}$ such that X is a symmetry of D (as in definition 3), and X is a symmetry of H, i.e. $L_X H = 0$. The basic 'Noether' type of result is the following: consider a vector field X on $\mathcal{X}$ which is a symmetry of $(\mathcal{X}, D, H)$ and assume that there exists a function $I : \mathcal{X} \to \mathbb{R}$ such that $(X, dI) \in D$ (we call X a Hamiltonian symmetry). Then I is a conserved quantity for $(\mathcal{X}, D, H)$, meaning that along every solution $x(t)$ of $(\mathcal{X}, D, H)$ we have $\frac{dI}{dt}(x(t)) = 0$. This follows directly from the property $D = D^\perp$, i.e.

$$\langle dI(x(t)), \dot{x}(t)\rangle + \langle dH(x(t)), X(x(t))\rangle = 0.$$

Furthermore, in (Blankenstein and van der Schaft, 1999*a*; Blankenstein and van der Schaft, 1999*b*) we describe a reduction procedure for implicit generalized Hamiltonian systems with symmetries and corresponding conserved quantities, generalizing the reduction theory for explicit Hamiltonian systems as well as constrained mechanical systems.

In general, adding external variables (inputs and outputs, or flows and efforts) breaks the symmetry of the system. In some cases it can be recovered by allowing for an external symmetry, which is a vector field on the space of external variables, representing a transformation on the external variables. If the resulting (mixed state-external) symmetry is Hamiltonian again, then the corresponding conserved quantity (in the case without external variables) will become a conservation law, i.e. its evolution is completely determined by the external variables of the system. Before giving the general theorem in section 4 we will give two examples to clarify the idea.

Example 2. Particle in $\mathbb{R}$. Consider a particle in $\mathbb{R}$ subject to an external force. The dynamics of the system is given by Newton's law and can be described by a port-controlled Hamiltonian system $(\mathcal{X}, \mathcal{F}, D, H)$, where $\mathcal{X} = T^*Q \simeq \mathbb{R}^2$ is the state space, $\mathcal{F} = \mathbb{R}$ is the space of external flows, $H(q,p) = \frac{p^2}{2}$ is the kinetic energy and

$$D := \{(X, f, \alpha, -e) \in \mathbb{R}^3 \times \mathbb{R}^3 \mid \begin{bmatrix} X_q \\ X_p \end{bmatrix} = \begin{bmatrix} 0 & 1 \\ -1 & 0 \end{bmatrix} \begin{bmatrix} \alpha_q \\ \alpha_p \end{bmatrix} + \begin{bmatrix} 0 \\ 1 \end{bmatrix} f, \quad e = \begin{bmatrix} 0 & 1 \end{bmatrix} \begin{bmatrix} \alpha_q \\ \alpha_p \end{bmatrix} \}$$

is the closed Dirac structure on $\mathcal{X} \times \mathcal{F}$. First consider the *autonomous* system (that is, taking $f = 0$ and leaving out e) given by $(\mathcal{X}, D_{\text{aut}}, H)$ where D_{aut} is the closed Dirac structure on $\mathcal{X}$ given by

$$D_{\text{aut}} = \{(X, \alpha) \in \mathbb{R}^2 \times \mathbb{R}^2 \mid \begin{bmatrix} X_q \\ X_p \end{bmatrix} = \begin{bmatrix} 0 & 1 \\ -1 & 0 \end{bmatrix} \begin{bmatrix} \alpha_q \\ \alpha_p \end{bmatrix} \}.$$

We see that $(\frac{\partial}{\partial q}, dp) \in D_{\text{aut}}$, which implies, by proposition 1, that $\frac{\partial}{\partial q}$ is a symmetry of $(\mathcal{X}, D_{\text{aut}}, H)$. Of course, $I(q,p) := p$ is a conserved quantity of the autonomous system since $\dot{I} = 0$ along solutions of $(\mathcal{X}, D_{\text{aut}}, H)$. Now, $\frac{\partial}{\partial q}$ is also a symmetry of D. Indeed, $(\frac{\partial}{\partial q}, d(I+P)) \in D$ with $P(q,p,y) := y$, a function on $\mathcal{X} \times \mathcal{F}$. However, in the presence of inputs, I is not a conserved quantity anymore. Indeed, along solutions of the system $(\mathcal{X}, \mathcal{F}, D, H)$, see (2),

$$\frac{dI}{dt}(q(t), p(t)) = -f(t).$$

We call I a conservation law.

Example 3. Particle in $\mathbb{R}^3$. Consider a particle in $\mathbb{R}^3$ subject to external forces. The dynamics can be described by a port-controlled Hamiltonian system (on $\mathbb{R}^6 \times \mathbb{R}^3$) of the form in example 2 by taking $q = (q_1, q_2, q_3)$, $p = (p_1, p_2, p_3)$, $f = (f_1, f_2, f_3)$, $e = (e_1, e_2, e_3)$ and replacing 1 by the 3×3 identity matrix and 0 by the 3×3 nulmatrix. For the autonomous system rotation around the $\mathfrak{e}_1$-axis is a symmetry. Indeed, $(X, dI) \in D_{\text{aut}}$, where

$$X(q,p) = -q_3 \frac{\partial}{\partial q_2} + q_2 \frac{\partial}{\partial q_3} - p_3 \frac{\partial}{\partial p_2} + p_2 \frac{\partial}{\partial p_3}$$

and $I(q,p) = \langle q \times p, \mathfrak{e}_1 \rangle$ is the corresponding conserved quantity representing the angular momentum around the $\mathfrak{e}_1$-axis. However, X is *not* a symmetry of the port-controlled Hamiltonian system $(\mathcal{X}, \mathcal{F}, D, H)$ (it is not a symmetry of D). This can be resolved by adding the vector field

$$T(y) = -y_3 \frac{\partial}{\partial y_2} + y_2 \frac{\partial}{\partial y_3}$$

on $\mathcal{F}$, which we call an *external symmetry*. It can easily be checked that

$$(X + T, d(I+P)) \in D$$

where $P(q,p,y) = -\langle q \times y, \mathfrak{e}_1 \rangle$, so by proposition 1 the vector field $X+T$ is a (mixed state-external) symmetry of D. Physically, the symmetry $X+T$ represents the fact that the system is invariant under (state-)rotation around the $\mathfrak{e}_1$-axis, as long as one also rotates the input (force) and the corresponding output in the same way. Note that the external symmetry in example 2 was trivial. The function I, which is a conserved quantity for the autonomous system, becomes a conservation law for the port-controlled Hamiltonian system. Along solutions of $(\mathcal{X}, \mathcal{F}, D, H)$

$$\begin{aligned}\frac{dI}{dt}(q(t), p(t)) &= \langle q(t) \times f(t), \mathfrak{e}_1 \rangle \\ &= \langle \Big(\int_0^t e(s) ds - e(0) \Big) \times f(t), \mathfrak{e}_1 \rangle.\end{aligned}$$

The derivative of I along solutions of the system is a function of the (integrated) external variables of the system.

4. MAIN RESULT

This section describes the main result regarding symmetries and conservation laws of implicit port-controlled generalized Hamiltonian systems. Due to space limitations we omit the proofs, they will appear elsewhere. We consider systems of the form (4)–(6), where $g(x) = [g_1(x) \dots g_m(x)]$ are the input vector fields, and $b(x) = [b_1(x) \dots b_\ell(x)]$ are the constraint vector fields. Assume without loss of generality that the vector fields $g_1, \dots, g_m, b_1, \dots, b_\ell$ are independent over $C^\infty(\mathcal{X})$. Denote with $B(x) = \text{Im}\, b(x)$ the distribution spanned by the constraint vector fields. The main assumption is that the input vector fields are *Hamiltonian*, that is,

$$g_i(x) = J(x) \frac{\partial G_i}{\partial x}(x), \quad i = 1, \dots, m \tag{7}$$

for some functions $G_i(x) \in C^\infty(\mathcal{X})$, $i = 1, \dots, m$. The generalized Dirac structure D on $\mathcal{X} \times \mathcal{F}$ corresponding to the system is given by

$$\begin{aligned} D = \{(X, f, \alpha, -e) \in T(\mathcal{X} \times \mathcal{F}) \oplus T^*(\mathcal{X} \times \mathcal{F}) \mid \\ X = J(x)\alpha + g(x) f + b(x)\lambda \\ e = g^T(x)\alpha, \\ 0 = b^T(x)\alpha \ \} \end{aligned}$$

and the autonomous (generalized) Dirac structure on $\mathcal{X}$ is given by

$$\begin{aligned} D_{\text{aut}} = \{(X, \alpha) \in T\mathcal{X} \oplus T^*\mathcal{X} \mid X = J(x)\alpha + b(x)\lambda \\ 0 = b^T(x)\alpha \ \}. \end{aligned}$$

Remark 2. It can be shown that if D is closed, then D_{aut} is closed (see (Blankenstein and van der Schaft, 1998)).

Let $\{\cdot, \cdot\}$ represent the generalized Poisson bracket with structure matrix J, and denote with X_G the Hamiltonian vector field, with Hamiltonian $G \in C^\infty(\mathcal{X})$, corresponding to this bracket. Finally, denote with ann B the annihilating codistribution of the distribution B.

Consider a Hamiltonian symmetry X of the autonomous system $(\mathcal{X}, D_{\text{aut}}, H)$, i.e $(X, dI) \in D_{\text{aut}}$ for some $I \in C^\infty(\mathcal{X})$. Then I is a conserved quantity of the system $(\mathcal{X}, D_{\text{aut}}, H)$. In general, it will not be a conserved quantity of the port-controlled system $(\mathcal{X}, \mathcal{F}, D, H)$, as can be seen in the previous examples. However, assume that there exists an external symmetry T(a vector field on $\mathcal{F}$) such that $X + T$ is a Hamiltonian symmetry of $(\mathcal{X}, \mathcal{F}, D, H)$, i.e. there exists a function $P \in C^\infty(\mathcal{X} \times \mathcal{F})$ such that

$$(X + T, d(I+P)) \in D \tag{8}$$

Then, by the next theorem, I will be a conservation law of the system $(\mathcal{X}, \mathcal{F}, D, H)$.

Theorem 2. Consider the implicit port-controlled generalized Hamiltonian system $(\mathcal{X}, \mathcal{F}, D, H)$ of the form (4)–(6), with Hamiltonian input vector fields g, see (7). Assume that $dG_i, d\{G_i, G_j\} \in \text{ann } B$, and $X_{\{G_i,G_j\}} \in B$, $i,j = 1,\ldots,m$. Let there be given a Hamiltonian symmetry X of $(\mathcal{X}, D_{\text{aut}}, H)$, with $(X, dI) \in D_{\text{aut}}$, and assume it can be extended, by adding an external symmetry T, to a Hamiltonian symmetry $X + T$ of $(\mathcal{X}, \mathcal{F}, D, H)$, i.e. (8) holds for some $P \in C^\infty(\mathcal{X} \times \mathcal{F})$. Furthermore, assume that $d\{G_i, I\} \in \text{ann } B$, $i = 1,\ldots,m$. Then I is a conservation law of $(\mathcal{X}, \mathcal{F}, D, H)$, i.e. the derivative of I along solutions of the system $(\mathcal{X}, \mathcal{F}, D, H)$ is a function of the (integrated) external variables of the system. Explicitly,

$$\frac{dI}{dt}(x(t)) = -\sum_{i=1}^{m}\left(\sum_{j=1}^{m}\left(\int_0^t e_j(s)ds + \sum_{r=1}^{m}\int_0^t c_{rj} f_r(s)ds + c'_j\right)\tilde{c}_{ij} + \hat{c}_i\right) f_i(t) \tag{9}$$

where $c_{ij}, c'_j, \tilde{c}_{ij}, \hat{c}_i \in \mathbb{R}$, $i,j = 1,\ldots,m$, are constants.

Remark 3. From (9) it follows immediately that I is a conserved quantity along solutions for which $f(t) = 0$, i.e. solutions of the autonomous system.

For mechanical systems described in example 1, $G_i(q,p) = G_i(q)$, $i = 1,\ldots,m$, which implies $\{G_i, G_j\} = 0$, $i,j = 1,\ldots,m$, and therefore the conditions $dG_i, d\{G_i, G_j\} \in \text{ann } B$, and $X_{\{G_i,G_j\}} \in B$, $i,j = 1,\ldots,m$, are automatically satisfied in this case.

In case the generalized Dirac structure D is closed, theorem 2 gets a particularly nice form.

Corollary 3. Consider the implicit port-controlled Hamiltonian system $(\mathcal{X}, \mathcal{F}, D, H)$ of the form (4)–(6) (with D closed), with Hamiltonian input vector fields g, see (7). Assume that $dG_i \in \text{ann } B$, $i = 1,\ldots,m$. Let there be given a Hamiltonian symmetry X of $(\mathcal{X}, D_{\text{aut}}, H)$, with $(X, dI) \in D_{\text{aut}}$, and assume it can be extended, by adding an external symmetry T, to a Hamiltonian symmetry $X+T$ of $(\mathcal{X}, \mathcal{F}, D, H)$, i.e. (8) holds for some $P \in C^\infty(\mathcal{X} \times \mathcal{F})$. Then I is a conservation law of $(\mathcal{X}, \mathcal{F}, D, H)$, i.e. the derivative of I along solutions of the system $(\mathcal{X}, \mathcal{F}, D, H)$ is a function of the (integrated) external variables of the system, see (9).

5. SYMMETRIES OF INTERCONNECTED SYSTEMS

In (van der Schaft and Maschke, 1997; Maschke and van der Schaft, 1997; Dalsmo and van der Schaft, 1999) it has been shown that implicit port-controlled generalized Hamiltonian systems are well suited for describing complex energy-conserving physical systems by using a "modular approach", that is, considering the system as a power-conserving interconnection of smaller subsystems. This section describes the symmetry properties of such interconnected systems.

For simplicity, consider two implicit port-controlled generalized Hamiltonian systems $(\mathcal{X}_i, \mathcal{F}_i, D_i, H_i)$, $i = 1,2$ (not necessarily of the form (4)–(6)). These two systems can be interconnected through their external variables using a power-conserving interconnection (van der Schaft and Maschke, 1997; Dalsmo and van der Schaft, 1999):

Definition 4. Consider two implicit port-controlled generalized Hamiltonian systems $(\mathcal{X}_i, \mathcal{F}_i, D_i, H_i)$, $i = 1,2$. A *power-conserving interconnection* of these two systems is defined as a (state-dependent) Dirac structure Γ on $\mathcal{F}_1 \times \mathcal{F}_2$. More precisely, for every $x \in \mathcal{X}_1 \times \mathcal{X}_2$, $\Gamma(x)$ defines a Dirac structure on $\mathcal{F}_1 \times \mathcal{F}_2$, not depending on $y \in \mathcal{F}_1 \times \mathcal{F}_2$ (i.e. $\Gamma(x,y) = \Gamma(x,y')$, $\forall (x,y),(x,y') \in \mathcal{X}_1 \times \mathcal{X}_2 \times \mathcal{F}_1 \times \mathcal{F}_2$).

(1) implies that

$$(f_1, f_2, e_1, e_2) \in \Gamma \Rightarrow \langle e_1(x), f_1(x)\rangle + \langle e_2(x), f_2(x)\rangle = 0,$$

$\forall x \in \mathcal{X}_1 \times \mathcal{X}_2$, which represents the conservation of power in the interconnection, that is, the incoming power $\langle e_1, f_1\rangle$ of the first system (see (3)) equals the outgoing power $-\langle e_2, f_2\rangle$ of the second system. The interconnected system will be energy conserving again, and is defined by the implicit generalized Hamiltonian system $(\mathcal{X}, D, H)$, where $\mathcal{X} = \mathcal{X}_1 \times \mathcal{X}_2$, $H(x_1,x_2) = H_1(x_1) + H_2(x_2)$ and D is given by

$$\begin{aligned} D := \{(X_1, X_2, \alpha_1, \alpha_2) \in T\mathcal{X} \oplus T^*\mathcal{X} \mid \\ \exists\, (f_1, f_2, e_1, e_2) \in \Gamma \text{ such that} \\ (X_i, f_i, \alpha_i, -e_i) \in D_i,\ i = 1,2\}, \end{aligned}$$

(van der Schaft and Maschke, 1997; Dalsmo and van der Schaft, 1999). Corresponding to D we define the *space of achievable flows and efforts*

$$\begin{aligned} E := \{(f_1, f_2, e_1, e_2) \mid \exists\, X_1, X_2, \alpha_1, \alpha_2 \\ \text{such that } (X_i, f_i, \alpha_i, -e_i) \in D_i,\ i = 1,2\}. \end{aligned}$$

Its orthogonal (in the sense of definition 1) is denoted by $E^\perp$ and is given by (van der Schaft and Maschke, 1997)

$$E^\perp = \{(f_1, f_2, e_1, e_2) \mid (0, f_i, 0, -e_i) \in D_i, i = 1,2\}.$$

Assume the system $(\mathcal{X}_i, \mathcal{F}_i, D_i, H_i)$ has a (mixed state-external) symmetry $\Delta_i(x_i, y_i) = X_i(x_i) +$

$T_i(y_i)$, $i = 1, 2$ (not necessarily Hamiltonian), as described in section 4. The vector field $\Delta = \Delta_1 + \Delta_2$ on $\mathcal{X} \times \mathcal{F}$ is called a *symmetry* of Γ (resp. $\Gamma \cap E$), if

$$(L_\Delta(f_1, f_2), -L_\Delta(e_1, e_2)) \in \Gamma \text{ (resp. } \Gamma \cap E)$$

for all $(f_1, f_2, e_1, e_2) \in \Gamma$ (resp. $\Gamma \cap E$), where L_Δ denotes the Lie derivative on $\mathcal{X} \times \mathcal{F}$ with respect to Δ. The following basic theorem gives necessary and sufficient conditions for the two symmetries Δ_1 and Δ_2 to result in a symmetry of the interconnected system. Note that the condition $E^\perp = 0$ is satisfied for systems of the form (4)–(6).

Theorem 4. Consider two implicit port-controlled generalized Hamiltonian systems $(\mathcal{X}_i, \mathcal{F}_i, D_i, H_i)$, $i = 1, 2$, interconnected by a power-conserving interconnection Γ. Assume that $E^\perp = 0$. Let $\Delta_i = X_i + T_i$ be a symmetry of $(\mathcal{X}_i, \mathcal{F}_i, D_i, H_i)$, $i = 1, 2$. Then the vector field $X = X_1 + X_2$ is a symmetry of the interconnected system $(\mathcal{X}, D, H)$ if and only if the vector field $\Delta = \Delta_1 + \Delta_2$ is a symmetry of Γ.

Corollary 5. (not necessarily $E^\perp = 0$) Consider two implicit port-controlled generalized Hamiltonian systems $(\mathcal{X}_i, \mathcal{F}_i, D_i, H_i)$, $i = 1, 2$, interconnected by a power-conserving interconnection Γ. Let $\Delta_i = X_i + T_i$ be a symmetry of $(\mathcal{X}_i, \mathcal{F}_i, D_i, H_i)$, $i = 1, 2$. Then the vector field $X = X_1 + X_2$ is a symmetry of the interconnected system $(\mathcal{X}, D, H)$ if the vector field $\Delta = \Delta_1 + \Delta_2$ is a symmetry of $\Gamma \cap E$.

Finally we remark that if Γ does not depend on the state x (i.e. $\Gamma(x) = \Gamma(x')$, $\forall x, x' \in \mathcal{X}$), then Δ being a symmetry of $\Gamma \cap E$ is implied by the vector field $T_1 + T_2$ being a symmetry of $\Gamma \cap E$.

6. CONCLUSIONS

In this paper we studied symmetries of implicit port-controlled generalized Hamiltonian systems (that is, systems with external variables). The main result describes the correspondence of Hamiltonian (mixed state-external) symmetries and conservation laws of these systems. This result generalizes the known ("Noether") correspondence between symmetries and conserved quantities in autonomous systems (i.e. systems without external variables). Secondly, the symmetry properties of interconnected implicit port-controlled generalized Hamiltonian systems are studied. A basic theorem is presented, giving necessary and sufficient conditions for two symmetries (of the two subsystems) to result in a symmetry of the interconnected system.

7. REFERENCES

Abraham, R. and J.E. Marsden (1978). *Foundations of Mechanics.* second ed.. Benjamin/Cummings Publishing Company.

Blankenstein, G. and A.J. van der Schaft (1998). Closedness of interconnected Dirac structures. In: *Preprints NOLCOS 1998.* pp. 381–386.

Blankenstein, G. and A.J. van der Schaft (1999*a*). Reduction of implicit Hamiltonian systems with symmetry. In: *Proceedings of the 5th European Control Conference, ECC'99.* Karlsruhe.

Blankenstein, G. and A.J. van der Schaft (1999*b*). Symmetry and reduction in implicit generalized Hamiltonian systems. Memorandum 1489. University of Twente. Faculty of Mathematical Sciences.

Bloch, A.M., P.S. Krishnaprasad, J.E. Marsden and R.M. Murray (1996). Nonholonomic mechanical systems with symmetries. *Arch. Rat. Mech.* **136**, 21–99.

Cantrijn, F., M. de Leon, J.C. Marrero and D. Martin de Diego (1999). Reduction of constrained systems with symmetries. *J. Math. Phys.* **40**(2), 795–820.

Courant, T. (1990). Dirac manifolds. *Trans. American Math. Soc.* **319**, 631–661.

Dalsmo, M. and A.J. van der Schaft (1999). On representations and integrability of mathematical structures in energy-conserving physical systems. *SIAM J. Cont. Opt.* **37**(1), 54–91.

Dorfman, I. (1993). *Dirac Structures and Integrability of Nonlinear Evolution Equations.* Chichester: John Wiley.

Marsden, J.E. and T. Ratiu (1986). Reduction of Poisson manifolds. *Letters in Mathematical Physics* **11**, 161–169.

Maschke, B.M. and A.J. van der Schaft (1997). Interconnected mechanical systems, part II: the dynamics of spatial mechanical networks. In: *Modelling and Control of Mechanical Systems.* Imperial College Press. pp. 17–30.

van der Schaft, A.J. (1981). Symmetries and conservation laws for Hamiltonian systems with inputs and outputs: A generalization of Noether's theorem. *System & Control Letters* **1**, 108–115.

van der Schaft, A.J. (1998). Implicit Hamiltonian Systems with Symmetry. *Rep. Math. Phys.* **41**, 203–221.

van der Schaft, A.J. and B.M Maschke (1997). Interconnected mechanical systems, part I: geometry of interconnection and implicit Hamiltonian systems. In: *Modelling and Control of Mechanical Systems* (A. Astolfi, D.J.N Limebeer, C. Melchiorri, A. Tornambè and R.B. Vinter, Eds.). Imperial College Press. pp. 1–15.

Copyright © IFAC Lagrangian and Hamiltonian Methods for
Nonlinear Control, Princeton, New Jersey, USA, 2000

ASYMPTOTIC STABILIZATION OF RELATIVE EQUILIBRIA WITH APPLICATION TO THE HEAVY TOP [1]

M. Egorov * T A. Posbergh *

* *Department of Aerospace Engineering and Mechanics, University of Minnesota, Minneapolis, Minnesota 55455.*

Abstract: This paper investigates the asymptotic stabilization of relative equilibria when motion is restricted to invariant manifolds. The motivating example is the heavy top. Three forms of feedback control laws are investigated. The first can be identified with double bracket dissipation, the second preserves energy but changes the momentum, and the third corresponds to a magnetic term in the closed loop equations. All three methods are derived and subsequently discussed. Numerical simulations are used to illustrate their performance. *Copyright ©2000 IFAC*

Keywords: Nonlinear control, Mechanical systems, Invariants, Stabilization, Asymptotic stability

1. INTRODUCTION

This paper investigates the asymptotic stabilization of relative equilibria when motion is restricted to invariant manifolds. In mechanical systems these invariant manifolds can be identified with conserved quantities and the corresponding symmetries. In this paper we investigate three types of feedback in the particular case of the heavy top. By design the closed loop dynamics with these feedback controls preserve one or more of the invariant quantities.

The control of the heavy top is of interest as it has features which are relevant to a number of practical applications. The heavy top is a classic dynamic model. In the context geometric mechanics relative equilibria for symmetric and asymmetric tops are investigated in (Lewis *et al.*, 1992). The heavy top motivates the study of integrability in (Audin, 1996) and optimal control on Lie groups in (Jurdjevic, 1997). For the heavy top two classes of relative equilibria are of interest. The first is the well known sleeping top. In this case the top spins steadily about its vertical axis. In the second case the spin axis is tipped from the vertical. For a symmetric top this motion is known as steady precession.

Stabilization of relative equilibria is the topic of a number of recent papers. Early work includes that of (Bloch and Marsden, 1990) and (Bloch *et al.*, 1992) where the energy-Casimir method is the basis of designing stabilizing uniform rotation of a rigid body and applied to control of a spacecraft by momentum wheels. In (Woolsey and Leonard, 1999) and related work the energy-Casimir method is used as the basis of stabilizing an underwater vehicle. The energy-momentum method is the basis of the robust stabilization of relative equilibria in (Zhao and Posbergh, 1993) and (Posbergh and Egorov, 1998). In (Jalnapurkar and Marsden, 1999) the energy-momentum method is used as the basis of stabilizing relative equilibria.

Feedback controls which do not asymptotically stabilize a relative equilibrium such as (Bloch and Marsden, 1990) or (Posbergh and Zhao, 1993) lead to systems with conserved quantities which are

[1] Research supported in part by NFS/CMS-9626287

not necessarily the same as those in the original systems but rather scaled and shifted versions of them. In contrast feedback control which leaves the original quantities invariant corresponds to the design of control systems on invariant manifolds in an ambient space. Early work in this regard includes that of (Brockett, 1973) who studied control systems designed on spheres. Such an example arises in the case of rigid body for which the angular momentum is to be preserved. Double bracket dissipation (Bloch *et al.*, 1996) is a well known example of the interesting dynamics which arise in such problems.

In the remainder of this paper three types of feedback control which preserve one or more invariant quantities are investigated. The motivating example is that of the heavy top. In Section 2 the model of an asymmetric heavy top along with the relevant notation, dynamics, and conserved quantities are introduced. In section 3 three types of feedback control which preserve one or more of these quantities are derived. Section 4 discusses and compares these feedback controls and presents numeric simulations. Section 5 is a summary the important points of research.

2. THE HEAVY TOP

The heavy top is a rigid body fixed at one point and free to move in a uniform gravitational field. In the following we follow the notation of (Marsden and Ratiu, 1994). A more extensive discussion of the top model in a geometric framework is found in (Lewis *et al.*, 1992). See also (Audin, 1996) and (Jurdjevic, 1997)

2.1 *Dynamics of the Top*

In describing the motion of the top it is convenient to distinguish two orthonormal coordinate systems. The basis vectors $(e_1,\ e_2,\ e_3)$ define a fixed, spatially referenced coordinate system. It is assumed that e_3 is vertical and $e_1,\ e_2$ lie in the horizontal plane. Denote the fixed inertia of the top by the inertia matrix $\mathbf{J}$. The basis vectors $(\boldsymbol{E}_1,\ \boldsymbol{E}_2,\ \boldsymbol{E}_3)$ are body fixed in the top and assumed to lie along the principle axes of $\mathbf{J}$.

The top has mass m and the vector from the fixed point to the center of mass lies along $\boldsymbol{E}_3$ and has length ℓ. Let $\boldsymbol{\Gamma}$ denote the body referenced unit vertical direction e_3 and $\boldsymbol{M} = mg\ell\boldsymbol{E}_3$ the body referenced mass vector scaled by g. Let $\boldsymbol{\Omega}$ be the body referenced angular velocity and $\boldsymbol{\Pi} = \mathbf{J}\boldsymbol{\Omega}$ the corresponding angular momentum.

For the heavy top the Hamiltonian is

$$H = \frac{1}{2}\boldsymbol{\Pi}\cdot\boldsymbol{\Omega} + \boldsymbol{\Gamma}\cdot\boldsymbol{M} \tag{1}$$

representing the sum of kinetic and potential energies. The phase space is coordinatized by $(\boldsymbol{\Gamma}, \boldsymbol{\Pi})$. The dynamics of the heavy top are given by

$$\dot{\boldsymbol{\Pi}} = \boldsymbol{\Pi}\times\boldsymbol{\Omega} + \boldsymbol{\Gamma}\times\boldsymbol{M}, \tag{2}$$

$$\dot{\boldsymbol{\Gamma}} = \boldsymbol{\Gamma}\times\boldsymbol{\Omega}. \tag{3}$$

These equations are derived in (Marsden and Ratiu, 1994). It is routine to show that the energy corresponding to H is constant along the solutions to (2) and (3). Denote by $\boldsymbol{X}_H = (\boldsymbol{X}_{\Pi}, \boldsymbol{X}_{\Gamma}) = (\boldsymbol{\Pi}\times\boldsymbol{\Omega} + \boldsymbol{\Gamma}\times\boldsymbol{M}, \boldsymbol{\Gamma}\times\boldsymbol{\Omega})$ the Hamiltonian vector field of (2) and (3). This can also viewed as a Lie bracket $[\boldsymbol{\mu}, \boldsymbol{\nu}] = (\boldsymbol{\mu}_1\times\boldsymbol{\nu}_1 + \boldsymbol{\mu}_2\times\boldsymbol{\nu}_2, \boldsymbol{\mu}_2\times\boldsymbol{\nu}_1)$ with $\boldsymbol{\mu} = (\boldsymbol{\mu}_1, \boldsymbol{\mu}_2) = (\boldsymbol{\Pi}, \boldsymbol{\Gamma})$ and $\boldsymbol{\nu} = (\boldsymbol{\nu}_1, \boldsymbol{\nu}_2) = (\boldsymbol{\Omega}, \boldsymbol{M})$. This bracket has a property which will be used later.

$$\boldsymbol{\rho}\cdot[\boldsymbol{\mu}, \boldsymbol{\nu}] = -\boldsymbol{\nu}\cdot[\boldsymbol{\mu}, \boldsymbol{\rho}]. \tag{4}$$

A top with distinct principal moments of inertia has a single S^1 invariance corresponding to rotations of the configuration about the vertical axis. This implies the existence of the two conserved quantities.

$$\pi_3(\boldsymbol{\Pi}, \boldsymbol{\Gamma}) = \boldsymbol{\Pi}\cdot\boldsymbol{\Gamma} \text{ and } \gamma(\boldsymbol{\Pi}, \boldsymbol{\Gamma}) = ||\boldsymbol{\Gamma}||. \tag{5}$$

Physically these quantities correspond to spatially referenced vertical angular momentum being conserved and the corresponding configuration a point on S^2.

Of interest in the remainder of this paper is the orbit space (Audin, 1996) for the heavy top.

$$\mathcal{O}_c = \{(\boldsymbol{\Gamma}, \boldsymbol{\Pi}) \mid \Gamma(\boldsymbol{\Pi},\ \boldsymbol{\Gamma}) = 1,\ \pi_3(\boldsymbol{\Pi},\ \boldsymbol{\Gamma}) = c\}. \tag{6}$$

This is identified with the invariant manifold in which π_3 and γ in (5) remain constant. One can check that $\boldsymbol{X}_H \in T(\mathcal{O}_c)$, the tangent space to $\mathcal{O}_c$, that is $\boldsymbol{X}_H$ is orthogonal to $\boldsymbol{\nabla}\gamma = (\boldsymbol{0}, \boldsymbol{\Gamma})$ and $\boldsymbol{\nabla}\pi_3 = (\boldsymbol{\Gamma}, \boldsymbol{\Pi})$.

3. FEEDBACK CONTROL OF THE TOP

For feedback control of the top the control $\boldsymbol{U} = (\boldsymbol{U}_1, \boldsymbol{U}_2)$ is added to the top equations (2) and (3),

$$\dot{\boldsymbol{\Pi}} = \boldsymbol{\Pi}\times\boldsymbol{\Omega} + \boldsymbol{\Gamma}\times\boldsymbol{M} + \boldsymbol{U}_1, \tag{7}$$

$$\dot{\boldsymbol{\Gamma}} = \boldsymbol{\Gamma}\times\boldsymbol{\Omega} + \boldsymbol{U}_2. \tag{8}$$

In general $\boldsymbol{U}$ will be a function of $(\boldsymbol{\Gamma}, \boldsymbol{\Pi})$

The objective is now to find $\boldsymbol{U}$ which will asymptotically stabilize (7) and (8) for a particular relative equilibrium while preserving one or more of the invariant quantities H, π_3 and Γ given in (1) and (5).

The general procedure followed is to restrict the vector field associated with the feedback to be tangent to the invariant manifold. Thus, the feedback will always lie in the kernel of the space spanned by the gradient vectors normal to our invariant manifolds. Three such feedbacks are considered in the following section.

3.1 *General Result*

The general methodology for asymptotic stabilization which preserves one or more invariants can now be described. Consider a nonlinear affine control system of the form

$$\dot{\boldsymbol{x}} = \boldsymbol{f}(\boldsymbol{x}) + \boldsymbol{u}(\boldsymbol{x}). \qquad (9)$$

Here $\boldsymbol{x} \in \mathbb{R}^n$ represents a local coordinate description of an appropriate phase space. The vector functions $\boldsymbol{f}$ and $\boldsymbol{u}$ are locally coordinatized smooth vector fields on the phase space and full state feedback is assumed. (See for example (Khalil, 1996))

Assume that there exists a set of k smooth functions

$$\Phi_1(\boldsymbol{x}),\ \Phi_2(\boldsymbol{x}),\ \dots,\ \Phi_k(\boldsymbol{x}),\ k \le n, \qquad (10)$$

which are independent and conserved along the dynamics of (9) in the case $\boldsymbol{u} \equiv \boldsymbol{0}$. For each conserved quantity the dynamics (9) admit symmetry. Relative equilibria represent solutions to (9) with $\boldsymbol{u} \equiv \boldsymbol{0}$ which also correspond to group actions. In the $(n-k)$-dimensional surface defined by (10) they may correspond to a point $\boldsymbol{x}_e$. (Note that $\boldsymbol{x}_e$ is the intersection of this surface and a solution to (9).)

The control problem is to find a smooth vector field $\boldsymbol{u}$ which stabilizes a point $\boldsymbol{x}_e$ in same neighborhood of $\boldsymbol{x}_e$ and which preserves one or more of the conserved quantities given in (10).

$$\Phi_{i_1}(\boldsymbol{x}),\ \Phi_{i_2}(\boldsymbol{x}),\ \dots,\ \Phi_{i_l}(\boldsymbol{x}),\ l \le k. \qquad (11)$$

Such $\boldsymbol{u}$ can be constructed under favorable assumptions on the convexity of one Φ_{i_0} of the quantities from (10) and which are not from (11), that is $i_0 \neq i_1,\ i_2,\ \dots,\ i_l$.

If Φ_{i_0} is positive definite in $\mathcal{N}_{\boldsymbol{x}_e}$, then choose

$$\boldsymbol{u} = -\alpha(\boldsymbol{x})(\boldsymbol{\nabla}\Phi_{i_0})|_{T\Phi_{i_1}, T\Phi_{i_2} \dots, T\Phi_{i_l}} \qquad (12)$$

where $\alpha(\boldsymbol{x})$ is a strictly positive scalar function in $\mathcal{N}_{\boldsymbol{x}_e}$ except $\boldsymbol{x}_e$ where $\alpha(\boldsymbol{x}_e) = 0$. Such choice of $\boldsymbol{u}$ preserves quantities in (11) and makes $\dot{\Phi}_{i_0} < 0$ in $\mathcal{N}_{\boldsymbol{x}_e}$ except at $\boldsymbol{x}_e$ where $\dot{\Phi}_{i_0} = 0$. By LaSalle's principle $\boldsymbol{u}$ from (12) asymptotically stabilizes (9) at $\boldsymbol{x}_e$ in $\mathcal{N}_{\boldsymbol{x}_e}$.

The procedure to construct $\boldsymbol{u}$ is based on the Gram-Schmidt procedure (Gantmacher, 1959) which uses $\boldsymbol{\nabla}\Phi_{i_1},\ \boldsymbol{\nabla}\Phi_{i_2},\ \dots,\ \boldsymbol{\nabla}\Phi_{i_l}$. That is

$$\boldsymbol{u} = \boldsymbol{\nabla}\Phi_{i_0} - \boldsymbol{k}_{i_1}(\boldsymbol{\nabla}\Phi_{i_0} \cdot \boldsymbol{k}_{i_1}) - \boldsymbol{k}_{i_2}(\boldsymbol{\nabla}\Phi_{i_0} \cdot \boldsymbol{i}_{i_2}) - \dots - \boldsymbol{k}_{i_l}(\boldsymbol{\nabla}\Phi_{i_0} \cdot \boldsymbol{i}_{i_l}), \qquad (13)$$

where $\boldsymbol{k}_{i_1}, \boldsymbol{k}_{i_2}, \dots, \boldsymbol{k}_{i_l}$ are orthonormal vectors such that

$$\text{span}[\boldsymbol{k}_{i_1}, \boldsymbol{k}_{i_2}, \dots, \boldsymbol{k}_{i_l}] = \text{span}[\boldsymbol{\nabla}\Phi_{i_1},\ \boldsymbol{\nabla}\Phi_{i_2}, \dots, \boldsymbol{\nabla}\Phi_{i_l}] \qquad (14)$$

3.2 *Dissipative Control of The Heavy Top*

By the *dissipative control* is meant a vector field $\boldsymbol{U}$ orthogonal to the Hamiltonian vector field $\boldsymbol{X}_H$, preserving $\mathcal{O}_c$ and diminishing the total energy H. If, in addition, $\boldsymbol{U}$ minimizes $\dot{H} = \boldsymbol{\Omega} \cdot \boldsymbol{U}_1 + \boldsymbol{\Gamma} \cdot \boldsymbol{U}_2$ for bounded $\|\boldsymbol{U}\| \le \delta$ at any instant of time, then such dissipation is called double bracket dissipation (Bloch *et al.*, 1996) and has the form

$$\boldsymbol{U}_1 = \boldsymbol{\Pi} \times (\boldsymbol{\Pi} \times \boldsymbol{\Omega} + \boldsymbol{\Gamma} \times \boldsymbol{M}) + \boldsymbol{\Gamma} \times (\boldsymbol{\Gamma} \times \boldsymbol{\Omega}), \qquad (15)$$

$$\boldsymbol{U}_2 = \boldsymbol{\Gamma} \times (\boldsymbol{\Pi} \times \boldsymbol{\Omega} + \boldsymbol{\Gamma} \times \boldsymbol{M}). \qquad (16)$$

One can show that such dissipative control is in the form of (12) as in the previous section and orthogonal to $\boldsymbol{X}_H$, that is

$$\boldsymbol{U}(\boldsymbol{\mu}) = -\alpha(\boldsymbol{\mu})\boldsymbol{\nabla}H^{\nu}(\boldsymbol{\mu}), \qquad (17)$$

where $\boldsymbol{\mu} = (\boldsymbol{\Pi}, \boldsymbol{\Gamma})$, $\alpha(\boldsymbol{\mu})$ is any positive function of $\boldsymbol{\mu}$ and $\boldsymbol{\nabla}H^{\nu}$ is the component of $\boldsymbol{\nabla}H$ in the orthogonal complement of span$[\boldsymbol{\nabla}\Gamma,\ \boldsymbol{\nabla}\pi_3,\ \boldsymbol{X}_H]$, that is in $\nu = \text{span}[\boldsymbol{\nabla}\Gamma,\ \boldsymbol{\nabla}\pi_3,\ \boldsymbol{X}_H]^{\perp} = \{\boldsymbol{\mu} \mid \boldsymbol{\mu} \in T(\pi_3) \bigcup T(\gamma)$ and $\boldsymbol{\mu} \perp \boldsymbol{X}_H\}$. So, $\boldsymbol{U}$ must satisfy the following

$$\boldsymbol{U}_2 \cdot \boldsymbol{\Gamma} = 0, \qquad (18)$$

$$\boldsymbol{U}_1 \cdot \boldsymbol{\Gamma} + \boldsymbol{U}_2 \cdot \boldsymbol{\Pi} = 0, \qquad (19)$$

$$\boldsymbol{U}_1 \cdot \boldsymbol{X}_{\Pi} + \boldsymbol{U}_2 \cdot \boldsymbol{X}_{\Gamma} = 0, \qquad (20)$$

$$(\boldsymbol{U}_1, \boldsymbol{U}_2) = \arg \min_{\|\boldsymbol{U}\| \le \delta} (\boldsymbol{U}_1 \cdot \boldsymbol{\Omega} + \boldsymbol{U}_2 \cdot \boldsymbol{M}). \qquad (21)$$

From (18) there is such vector $\boldsymbol{A}_1$ that

$$\boldsymbol{U}_2 = \boldsymbol{\Gamma} \times \boldsymbol{A}_1, \qquad (22)$$

with this relation (19) becomes $(\boldsymbol{U}_1 - \boldsymbol{\Pi} \times \boldsymbol{A}_1) \cdot \boldsymbol{\Gamma} = 0$ and

$$\boldsymbol{U}_1 = \boldsymbol{\Pi} \times \boldsymbol{A}_1 + \boldsymbol{\Gamma} \times \boldsymbol{A}_2. \qquad (23)$$

Substituting (23) and (22) into (20) and (21) gives

$$\boldsymbol{A}_1 \cdot (\boldsymbol{\Pi} \times \boldsymbol{X}_{\Pi} - \boldsymbol{\Gamma} \times \boldsymbol{X}_{\Gamma}) + \boldsymbol{A}_2 \cdot (\boldsymbol{\Gamma} \times \boldsymbol{X}_{\Gamma}) = 0, \qquad (24)$$

$$\boldsymbol{A} = \arg \min_{\|\boldsymbol{A}\| \le \delta_U} (-\boldsymbol{A}_1 \cdot \boldsymbol{X}_{\Pi} - \boldsymbol{A}_2 \cdot \boldsymbol{X}_{\Gamma}), \qquad (25)$$

where $\boldsymbol{A} = (\boldsymbol{A}_1, \boldsymbol{A}_2)$ and δ_U is derived from $\|\boldsymbol{U}\| \le \delta$, (22) and (23).

These two relations can be rewritten

$$\boldsymbol{\alpha} \cdot [\boldsymbol{\mu}, \boldsymbol{\nu}] = 0, \qquad (26)$$

$$-(\boldsymbol{\alpha} \cdot \boldsymbol{\nu}) < 0, \qquad (27)$$

where $\boldsymbol{\alpha}(\boldsymbol{A}_1, \boldsymbol{A}_2)$, $\boldsymbol{\mu} = (\boldsymbol{\Pi}, \boldsymbol{\Gamma})$, and $(\boldsymbol{\nu} = \boldsymbol{X}_{\Pi}, \boldsymbol{X}_{\Gamma})$. Relation (4) tells an obvious choice is $\boldsymbol{\alpha} = \boldsymbol{\nu}$, that is $\boldsymbol{A}_1 = \boldsymbol{X}_{\Pi}$ and $\boldsymbol{A}_2 = \boldsymbol{X}_{\Gamma}$, which proves (15) and (16) up to the scaling function $\alpha(\boldsymbol{\Pi}, \boldsymbol{\Gamma})$.

3.3 *Momentum Management of The Heavy Top*

Another qualitatively different type of control will be called *momentum management.* It is a vector field $\boldsymbol{U}$ preserving γ and H and extremizing $\dot{\pi}_3 = \boldsymbol{\Gamma}\cdot\boldsymbol{U}_1 + \boldsymbol{\Pi}\cdot\boldsymbol{U}_2$ for bounded $||\boldsymbol{U}|| \le \delta$ at any instant of time.

Again following (12) the momentum management is given in the following form

$$\boldsymbol{U}(\boldsymbol{\mu}) = -\alpha(\boldsymbol{\mu})\boldsymbol{\nabla}\pi_3^{\nu}(\boldsymbol{\mu}) \tag{28}$$

where $\alpha(\boldsymbol{\mu})$ is some positive function of $\boldsymbol{\mu}$ and $\boldsymbol{\nabla}\pi_3^{\nu}$ is the component of $\boldsymbol{\nabla}\pi_3$ in the orthogonal complement of span$[\boldsymbol{\nabla}\Gamma,\ \boldsymbol{\nabla}H]$, that is in $\nu =$ span$[\boldsymbol{\nabla}\Gamma,\ \boldsymbol{\nabla}H]^{\perp} = \{\boldsymbol{\mu} \mid \boldsymbol{\mu} \in T(H)\bigcup T(\gamma)\}$. So, $\boldsymbol{U}$ has to necessarily satisfy the following relations

$$\boldsymbol{U}_2\cdot\boldsymbol{\Gamma} = 0, \tag{29}$$

$$\boldsymbol{U}_1\cdot\boldsymbol{\Omega} + \boldsymbol{U}_2\cdot\boldsymbol{M} = 0, \tag{30}$$

$$(\boldsymbol{U}_1, \boldsymbol{U}_2) = \arg\max_{||\boldsymbol{U}||\le\delta}(\boldsymbol{U}_1\cdot\boldsymbol{\Gamma}+\boldsymbol{U}_2\cdot\boldsymbol{\Omega}). \tag{31}$$

From (18)

$$\boldsymbol{U}_2 = \boldsymbol{A}_2\times\boldsymbol{\Gamma}, \tag{32}$$

and with this relation (30) becomes $\boldsymbol{U}_1\cdot\boldsymbol{\Omega}+(\boldsymbol{A}_2\times\boldsymbol{\Gamma})\cdot\boldsymbol{M} = 0$ from which

$$\boldsymbol{U}_1 = -\boldsymbol{\Omega}\frac{(\boldsymbol{A}_2\times\boldsymbol{\Gamma})\cdot\boldsymbol{M}}{||\boldsymbol{\Omega}||^2} + \boldsymbol{A}_1\times\boldsymbol{\Omega}. \tag{33}$$

Substituting (32) and (33) into (31) gives

$$\boldsymbol{A} = \arg\max_{||\boldsymbol{A}||\le\delta_U}\{[(\frac{(\boldsymbol{\Gamma}\cdot\boldsymbol{\Omega})}{||\boldsymbol{\Omega}||^2}\boldsymbol{M} - \boldsymbol{\Pi})\times\boldsymbol{\Gamma}]\cdot\boldsymbol{A}_2 + (\boldsymbol{\Omega}\times\boldsymbol{\Gamma})\cdot\boldsymbol{A}_1\}, \tag{34}$$

where $\boldsymbol{A} = (\boldsymbol{A}_1, \boldsymbol{A}_2)$ and and δ_U is derived from $||\boldsymbol{U}|| \le \delta$, (32) and (33). From which

$$\boldsymbol{A}_1 = \boldsymbol{\Omega}\times\boldsymbol{\Gamma} \tag{35}$$

$$\boldsymbol{A}_2 = (\frac{(\boldsymbol{\Gamma}\cdot\boldsymbol{\Omega})}{||\boldsymbol{\Omega}||^2}\boldsymbol{M} - \boldsymbol{\Pi})\times\boldsymbol{\Gamma} \tag{36}$$

and $\boldsymbol{U} = (\boldsymbol{U}_1, \boldsymbol{U}_2)$ is given by (32) and (33)

3.4 *Magnetic Control of The Heavy Top*

This control preserves Γ, π_3 and H. Such control will be called *magnetic control* because its vector field lies in the tangent to the space of γ and π_3 and can be characterized by symplectic form on $\mathcal{O}_c$, see (Marsden and Ratiu, 1994). Magnetic control preserves

$$\gamma,\ \pi_3 \text{ and } H \tag{37}$$

and therefore can not stabilize the system to an isolated vertical equilibrium (sleeping top) which is an intersection of hyperserfaces of (37) with different values. What magnetic control can do is to stabilize the system to a part of the vertical equilibrium which is known as steady precession of the heavy top about the vertical, see (Lewis *et al.*, 1992).

Such a control will require vector fields which span the tangent space to (37). One can check that

$$\boldsymbol{A}_1 = \{\boldsymbol{\Omega}\times\boldsymbol{\Gamma}, \boldsymbol{0}\} \tag{38}$$

$$\boldsymbol{A}_2 = \{\boldsymbol{L}\times(\boldsymbol{\Gamma}-\boldsymbol{\Omega}), (\boldsymbol{\Pi}-\boldsymbol{M})\times\boldsymbol{\Gamma}\} \tag{39}$$

where

$$\boldsymbol{L} = -(\boldsymbol{M}\times\boldsymbol{\Pi})\cdot\boldsymbol{\Gamma}\frac{\boldsymbol{\Gamma}\times\boldsymbol{\Omega}}{||\boldsymbol{\Gamma}\times\boldsymbol{\Omega}||^2}, \tag{40}$$

are orthogonal to $\boldsymbol{\nabla}\gamma$, $\boldsymbol{\nabla}\pi_3$ and $\boldsymbol{\nabla}H$, that is lie in the tangent space to manifold (37), and also that $\boldsymbol{A}_1$, $\boldsymbol{A}_2$, $\boldsymbol{X}_H$ are linearly independent. Since the motion of the heavy top preserves three quantities (37), the trajectory of the heavy top lies in some three dimensional manifold. In this manifold there is an imbeded submaifold of relative equilibria corresponding to steady precession which intersects the set of vertical equilibria. So, deflecting $\boldsymbol{X}_H$ to the vertical equilibria should stabilize the system to the intersection.

The control is

$$\boldsymbol{U}(\boldsymbol{\mu}) = -\alpha(\boldsymbol{\mu})\boldsymbol{K}(\boldsymbol{\mu}), \tag{41}$$

where $\boldsymbol{K}(\boldsymbol{\mu})$ is the projection of $\boldsymbol{\mu} - \boldsymbol{\mu}_e$ on the orthogonal to $\boldsymbol{X}_H$ subspace tangent to (37) and $\boldsymbol{\mu} = (\boldsymbol{\Pi}, \boldsymbol{\Gamma})$ and $\boldsymbol{\mu}_e$ is a vertical equilibrium corresponding $\pi_3(\boldsymbol{\Pi}, \boldsymbol{\Gamma})$ for a different value of H.

Orthogonalizing $\boldsymbol{A}_1$, $\boldsymbol{A}_1$ and $\boldsymbol{X}_H$

$$\boldsymbol{k}_1 = \frac{\boldsymbol{X}_H}{||\boldsymbol{X}_H||}, \tag{42}$$

$$\boldsymbol{k}_2 = \frac{\boldsymbol{A}_1 - \boldsymbol{k}_1(\boldsymbol{A}_1\cdot\boldsymbol{k}_1)}{||\boldsymbol{A}_1 - \boldsymbol{k}_1(\boldsymbol{A}_1\cdot\boldsymbol{k}_1)||}, \tag{43}$$

$$\boldsymbol{k}_3 = \frac{\boldsymbol{A}_3 - \boldsymbol{k}_1(\boldsymbol{A}_3\cdot\boldsymbol{k}_1) - \boldsymbol{k}_2(\boldsymbol{A}_3\cdot\boldsymbol{k}_2)}{||\boldsymbol{A}_3 - \boldsymbol{k}_1(\boldsymbol{A}_3\cdot\boldsymbol{k}_1) - \boldsymbol{k}_2(\boldsymbol{A}_3\cdot\boldsymbol{k}_2)||}, \tag{44}$$

one finds that $\boldsymbol{K}$ is given by

$$\boldsymbol{K} = \boldsymbol{k}_2((\boldsymbol{\mu}-\boldsymbol{\mu}_0)\cdot\boldsymbol{k}_2) + \boldsymbol{k}_3((\boldsymbol{\mu}-\boldsymbol{\mu}_0)\cdot\boldsymbol{k}_3) \tag{45}$$

4. NUMERIC EXAMPLES

Derivations were given for three control schemes in the previous section. Double bracket dissipation preserves γ and π_3 and is orthogonal to $\boldsymbol{X}_H$, Momentum management preserves γ and H and is orthogonal to $\boldsymbol{X}_H$, finally, magnetic control preserves γ, π_3 and H and is orthogonal to $\boldsymbol{X}_H$.

To demonstrate the stabilizing property of dissipative control, momentum management and magnetic control (7) and (8) were integrated with $\boldsymbol{U}$ given by (17), (28) and (41). For parameters of the heavy top we took $\mathbf{J} = \text{diag}(2,3,4)$, $\boldsymbol{M} = mlg\boldsymbol{E}_3$, where mass $m = 3$, $l = 3$ and $g = 10$. For initial conditions we took $\boldsymbol{\Pi}(0) = \mathbf{J}\boldsymbol{\Omega}(0)$, $\boldsymbol{\Omega}(0) = (3, 0, 15)$ and $\boldsymbol{\Gamma}(0) = (\sin 50^\circ\cos 30^\circ, \sin 50^\circ\sin 30^\circ, \cos 50^\circ)$.

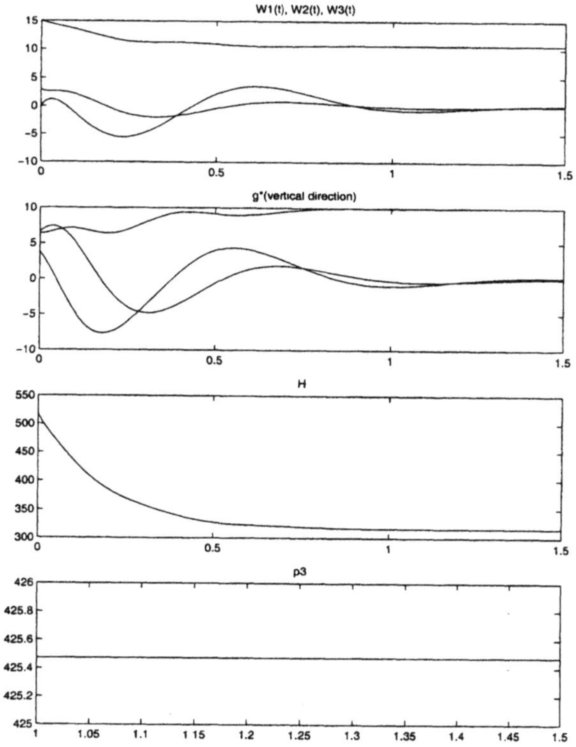

Fig. 1. Dissipative Control Performance

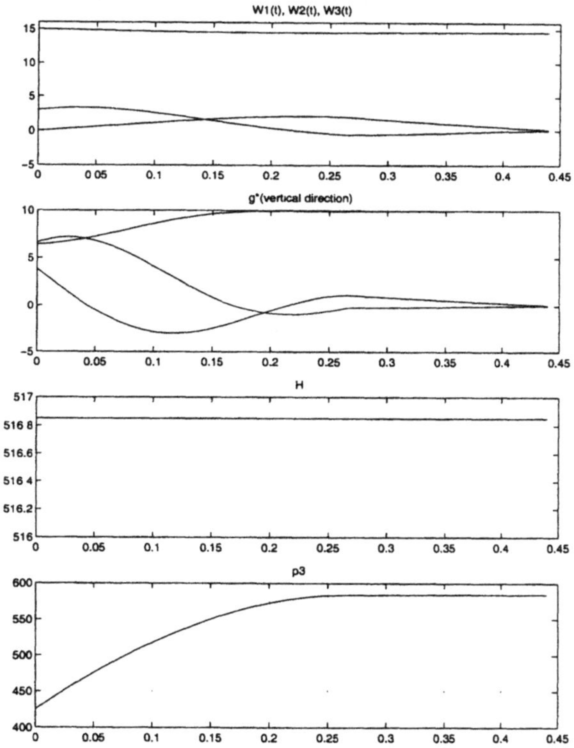

Fig. 2. Momentum Management Control Performance

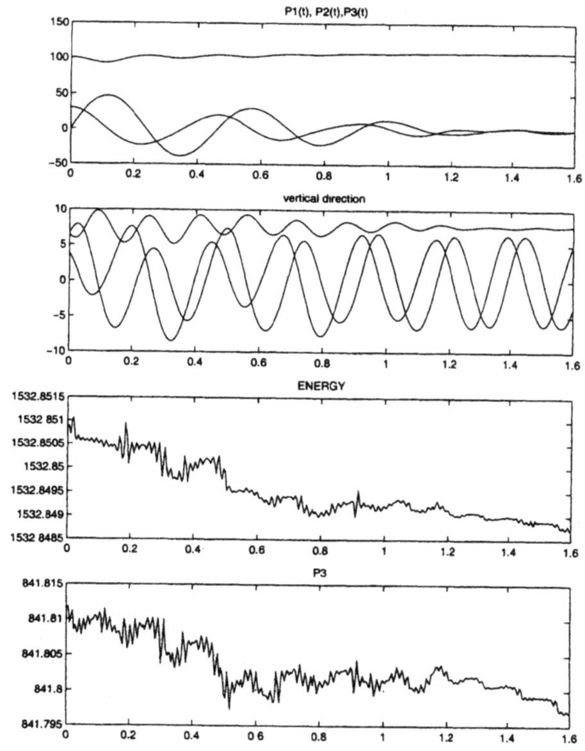

Fig. 3. Magnetic Control Performance

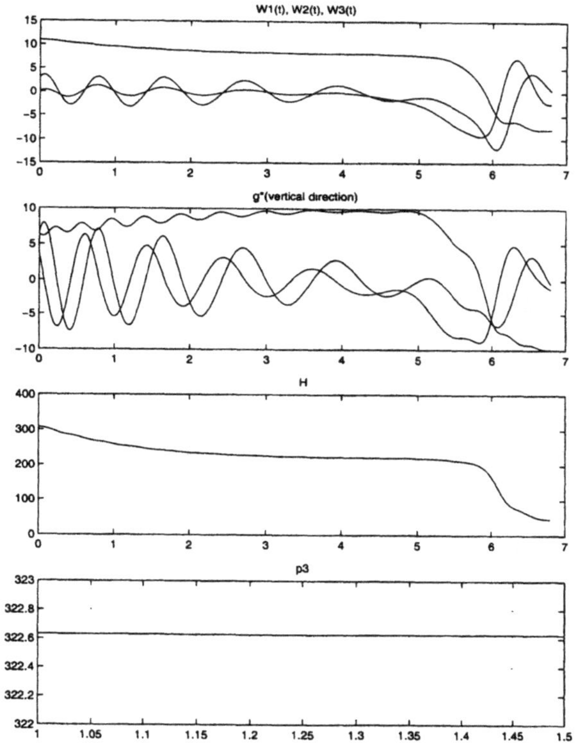

Fig. 4. Failure of Dissipative Control

The results are plotted on figures 1, 2 and 3. As can seen in the figures dissipative and momentum controls asymptotically stabilize vertical equilibrium, preserving π_3 and γ correspondingly.

On the figure 4, which is for the initial condition with slower vertical rotation, one can see that dissipative control fails to stabilize the system. First the system is being stabilized asymptotically moving $\mathbf{\Gamma}$ toward the vertical, then the system becomes unstable.

5. CONCLUSIONS

Three qualitatively different feedback laws have been considered which asymptotically stabilize the relative equilibria corresponding to vertical rotation and steady precession of the heavy top. The vertical rotation can be stabilized by dissipative control and momentum management.

Dissipative control is feasible if the initial conditions have sufficiently large π_3 for which the vertical rotation is stable. This means that for insufficient π_3, dissipative control can destabilize the system.

For the momentum management feedback if the initial conditions of the perturbed motion correspond to stable motion then momentum management does not destabilize the system.

Magnetic control can only stabilize the system to the relative equilibria corresponding to the steady precession which can be considered as a part of the relative equilibria corresponding to the vertical rotation.

The controls which were discussed are not realizable physically because $\boldsymbol{U}_2$ is not accessible. A realizable control would be $\boldsymbol{U} = (\boldsymbol{U}_1, \mathbf{0})$. To this control one can also apply the above technique which gives for the dissipative and momentum management controls

$$\boldsymbol{U}_1 = \mathbf{\Gamma} \times (\mathbf{\Gamma} \times \mathbf{\Omega}), \tag{46}$$

$$\boldsymbol{U}_1 = \mathbf{\Omega} \times (\mathbf{\Gamma} \times \mathbf{\Omega}). \tag{47}$$

This will be done in the forthcoming paper to asymptotically stabilize the heavy top with parametric uncertainties in the context (Posbergh and Egorov, 1998).

6. REFERENCES

Audin, M. (1996). *Spinning Tops.* number 51 In: *Cambridge Studies in Advanced Mathematics.* Cambridge. Cambridge.

Bloch, A. M. and J. E. Marsden (1990). Stabilization of rigid body dynamics by the energy-Casimir method. *Systems & Control Letters* **14**, 341–346.

Bloch, A. M., P. S. Krishnaprasad, J. E. Marsden and G. Sánchez de Alverez (1992). Stabilization of rigid body dynamics by internal and external torques. *Automatica* **28**, 745–756.

Bloch, A. M., P. S. Krishnaprasad, J. E. Marsden and T. S. Ratiu (1996). The Euler-Poincaré equations and double bracket dissipation. *Comm. Math. Phys.* **175**, 1–42.

Brockett, R. W. (1973). Lie theory and control systems defined on spheres. *SIAM J. Appl. Math.* **25**(2), 213–225.

Gantmacher, F. R. (1959). *The Theory of Matrices, Volume II.* New York. Chelsea.

Jalnapurkar, S. M. and J. E. Marsden (1999). Stabilization of relative equilibria II. *Reg. and Chaotic Dyn.* **3**, 161–179.

Jurdjevic, V. (1997). *Geometric Control Systems.* Cambridge Studies in Advanced Mathematics. Cambridge. Cambridge.

Khalil, H. K. (1996). *Nonlinear Systems.* second ed.. Prentice-Hall. Upper Saddle River.

Lewis, D., T. Ratiu, J. C. Simo and J. E. Marsden (1992). The heavy top: A geometric treatment. *Nonlinearity* **5**, 1–48.

Marsden, J. E. and T. Ratiu (1994). *Introduction to Mechanics and Symmetry.* number 17 In: *Text in Applied Mathematics.* Springer. New York.

Posbergh, T. A. and R. Zhao (1993). Stabilization of the uniform rotation of a rigid body by the energy-momentum method. In: *Dynamics and Control of Mechanical Systems* (M.. Enos, Ed.). Vol. 1 of *Fields Insitute Communications.* pp. 263–280. AMS. Providence.

Posbergh, T.A. and M.A. Egorov (1998). Robust stabilization of relative equilibria with application to a heavy top.. *IMA preprint series, No. 1590.*

Woolsey, C. A. and N. E. Leonard (1999). Underwater vehicle stabilization by internal rotors. *Proceedings of the 1999 Amer. Contr. Conf.* pp. 3417–3421.

Zhao, R. and T. A. Posbergh (1993). Robust stabilization of uniformly rotating rigid body. *IMA Preprints Series, No. 1185.*

NONLINEAR CONTROL OF UNDERACTUATED MULTI-BODY SPACE VEHICLES

N. Harris McClamroch[1]

Department of Aerospace Engineering
University of Michigan
Ann Arbor, Michigan 48109-2140
nhm@umich.edu

Abstract: This paper provides an overview of dynamics and control issues for a class of multi-body space vehicles, that is space vehicles consisting of an interconnection of a rigid base body and a finite number of connected bodies that define the shape of the vehicle. A summary of multi-body space vehicle dynamics is provided that describes the coupling between the dynamics of the base body and the shape dynamics. Several categories of underactuated multi-body space vehicle models are introduced and it is shown that they are in a standard normal form for underactuated mechanical control systems. A number of control problems are posed and references to available literature are given. The development is motivated by rest-to-rest maneuvering problems for underactuated multi-body space vehicles.

Keywords: nonlinear control, nonlinear systems, space vehicles, dynamic modeling, mechanical systems.

1. INTRODUCTION

In this paper, models for multi-body space vehicles are proposed; the models arise from the assumption that the vehicle consists of a rigid base body and multiple interconnected bodies whose relative motion with respect to the base body defines the shape of the space vehicle. The perspective of the paper reflects an interest in formulating interesting and challenging multi-body space vehicle rest-to-rest maneuvering problems as nonlinear control problems. Many details are omitted, but references are made to the available literature. A number of new research problems are identified that have not been studied in the literature.

The multi-body space vehicle models presented in this paper are developed in (Cho, *et al.*, 2000).

[1]Support from the National Science Foundation, Grant ECS-9625173 and ECS-9906018 are gratefully acknowledged.

Related publications on underactuated (or equivalently superarticulated) mechanical control systems are (Baillieul, 1999), (Bullo, 2000), (Bullo, *et al.*, 2000), (Ostrowski, 1999), (Reyhanoglu, *et al.*, 1999) and (Seto and Baillieul, 1994).

2. EQUATIONS OF MOTION

Gravity effects are ignored, so zero potential energy is assumed. Following the development in (Cho, *et al.*, 2000), the equations of motion of the space vehicle are expressed in terms of the inertial position vector x of the center of mass of the base body and the attitude defined by a rotation matrix R of the base body, where $(x,R) \in SE(3)$; the space vehicle base body translational velocity vector $V \in R^3$ and translational momentum vector $P \in R^3$ and the space vehicle base body angular velocity vector $\omega \in R^3$ and angular momentum vector $\Pi \in R^3$, all expressed in base body coordinates, and a vector of shape coordinates $r \in R^n$, $n \geq 1$, that describes the shape or mass distribution of the space vehicle defined with respect to the base body. In these coordinates, the equations of motion for the space vehicle base body can be shown to be given by:

$$\begin{aligned} \dot{R} &= R\hat{\omega}, \\ \dot{x} &= RV, \\ \dot{\Pi} &= \hat{\Pi}\omega + \hat{P}V + B_a\tau_a, \\ \dot{P} &= \hat{P}\omega + B_t\tau_t, \end{aligned} \qquad (1)$$

where the momenta are given by

$$\begin{bmatrix} P \\ \Pi \end{bmatrix} = J(r) \begin{bmatrix} V + A_{ts}(r)\dot{r} \\ \omega + A_{as}(r)\dot{r} \end{bmatrix}. \qquad (2)$$

The equations of motion for the shape coordinates can be shown to be given by:

$$\begin{aligned} m(r)\ddot{r} + F(r,\dot{r},P,\Pi) &= B_s(r)\tau_s \\ &- A'_{as}(r)B_a\tau_a - A'_{ts}(r)B_t\tau_t. \end{aligned} \qquad (3)$$

The function $F(r,\dot{r},P,\Pi)$ is quadratic in $(\dot{r},P,\Pi)$ and satisfies $F(r,0,0,0)=0$ for all r. The shape dependent functions $A_{as}(r), A_{ts}(r)$ are mechanical connections that summarize the coupling between the rate of change of the shape and the angular velocity of the base body and between the rate of change of the shape and the translational velocity of the base body, respectively. The input matrices $B_a, B_t, B_s(r)$ are assumed to be full rank for all r. The hat over a vector denotes the skew symmetric matrix formed from that vector; a transpose denotes matrix transpose.

The generalized forces and moments on the space vehicle are assumed to consist of control inputs which can be partitioned into three parts: $\tau_a \in R^{r_a}$ (typically from symmetric rotors, reaction wheels, and thrusters) is the vector of generalized control moments that act on the space vehicle base body, $\tau_t \in R^{r_t}$ (typically from thrusters) is the vector of generalized control forces that act on the space vehicle base body, and $\tau_s \in R^{r_s}$ is the vector of generalized control forces and moments (typically from shape change actuators) that directly control the space vehicle shape dynamics.

Terminology is now introduced that is useful in describing a conceptual classification of multi-body space vehicle maneuvering problems. The classification is based on the properties of the control inputs that can be used to achieve specific vehicle maneuvering objectives.

The assumption of full base body attitude actuation corresponds to $r_a = 3$, the assumption of full base body translational actuation corresponds to $r_t = 3$, and the assumption of full shape actuation corresponds to $r_s = n$. Many of the full actuation cases are relatively simple and have been studied in the control literature.

Underactuated space vehicles corresponding to the assumption $r_a + r_t + r_s < 6 + n$ are now considered; that is the total number of vehicle controls is less than the total number of vehicle degrees of freedom. The following terminology is useful: partial actuation of the base body attitude means that $1 \le r_a < 3$, partial actuation of the base body translation means that $1 \le r_t < 3$, and partial shape actuation means that $1 \le r_s < n$; unactuated base body attitude means that $r_a = 0$, unactuated base body translation means that $r_t = 0$, unactuated shape means that $r_s = 0$.

If there is no actuation, that is all input forces and moments are identically zero or equivalently $r_a = 0$, $r_t = 0$, $r_s = 0$, then there is an equilibrium manifold described by

$$\Omega = 0, V = 0, \dot{r} = 0 \qquad (4)$$

or equivalently by

$$\Pi = 0, P = 0, \dot{r} = 0. \qquad (5)$$

In other words, the space vehicle remains in equilibrium at any fixed position, attitude, and shape.

3. CONTROL PHILOSOPHY

The traditional approach to control problems for rest-to-rest maneuvering of *fully actuated* multi-body space vehicles is based on the following:

- Base body control moments are chosen to control the base body attitude dynamics only; the control moments should attenuate spillover from the base body translational dynamics and from the shape dynamics while not exciting the base body translational dynamics or the shape dynamics;
- Base body control forces are chosen to control the base body translational dynamics only; the control forces should attenuate spillover from the base body attitude

dynamics and from the shape dynamics while not exciting the base body attitude dynamics or the shape dynamics;

- Shape control forces and moments are chosen to control the shape dynamics only; the shape control forces and moments should attenuate spillover from the base body attitude dynamics and from the base body translational dynamics while not exciting the base body attitude dynamics or the base body translational dynamics.

That is, traditional control approaches ignore coupling and control spillover between the base body attitude dynamics, the base body translational dynamics, and the shape dynamics, thus leading to more conservative, reduced performance space vehicle systems. In contrast to this traditional approach, it is important to recognize that space vehicle shape change features are often required as a part of the space mission requirements. An integrated vehicle control system should exploit the coupling between the base body attitude dynamics, the translational dynamics, and the vehicle shape dynamics to achieve improved performance at the space vehicle system level. These issues provide motivation for the perspective espoused in this paper.

In the subsequent sections of the paper, several classes of multi-body space vehicles are studied. Specific underactuation assumptions are made, and the equations of motion are expressed in a normal form for underactuated mechanical systems that has been studied in (Reyhanoglu, *et al.*, 1999) and elsewhere. In this normal form, the accelerations of the actuated degrees of freedom are transformed control inputs, and the accelerations of the unactuated degrees of freedom are affine functions of the transformed control inputs. In addition, many vehicle maneuvering problems are suggested in subsequent sections.

4. FULLY ACTUATED BASE AND UNACTUATED SHAPE

If both the base body attitude and translation are fully actuated, that is $r_a = 3$, $r_t = 3$, and the shape is unactuated, that is $r_s = 0$, it is easy to show the rotational and translational dynamics of the base body can be "effectively" made linear and controllable by feedback transformations to obtain the base body attitude dynamics

$$\dot{R} = R\hat{\Omega}, \quad \dot{\Omega} = u_a, \qquad (6)$$

the base body translational dynamics

$$\ddot{x} = Ru_t, \qquad (7)$$

and the resulting shape dynamics

$$\ddot{r} + F_s(r,\dot{r},P,\Pi) = C_{sa}(r)u_a + C_{st}(r)u_t. \quad (8)$$

The function $F_s(r,\dot{r},P,\Pi)$ is quadratic in $(\dot{r},P,\Pi)$ and satisfies $F_s(r,0,0,0) = 0$. The base body acceleration control variables u_a, u_t can be expressed in terms of the torque and force control variables τ_a, τ_t. These equations are in the normal form for underactuated mechanical systems presented in (Reyhanoglu, *et al*, 1999). Such models can be used to define control problems that correspond to use of base body controls to control both the base body dynamics and the shape dynamics.

Several classes of space vehicle rest-to-rest maneuvering problems can be defined by the requirements that:

- An initial equilibrium defined by a base body attitude and position and vehicle shape be transferred to a final equilibrium defined by a base body attitude and position and vehicle shape, in a small time period, using only base body controls;
- An initial equilibrium defined by a base body attitude and position be transferred to a final equilibrium defined by a base body attitude and position, without excitation of the shape dynamics, in a small time period, using only base body controls.

These problems can be formulated either as open loop control problems or as feedback stabilization problems.

There are several variations on possible problem formulations. If the base body attitude is fully actuated, then control problems can be defined solely for the base body attitude and vehicle shape, modulo the base body translation. If the base body translation is fully actuated, then control problems can be defined solely for the base body translation and the vehicle shape, modulo the base body attitude. These models can also be expressed in the normal form for underactuated mechanical systems.

5. UNACTUATED BASE AND FULLY ACTUATED SHAPE

Here the base body is assumed to be unactuated while the shape dynamics are fully actuated, that is $r_a = 0$, $r_t = 0$, $r_s = n$. In this case it is easy to show the shape dynamics can be made linear and controllable by a feedback transformation to obtain the shape dynamics

$$\ddot{r} = u_s \quad , \tag{9}$$

the attitude dynamics of the base body

$$\begin{aligned} \dot{R} &= R\hat{\omega}, \\ \dot{\omega} &= F_a(r, \dot{r}, P, \Pi) + C_{as}(r)u_s, \end{aligned} \tag{10}$$

and the translational dynamics of the base body

$$\ddot{x} = R\big(F_t(r, \dot{r}, P, \Pi) + C_{ts}(r)u_s\big). \tag{11}$$

The two functions $F_a(r, \dot{r}, P, \Pi)$, $F_t(r, \dot{r}, P, \Pi)$ are quadratic in $(\dot{r}, P, \Pi)$ and they satisfy $F_a(r, 0, 0, 0) = 0$, $F_t(r, 0, 0, 0) = 0$ for all r. The shape acceleration control variables u_s can be expressed in terms of the torque/force control variable τ_s.

Such models can be used to define control problems that correspond to use of shape controls to control the base body dynamics and the shape dynamics or to control the base body dynamics only.

Several classes of space vehicle rest- to-rest maneuvering problems can be defined by the requirements that:

- An initial equilibrium defined by a base body attitude and position and vehicle shape be transferred to a final equilibrium defined by a base body attitude and position and vehicle shape, in a small time period, using only shape controls;
- An initial equilibrium defined by a vehicle shape be transferred to a final equilibrium defined by a vehicle shape, without excitation of the base body attitude or translational dynamics, in a small time period, using only shape controls;
- An initial equilibrium base body attitude and position be transferred to a specified final equilibrium base body attitude and position, in a small time period, using only shape controls; in this case the vehicle shape is free.

These problems can be formulated either as open loop control problems or as feedback stabilization problems.

There are again several variations on possible problem formulations in this category. If only the base body attitude is to be controlled, then control problems can be defined solely for the base body attitude and vehicle shape, modulo the base body translation. If only the base body translation is to be controlled, then control problems can be defined solely for the base body translation and the vehicle shape, modulo the base body attitude. These models can also be expressed in the normal form for underactuated mechanical systems.

6. OTHER MULTI-BODY SPACE VEHICLE CLASSIFICATIONS

The models and problems considered in Sections 4 and 5 correspond to specific actuation assumptions. There are many other actuation assumptions that could be made, including:

- Fully actuated base and partially actuated shape;
- Partially actuated base and fully actuated shape;
- Partially actuated base and partially actuated shape;
- Partially actuated base and unactuated shape;
- Partially actuated shape and unactuated base.

Many interesting and challenging control problems can be formulated for these classes of multi-body space vehicles.

7. MANEUVER EXAMPLES FOR MULTI-BODY SPACE VEHICLES

Relatively few three dimensional rest-to-rest maneuvering problems for underactuated multi-body space vehicles have been studied in detail. One area that has received attention is simultaneous attitude and shape control, assuming zero angular momentum of the space vehicle. Results, corresponding to different actuation and vehicle assumptions, have been developed in (Rui, *et al.*, 1996), (Rui, *et al.*, 1997), (Rui, *et al.*, 1998), and (Rui, *et al.*, 2000).

The development in Sections 3, 4, 5 is easily modified if it assumed that the maneuvers are planar, that is the models can be expressed in terms of the base body inertial position vector x and rotation matrix R, where $(x,R) \in SE(2)$ and the planar shape coordinates $r \in R^n$, $n \geq 1$. A brief review of results for this class of problems is given.

Models for planar two-body vehicles, using a base and shape variable formulation, have been presented. (Baillieul, 1999) treats a planar body with a pendulum attachment, and (Ostrowski, 1999) treats "Elroy's beanie."

Control problems for planar maneuvers of multi-body space vehicles with fully actuated base and unactuated shape have been studied in (Reyhanoglu, *et al.*, 1999). A space vehicle with a single unactuated shape degree of freedom is shown to be controllable; specific controller construction procedures have been suggested in (Reyhanoglu, 1998), (Reyhanoglu, *et al.*, 1999) and (Reyhanoglu, *et al.*, 2000).

Control problems for unactuated base and fully actuated shape dynamics of planar multi-body space vehicles have received some attention. Results for simultaneous control of base body attitude and shape have been studied for a number of different cases by (Kolmanovsky and McClamroch, 1993), (Kolmanovsky, *et al.*, 1995), (Reyhanoglu and McClamroch, 1992) and by (Reyhanoglu and McClamroch, 1993). Preliminary control results in (Shen, McClamroch, 2000) demonstrate that the base body position and attitude are controllable if there are two or more shape change degrees of freedom. All of the results cited assume that the space vehicle momenta is zero, so that the models can be simplified and expressed in a nonlinear drift free control form that is easier to analyze.

8. CONCLUDING REMARKS

An overview of maneuver problems for underactuated multi-body space vehicles has been presented. Remarks are made about the available control theoretical results, possible control construction approaches, and the need for continuing research into this class of problems.

As suggested in the previous sections, control problems for multi-body space vehicles are very challenging since traditional linear and smooth nonlinear control approaches generally do not suffice. In terms of available theory, there are relatively few theoretical control results available for second order mechanical systems in general or for multi-body space vehicle systems in particular. Results in (Reyhanoglu, *et al.*, 1999) provide Lie algebraic conditions for controllability of underactuated mechanical systems of the form presented in this paper; it is also demonstrated there that any equilibrium is not asymptotically stabilizable by time invariant smooth static feedback. Several results are given in (Baillieul, 1999) that characterize when underactuated mechanical systems can be expressed in a special "velocity controlled" form.

Procedures for construction of controllers, either in open loop or feedback form, are complicated by the fact that traditional linear or smooth nonlinear control methods are not applicable. In addition, the above models are not drift free, so that the simplest nonlinear control constructions are not applicable. Two approaches that have been used successfully involve use of oscillatory control inputs, see (Baillieul, 1999), and use of non-smooth transformations, see (Reyhanoglu, *et al.*, 2000). In some classes of problems, symmetry properties, backstepping approaches, or approximations can yield simplified nonlinear control models that are more tractable. As demonstrated in (Rui, *et al.*, 2000) and other references previously cited, momenta conservation provides an effective simplification for multi-body space vehicle control problems involving an unactuated base body and fully actuated shape. This general area of control construction for multi-body space vehicles remains a rich source of research challenges.

There are a number of important dynamics and control topics for underactuated multi-body space vehicles that have received little or no attention; the following is a list of areas where future research should be concentrated:

- Control theory for underactuated mechanical systems, specifically for classes of underactuated multi-body space vehicles;
- Control construction methods for open loop maneuvers and for feedback stabilization;
- Computational approaches, using both symbolic and numerical tools, for dynamics and control of underactuated multi-body space vehicles;
- Continued study of special classes and specific examples of underactuated multi-body space vehicles dynamics and control problems.

It is hoped that the overview provided in this paper will stimulate further research on control of multi-body space vehicles.

9. ACKNOWLEDGEMENTS

Many colleagues have contributed to the development of the ideas presented in this paper: A. M. Bloch, M. Reyhanoglu, H. Krishnan, I. Kolmanovsky, C. Rui, S. Cho, and J. Shen. Their help is gratefully acknowledged.

10. REFERENCES

Baillieul, J. (1999), The Geometry of Controlled Mechanical Systems. In: *Mathematical Control Theory*. (J. Baillieul and J. C. Willems, Ed), 323-354.

Bullo, F. (2000), Stabilization of Relative Equilibria for Underactuated Systems on Riemannian Manifolds, to appear in *Automatica*.

Bullo, F., N. E. Leonard, and A. D. Lewis (2000), Controllability and Motion Algorithms for Underactuated Lagrangian Systems on Lie Groups, to appear in *IEEE Transactions on Automatic Control*.

Cho, S., N. H. McClamroch, and M. Reyhanoglu (2000), On Dynamics and Control of Multibody Vehicle Systems with SE(3) Symmetries, American Control Conference.

Kolmanovsky, I. and N. H. McClamroch (1993), Planar Reorientation of a Free-Free Beam in Space using Embedded Electromechanical Actuators, Proceedings of SPIE Conference on Smart Structures and Materials '93, Albuquerque, NM, 260-270.

Kolmanovsky, I., N. H. McClamroch, and V. T. Coppola (1995), New Results on Control of Multibody Systems which Conserve Angular Momentum, *Journal of Dynamical and Control Systems*, **1**, No. 4, 447-462.

Ostrowski, J. P. (1999), Computing Reduced Equations for Robotic Systems with Constraints and Symmetries, *IEEE Transactions on Robotics and Automation*, **15**, No. 1, February, 111-123.

Reyhanoglu, M. and N. H. McClamroch (1992), Reorientation Maneuvers of Planar Multibody Systems in Space using Internal Controls, AIAA *Journal of Guidance, Control and Dynamics*, **15**, No. 6, 1475-1480.

Reyhanoglu, M. and N. H. McClamroch (1993), Nonlinear Attitude Control of Planar Structures in Space using only Internal Controls, *Fields Institute Communications*, **2**, American Mathematical Society, 91-100.

Reyhanoglu, M., S. Cho, N. H. McClamroch, and I. Kolmanovsky (1998), Discontinuous Feedback Control of a Planar Rigid Body with an Unactuated Degree of Freedom, Proceedings of 37th IEEE Conference on Decision and Control, December, 433-438.

Reyhanoglu, M., S. Cho, and N. H. McClamroch (1999), Feedback Control for Planar Maneuvers of an Aerospace Vehicle with an Unactuated Internal Degree of Freedom, Proceedings of American Control Conference, 3432-3436.

Reyhanoglu, M., S. Cho, and N. H. McClamroch (2000), Discontinuous Feedback Control of a Special Class of Underactuated Mechanical Systems, to appear in *International J. of Robust and Nonlinear Control*.

Reyhanoglu, M., A. van der Schaft, I. Kolmanovsky, and N. H. McClamroch (1999), Dynamics and Control of a Class of Underactuated Mechanical Systems, *IEEE Transactions on Automatic Control*, **44**, September, 1999, 1663-1671.

Rui, C., I. Kolmanovsky, and N. H. McClamroch (1996), Feedback Reorientation of Underactuated Multibody Spacecraft, Proceedings of 35th IEEE Conference on Decision and Control, Kobe, Japan, December, 489-494.

Rui, C., N. H. McClamroch, and A. M. Bloch (1997), Three Dimensional Reorientation of a Spacecraft Containing a Single Reaction Wheel and a Single Movable Appendage, Proceedings of 36th IEEE Conference on Decision and Control, December, 4844-4849.

Rui, C., I. Kolmanovsky, and N. H. McClamroch (1998), Three Dimensional Attitude and Shape Control of Spacecraft with Appendages and Reaction Wheels, Proceedings of 37th IEEE Conference on Decision and Control, December, 4176-4181.

Rui, C., I. Kolmanovsky, and N. H. McClamroch (2000), "Nonlinear Attitude and Shape Control of Spacecraft with Appendages and Reaction Wheels," to appear in *IEEE Transactions on Automatic Control*.

Seto, D. and J. Baillieul (1994), Control Problems in Superarticulated Mechanical Systems, *IEEE Transactions on Automatic Control*, **39**, 2442-2453.

Shen, J. and N. H. McClamroch, Control of Spacecraft Planar Motions via Shape Change using Linear Proof Mass Actuators, preprint.

GENERALIZED KIRCHHOFF EQUATIONS

Alexander R. Galper *

* *Faculty of Engineering, Tel-Aviv University, Israel 69978*

Abstract: Existing theories of control of autonomous underwater vehicle are based mainly on a potential framework for the surrounding fluid. It leads to ODE governing the motion of a vehicle. On the other hand an influence of more realistic rheology accounted for a different types of vorticity shedding can critically change the "potential" stability criteria. In the sequel some possible generalizations of the Kirchhoff equations are suggested based on extended potential formalism in order to account for the vorticity shedding phenomena. A linear stability of steadily translating axisymmetric bodies with a motion-dependent shape is treated and some applications for a stability control are discussed. *Copyright ©2000 IFAC*

Keywords: Autonomous vehicles, Control, Stability analysis

1. THE KIRCHHOFF EQUATIONS

The classical hydrodynamic problem of the motion of submerged rigid (deformable) bodies (Lamb, 1945 §3-5) has recently received renewed interest stimulated by the rapidly growing industry of *autonomous underwater vehicles* (denoted further as AUV) for deep sea explorations.

The problem of the non-linear control of AUV is an extremely difficult problem both of practical and fundamental significance. In its full size the problem should couple the following different areas:

(I). Navier-Stokes equation for the ambient flow field hydrodynamics (including turbulence phenomena) in a domain bounded by rigid or free-surface,

(II). The Euler dynamic equation for the motion of a 3-*D* body (AUV).

(III). Modern strategy for active or internal control of a chaos exhibited by a moving AUV.

[1] The author acknowledges the support of the BSF Foundation, Contract No. 94-00287.

The problem of the motion of an AUV is strongly simplified (still remains non-trivial) for the case when a *potential framework* for the hydrodynamical circumstances can be justified. A potential framework allows the unique opportunity for an effective *analytical* treatment of the problem of an AUV controlling. In a *perfect* unbounded fluid treated within a potential framework the equations of motion of a rigid AUV with a translational velocity $\mathbf{U}$ and an angular velocity $\boldsymbol{\Omega}$ reduce to a system of ordinary differential equations (the, so-called; Kirchhoff equations (Lamb, 1945 §6)). The classical Kirchhoff equations can be written in the Lagrangian form, given by

$$\frac{d}{dt}\left(\frac{\partial E_{tot}}{\partial \mathbf{U}}\right) + \boldsymbol{\Omega} \wedge \frac{\partial E_{tot}}{\partial \mathbf{U}} = 0, \qquad (1)$$

$$\frac{d}{dt}\left(\frac{\partial E_{tot}}{\partial \boldsymbol{\Omega}}\right) + \boldsymbol{\Omega} \wedge \frac{\partial E_{tot}}{\partial \boldsymbol{\Omega}} + \mathbf{U} \wedge \frac{\partial E_{tot}}{\partial \mathbf{U}} = 0.(2)$$

Here E_{tot} is a kinetic energy of the body and the surrounding fluid, $\mathbf{U}$ and $\boldsymbol{\Omega}$ in (1) and (2) are referred instantaneously to a frame of coordinates moving with the body and we use "$\wedge$" for a vector product. Note, that also a corresponding non-canonical Hamiltonian theory for the body dynamics governing by (1) and (2) can be created (Aref and Jones, 1993).

It is important to emphasize that the general form (1) and (2) of the Kirchhoff equations holds also for the motion of a deformable body in an ambient non-uniform potential stream. An interesting and not obvious fact that the conserve integral of motion (for an autonomous UV) $E_{tot} > 0$ can serve as a Lagrangian for the motion of a body itself was justified by Birkhoff (1960) for some particular cases. Thus, it can be shown that this first integral is not independent (and hence, additional) first integral of motion.

The main characteristic property of the motion of AUV governed by the Kirchhoff equations and moving in an otherwise quiescent fluid (or, more generally, in an arbitrary *potential* nonuniform bounded stream) is its high shape-sensitivity. This fact manifests itself in a chaotization of the motion for nearly all shapes (Holmes, et al., 1998).

The problem of a control of a *rigid* AUV has been extensively and rigorously elaborated in recent years by a number of investigators. Their research treats mainly the *potential* hydrodynamic model applied in an unbounded or bounded space. The methodology is based on the Hamiltonian structure of the dynamical equations governing the body motion within the potential framework (see Galper and Miloh, 1995). The stability problem is then effectively treated using the, so-called, Energy-Casimir method (Galper and Miloh, 1998, 1999; Leonard, 1996, 1997a, 1997b; Leonard and Marsden, 1997 and Holmes, et al., 1998).

2. POTENTIAL CONTROL STRATEGIES

Thus, it appears, that the simple non-coincident centers of gravity and buoyancy can lead to a stabilization of the motion of the AUV due to the additionally induced rotation (Galper and Miloh, 1998, Leonard and Marsden, 1997) for the model case of a potential framework for the surrounding liquid.

The alternative avenue, seeks to realize the motion control through active modification of the moment of inertia and center of mass (Aref, 1993, 1994; Aref, et al., 1998), which appears as coefficients in the Kirchhoff equation. The key physical reason for this is *sharp transitions* between regimes of steady motion, periodic oscillation, tumbling and chaotic motions (the phenomena also found in many experiments (Weiss, 1998) in real viscous fluid). Only by designing and operating the body close to these sharp transition boundaries, some finite $(0(1))$ change in the motion of the body can be realized with an infinitesimal $0(\epsilon)$ change in the controlling parameters.

Given a body in motion, under specific external forces such as propulsion and gravitational/ buoyancy forces, the Kirchhoff equation can be used to identify the critical transition parameters.

Also an active control of an underwater vehicle can be performed by small prescribed periodic changes of its surface (e.g. Miloh and Galper, 1993) using some special mechanisms for this goal (wings, elastic shells, etc.) . The possibility to stabilize the motion of the underwater vehicle in a non-uniform ambient potential stream is based on a parametric resonant interaction between the surface pattern and the flow field non-uniformity, as pointed out by Galper and Miloh (1995). The controlling parameters become now the periodically changing added-masses and the deformation Kelvin impulse (deformation Kelvin impulse-couple), which are induced by the surface deformation motion of the vehicle. Some properly chosen surface deformations are used in this case in order to absorb energy from an ambient spatially non-uniform stream. Even in an otherwise quiescent fluid surface deformations can be used for creating the so-called "self-propulsion" phenomena (Miloh and Galper, 1993), which is especially efficient for a vehicle with non-zero initial kinetic energy.

3. LIFTING FORCES AND BRAYAN EQUATION

The stability analysis of the motion of AUV, which is based only on the potential framework, ceases to be valid for the case where the added-mass concept is of a less importance role compared with that of the lifting forces. These lifting forces are essentially important for slender bodies, like the hull of a submarine or a UV with a typical submarine shape. Also for a deformable symmetric body even small deformations can lead to a non-symmetrical flow which drastically changes the type of solution.

The classical example of a wing with a flap where the deviation of a flap induces lift must be kept in mind.

The main advantage of the Kirchoff equations is that they compose a self-consistent system. They also correctly account for the effects of fluid inertia. On the other hand, the Kirchhoff equations do not include the important forces and moments from viscous forces, or vortex shedding, and do not include forces due to propulsion.

In general, a kinetic energy of the "body + fluid" system E_{tot} depends functionally on the whole motion history of the body, i.e. $E_{tot}(t) = E_{tot}[\mathbf{U}(\tau), \mathbf{\Omega}(\tau)]$, for all τ, where $0 < \tau \leq t$. Using again E_{tot} as a Lagrangian for the corresponding

Lagrangian formalism one obtains the following generalized Kirchhoff equation, expressed in the terms of variational derivatives;

$$\frac{d}{dt}\left(\frac{\delta E_{tot}}{\delta \mathbf{U}(t)}\right) + \mathbf{\Omega}(t) \wedge \frac{\delta E_{tot}}{\delta \mathbf{U}(t)} = 0, \quad (3)$$

$$\frac{d}{dt}\left(\frac{\delta E_{tot}}{\delta \mathbf{\Omega}(t)}\right) + \mathbf{\Omega}(t) \wedge \frac{\delta E_{tot}}{\delta \mathbf{\Omega}(t)} + \mathbf{U}(t) \wedge \frac{\delta E_{tot}}{\delta \mathbf{U}(t)} = 0. \quad (4)$$

When the force and moment vectors, $\mathbf{F}[\mathbf{U}(\tau), \mathbf{\Omega}(\tau))]$ and $\mathbf{T}[\mathbf{U}(\tau), \mathbf{\Omega}(\tau)]$, representing the additional viscous effects are added to the RHS of the generalized Kirchoff equations (3) and (4), one obtains the following generalized *variational* Brayan equation

$$\frac{d}{dt}\left(\frac{\delta E_{tot}}{\delta \mathbf{U}(t)}\right) + \mathbf{\Omega}(t) \wedge \frac{\delta E_{tot}}{\delta \mathbf{U}(t)} = \mathbf{F}[\mathbf{U}(\tau), \mathbf{\Omega}(\tau)], \quad (5)$$

$$\frac{d}{dt}\left(\frac{\delta E_{tot}}{\delta \mathbf{\Omega}(t)}\right) + \mathbf{\Omega}(t) \wedge \frac{\delta E_{tot}}{\delta \mathbf{\Omega}(t)} + \mathbf{U}(t) \wedge \frac{\delta E_{tot}}{\delta \mathbf{U}(t)} = \mathbf{T}[\mathbf{U}(\tau), \mathbf{\Omega}(\tau)]. \quad (6)$$

If one neglects the variational dependence of all terms in (5) and (6) on the whole history of the process the generalized Brayan equations cast to the ordinary ***differential*** Brayan equations (see, for example, Abzug and Larrabee, 1997)

$$\frac{d\mathbf{p}}{dt} + \mathbf{\Omega} \wedge \mathbf{p} = \mathbf{F}, \quad (7)$$

$$\frac{d\mathbf{M}}{dt} + \mathbf{\Omega} \wedge \mathbf{M} + \mathbf{U} \wedge \mathbf{M} = \mathbf{T}. \quad (8)$$

Clearly, even these simplified Brayan equations are not self-consistent, since the effects of the fluid motion, such as vortex shedding, on the liner and angular velocities of the body must be parameterized in terms of the lift (C_L), drag (C_D) and moments (C_M^x, C_M^y, C_M^z) coefficients. In general, the back-effect of the motion of the body on the external fluid flow cannot be parameterized in this simple way (Aref, et al., 1998). Nevertheless, the ODE form of the Bryan's equation is an advantage for an analytical treatment of the stability and control of AUV.

4. VORTICITY SHEDDING

The Kirchhoff equations correctly account for the effects of fluid inertia, but do not include the important forces and moments from viscous forces or vortex shedding.

While the Kirchhoff equations ((1) and (2)) provide a rigorous, self-consistent model of the motion of a body in an ideal fluid, the Bryan's equations ((7) and (8)) are generally not self-consistent. In contrast, we propose to account for a vortex shedding effect, by still using the self-consistent Kirchhoff equations (which is a great advantage for applications). For this goal we will consider a rigid body together with a boundary layer around it. One can speak now about the combined shape of a body composed by the body itself and by its own boundary layer (Wu, 1995). The thickness of the boundary layer depends on the bodies velocities. In order to estimate the amount of vorticity shedding, a number of already existing models can be chosen (Wu and Wu, 1997). This combined deformable (non-homogeneous) body with a motion-dependent shape can be treated within a certain Hamiltonian theory. Such an approach, using also the Energy-Casimir stability method, leads directly to some stability criteria of the translational motion (in a vortical realm !). The corresponding treatment is similar to the stability problem elaborated in Galper and Miloh (1999). Clearly, the derived stability criteria will be in accordance with the chosen model for the vorticity shedding. It is expected that these stability criteria will depend only on the qualitative character of the motion-combined-shape dependence. It allows to overcome the difficulty arising from the lack of a precise analytical expression for the boundary layer around a moving body.

In order to illustrate a possible influence of a vorticity shedding phenomena on a stability of a solid motion one has to derive the generalized Kirchhoff equations for the motion dependent shape S, given by

$$S = S(\mathbf{U}, \mathbf{\Omega}). \quad (9)$$

The motion-dependent shape (9) can be used also for the treating of the stability problem of motion for elastic containers (carrying water) in a sea environment and for bubbles or drops motion. The deformational velocity V_d of the shape (9) is given by $V_d = \dot{S}/|\nabla S|$. Hence, the corresponding Lagrangian L which is equal to the total kinetic energy of the system, i.e, $L = E_{tot}(\mathbf{U}, \mathbf{\Omega}, \dot{\mathbf{U}}, \dot{\mathbf{\Omega}})$ leads to the theories with higher derivatives (see, for example, Gitman and Tyutin, 1990, §7). It can be shown that the generalized Kirchhoff equation takes next the form

$$\frac{d\mathbf{p}}{dt} + \mathbf{\Omega} \wedge \mathbf{p} = 0, \quad (10)$$

$$\frac{d\mathbf{m}}{dt} + \mathbf{\Omega} \wedge \mathbf{m} + \mathbf{U} \wedge \mathbf{p} = 0, \quad (11)$$

where the generalized impulses $\mathbf{p}$ and $\mathbf{m}$ are given by

$$\mathbf{p} \equiv \frac{\partial E_{tot}}{\partial \mathbf{U}} - \frac{d}{dt}\left(\frac{\partial E_{tot}}{\partial \dot{\mathbf{U}}}\right), \tag{12}$$

$$\mathbf{m} \equiv \frac{\partial E_{tot}}{\partial \mathbf{\Omega}} - \frac{d}{dt}\left(\frac{\partial E_{tot}}{\partial \dot{\mathbf{\Omega}}}\right). \tag{13}$$

Two first integrals of motion can be immediately found for (10) and (11), namely (compare with Aref and Jones, 1993):

$$\mathbf{p}^2 = \text{const}, \quad \mathbf{p}\cdot\mathbf{m} = \text{const}. \tag{14}$$

5. SPHERE WITH A BOUNDARY LAYER

One can pose now the following problem: if the steady translation of a rigid sphere (accounting for a vorticity shedding) is linearly stable (being of a neutral stability for a purely potential framework)? The combined motion-dependent (pseudo) body (i.e., a rigid sphere plus its boundary layer) is, clearly, a body of revolution (similar to a prolate spheroidal shape, shrunk in the direction orthogonal to a velocity). Note, that a translation of a (rigid) prolate spheroidal body along its longest axis of symmetry is proven to be linearly unstable (Lamb, 1945). Due to a high symmetry of a sphere, the combined shape depends on a module of a translational velocity and we assume also that it is independent from the sphere's angular velocity, i.e., $S = S(|\mathbf{U}|)$. Note, further, that based on the construction of a pseudo-body the angular velocity of the combined pseudo- body is directly connected with its translational velocity by the following relationship

$$\mathbf{\Omega} = \frac{d}{dt}\left(\frac{\mathbf{U}}{|\mathbf{U}|}\right). \tag{15}$$

Hence, only the first (momentum conservation) equation (10) should be taken into account for its dynamics. Finally, the corresponding generalized Kirchhoff equation of motion for a translating sphere (accounting the separation phenomena) is given now by

$$\frac{d}{dt}\left(\frac{\partial E_{tot}}{\partial \mathbf{U}}\right) - \frac{d^2}{dt^2}\left(\frac{\partial E_{tot}}{\partial \dot{\mathbf{U}}}\right)$$
$$+\frac{d}{dt}\left(\frac{\mathbf{U}}{|\mathbf{U}|}\right)\wedge\left(\frac{\partial E_{tot}}{\partial \mathbf{U}} - \frac{d}{dt}\left(\frac{\partial E_{tot}}{\partial \dot{\mathbf{U}}}\right)\right) = 0 \tag{16}$$

Next, a direct calculation for a total kinetic energy for the system "pseudo-body + surrounding fluid" leads to

$$E_{tot} = \frac{1}{2}\hat{T}(|\mathbf{U}|)\mathbf{U}\cdot\mathbf{U} + k(|\mathbf{U}|)\mathbf{U}\cdot\dot{\mathbf{U}}$$
$$+\frac{1}{2}\hat{\Pi}(|\mathbf{U}|)\dot{\mathbf{U}}\cdot\dot{\mathbf{U}}. \tag{17}$$

In (17) a symmetric tensor $\hat{T}$ has a meaning of a translational added-mass tensor for a pseudo-body S, the tensor $\hat{\Pi}$ is a symmetric and positively defined tensor whereas $k(|\mathbf{U}|)$ plays a role of a deformation Kelvin impulse. In derivation of (17) we use the axisymmetricity of the pseudo-shape. Note further, that the term $k(|\mathbf{U}|)\mathbf{U}\cdot\dot{\mathbf{U}}$ can be neglected in the Lagrangian as a full time derivative. After the substitution of (17) into (16) one obtains, for example

$$\frac{d}{dt}\left(\frac{\partial E_{tot}}{\partial \mathbf{U}}\right) = \hat{T}\dot{\mathbf{U}} + \frac{\partial \hat{T}}{\partial |\mathbf{U}|}\mathbf{U}\left(\dot{\mathbf{U}}\cdot\mathbf{e_U}\right)$$
$$+\frac{\partial T_{ij}}{\partial |\mathbf{U}|}\dot{U}_i U_j \mathbf{e_U} + \frac{1}{2}\frac{\partial T_{ij}}{\partial |\mathbf{U}|}U_i U_j \dot{\mathbf{e}}_\mathbf{U}$$
$$+\frac{1}{2}\frac{\partial^2 T_{ij}}{\partial |\mathbf{U}|^2}U_i U_j\left(\dot{\mathbf{U}}\cdot\mathbf{e_U}\right)\mathbf{e_U}$$
$$+\frac{1}{2}\frac{d}{dt}\left(\frac{\partial \Pi_{ij}}{\partial |\mathbf{U}|}\dot{U}_i\dot{U}_j\mathbf{e_U}\right), \tag{18}$$

where is denoted

$$\mathbf{e_U} \equiv \frac{\mathbf{U}}{|\mathbf{U}|} \tag{19}$$

. The similar calculations should be done also for

$$\frac{d^2}{dt^2}\left(\frac{\partial E_{tot}}{\partial \dot{\mathbf{U}}}\right) = \frac{d^2}{dt^2}\left(\Pi\dot{\mathbf{U}}\right), \tag{20}$$

and for other terms in (16). One can investigate now a linear stability of a stationary translational motion of a sphere

$$\mathbf{U}_0 = (U_1, 0, 0) = const. \tag{21}$$

The direct calculations (omitted here) lead to the following dispersion relation for the perturbations orthogonal to $\mathbf{U}_0$

$$\left(B - \Pi_2\omega^2\right)^2 + A^2 = 0, \tag{22}$$

where it is denoted

$$A \equiv \left(T_1 + \frac{1}{2}\frac{\partial T_1}{\partial |\mathbf{U}|}|\mathbf{U}|\right)\Bigg|_{\mathbf{U}=\mathbf{U}_0}, \tag{23}$$

$$B \equiv \left(T_2 + \frac{1}{2}\frac{\partial T_1}{\partial |\mathbf{U}|}|\mathbf{U}|\right)\Bigg|_{\mathbf{U}=\mathbf{U}_0}. \tag{24}$$

Here $\hat{T} = diag(T_1, T_2, T_3)$, $\hat{\Pi} = diag(\Pi_1, \Pi_2, \Pi_3)$ and $T_2 = T_3$, $\Pi_2 = \Pi_3$. Clearly, there is a solution of (22) with a positive real part of ω Hence, the translational motion of a sphere within a potential framework accounted for a separation phenomena is *linearly unstable* - a quite surprising result!

6. AXISYMMETRIC BODY WITH SEPARATION

To understand a critical influence on a stability problem makes by a velocity shape dependence one should consider the generalized Kirchhoff equations applied to a deformable body with a motion dependent shape $S = S(|\mathbf{U}|)$, which is chosen for the simplicity to be axisymmetric. The corresponding total energy reads now (excluding the full time derivatives) as

$$E_{tot} = \frac{1}{2}\hat{T}(|\mathbf{U}|)\mathbf{U}\cdot\mathbf{U} + \frac{1}{2}\hat{R}(|\mathbf{U}|)\mathbf{\Omega}\cdot\mathbf{\Omega} + \frac{1}{2}\hat{\Pi}(|\mathbf{U}|)\dot{\mathbf{U}}\cdot\dot{\mathbf{U}}, \quad (25)$$

where $\hat{R}$ denotes a rotational added-mass tensor. The corresponding generalized Kirchhoff equations lead to

$$\frac{d}{dt}\left(\frac{\partial E_{tot}}{\partial \mathbf{U}} - \frac{d^2}{dt^2}\left(\frac{\partial E_{tot}}{\partial \dot{\mathbf{U}}}\right)\right) + \mathbf{\Omega}\wedge\left(\frac{\partial E_{tot}}{\partial \mathbf{U}} - \frac{d}{dt}\left(\frac{\partial E_{tot}}{\partial \dot{\mathbf{U}}}\right)\right) = 0, \quad (26)$$

$$\frac{d}{dt}\frac{\partial E_{tot}}{\partial \mathbf{\Omega}} + \mathbf{\Omega}\wedge\frac{\partial E_{tot}}{\partial \mathbf{\Omega}} + \mathbf{U}\wedge\left(\frac{\partial E_{tot}}{\partial \mathbf{U}} - \frac{d}{dt}\left(\frac{\partial E_{tot}}{\partial \dot{\mathbf{U}}}\right)\right) = 0. \quad (27)$$

One can check now a linear stability of a relative equilibria (a translation without rotation) of (26), (27) given by

$$\mathbf{U} = \mathbf{U}_0 = const, \quad \mathbf{\Omega} = 0, \quad (28)$$

where $\mathbf{U}_0$ is aligned along the eigendirection of the added-mass tensor $\hat{T}(|\mathbf{U}_0|)$. The direct calculations applied to a spectral analysis of the linearized equations leads to the following dispersion relation

$$\Pi_2 R_2 \omega^4 + U_1\Pi_2 A - R_2 B)\omega^2 + AU_1(T_2 - T_1) = 0. \quad (29)$$

Here, due to the axisymmetry of the shape all tensors can be simultaneously diagonalized, i.e., $\hat{T} = diag(T_1, T_2, T_3)$, $\hat{R} = diag(R_1, R_2, R_3)$ and also $\hat{\Pi} = diag(\Pi_1, \Pi_2, \Pi_3)$ whereas $\mathbf{U}_0 = (U_1, 0, 0)$.

For a rigid shape (i.e., no motion-dependence) $\hat{\Pi} = 0$ and one obtains a standard instability criterion i.e, the instability of a short-side motion (see, Lamb, 1945), given by

$$T_2 > T_1. \quad (30)$$

For a non-zero Π the instability criterion is qualitatively changed, leading to the negativity of the corresponding determinant and resulting in

$$(AU_1\Pi_2 - R_2B)^2 < 4AU_1(T_2 - T_1). \quad (31)$$

It is clear that (31) cannot be satisfied for a high enough velocity U_1. Thus an increase of a velocity can stabilize the motion of a body with a *motion-dependent* shape, even for the cases known as unstable translations for a *rigid* body.

7. SUMMARY

A number of generalization of the existing Kirchhoff dynamic model for the motion of autonomous underwater vehicle is suggested towards more realistic hydrodynamic rheology of a surrounding fluid compared with a purely potential framework. It is found that an accounting of a vorticity shedding phenomena can strongly change the "potential" stability criteria, leading, for example, to an instability of a steady translation of a sphere. It is shown also that a motion-dependence of a solid shape can serve for a stabilization of some types of translational motion.

8. REFERENCES

Abzug, M. J. & Larrabee, E. E. 1997 *Airplane Stability and Control* Cambridge University Press

Aref, H., & Jones S., W. 1993 Chaotic motion of a solid through ideal fluid. *Physics of Fluids A* 5 **12** 3026-3028

Aref, H., & Jones S., W. 1994 Motion of a solid through ideal fluid. Proc. *DCAMM 25th Anniversary meeting*, Vedbak, Denmark

Aref, H., Aluru, N., Balachandar, S., Liu C. Selig M. & Bar-Cohen Y. 1998 Control of Unmanned Vehicles Using Adaptive Internal Actuation *Technical Proposal*

Birkhoff, G. 1960 *Hydrodynamics -A Study in Logic, Fact and Similitude. Revised Version,* Princeton University Press, Ch. 6

Galper A. R. and Miloh T, 1995 Dynamical equations for the motion of a deformable body in an arbitrary potential non-uniform flow field, *J. Fluid Mech.* **295**,

Galper A. R. & Miloh T. 1998 Motion stability of a deformable body in an ideal fluid with applications to the N spheres problem *Phys. of Fluids* , Vol. 1, **10**, 119-131.

Galper A. R. & Miloh T. 1999 Hydrodynamics and stability of a deformable body moving in the proximity of interfaces *Phys. of Fluids* , Vol. 3, **11**.

Gitman D., & Tyutin I. 1990 *Quantization of Fields with Constraints* Springer - Verlag

Holmes Ph., Jenkins J., & Leonard N. 1998 Dynamics of the Kirchhoff Equations I. *Physica D.* **118**, 311-342.

H. Lamb, 1945 *Hydrodynamics.* (New York, Dover).

N. E. Leonard 1996 "Stabilization of Steady Motion of an Underwater Vehicle" Proceeding of the 35th Conference on Decision and Control, Kobe, Japan

N. E. Leonard 1997a "Stabilization of Underwater Vehicle Dynamics with Symmetry-Breaking Potentials" *Systems and Control Letters* 32: (1) 35-42 OCT 26

N. E. Leonard 1997b "Stability of a Bottom-heavy Underwater Vehicle" *Automatic* **33**, pp. 331-346.

N. E. Leonard and J. E. Marsden, 1997 Stability and drift of underwater vehicle dynamics: mechanical systems with rigid motion symmetry, *Physica D* **105**, 130

T. Miloh and A. R. Galper, 1993 Self-propulsion of a maneuvering deformable body in a perfect fluid, *Proc. Roy. Soc. London* A **442**, 273

Proceedings of the Symposium on Autonomous Underwater Vehicle Technology 1996 IEEE New York, NY, USA

Weiss P., 1998 The puzzle of flutter and tumble *Science News* **154**, 285-87

Wu J. Z. 1994 A theory of three-dimensional interfacial vorticity dynamics. *Phys. of Fluids* **7**, 2375-2395

Wu J. Z. & Wu J. M. 1996 Vorticity dynamics on boundaries. *Advanced in Applied Mechanics.* **32**

TIME-OPTIMAL CONTROL FOR UNDERWATER VEHICLES

M. Chyba *,[1] N.E. Leonard *,[1] E.D. Sontag **,[2]

* Department of Mechanical and Aerospace Engineering, Princeton University, Princeton NJ 08544
** Department of Mathematics, Rutgers University, New Brunswick NJ 08903

Abstract: This paper addresses time-optimal control problems for a special class of controlled mechanical systems, underwater vehicles. Lie algebras associated to mechanical systems enjoy certain very special properties, which, together with the maximum principle, allow the deduction of information regarding the structure of singular extremals, and in particular of time-optimal trajectories. We apply the general theory to a model of an underwater vehicle and illustrate our results with some simulations. We consider the fully actuated and the underactuated situations.

Keywords: Time-Optimal Control, Controlled Mechanical Systems, Underwater Vehicle

1. INTRODUCTION

We consider the time-optimal problem for controlled mechanical systems, studying both the fully actuated and underactuated cases. Fully actuated systems were studied in (Sontag and Sussmann, 1986; Sontag, 1989), with the 2-link manipulator as a motivating example. As in that work, we base our study on the Pontryagin maximum principle, which gives necessary conditions for trajectories of control systems to be optimal with respect to a criterion such as energy or time. One may associate, to any given controlled mechanical system, a set of vector fields describing a control system. Then, from the special form of the Lie algebra generated by these vector fields, one can extract from the maximum principle information on the structure of the optimal trajectories. More precisely, according to the maximum principle, a time-optimal trajectory can be lifted to the cotangent bundle of its phase space as a trajectory of a constrained Hamiltonian system. This trajectory, combined with the corresponding control, is called an extremal. When the pointwise constraints in the maximum principle are nontrivial, one has nonsingular trajectories. These lead to boundary-valued controls, determined by the signs of the associated switching functions. However, it is well-known that an optimal trajectory may well be singular; that is, switching functions may vanish identically along the trajectory. The characterization of such trajectories, which is the question addressed in this paper, is in general a highly nontrivial problem.

Our study is motivated by the problem of motion planning for underwater vehicles. The optimality of a trajectory with respect to a given criterion is in many ways interesting. For underwater vehicles, it is not only worthwhile to minimize the energy expended to realize the desired motion, but one is often also concerned with the continuous power consumption of all devices on board, such as sensors and computers. Because of this last quantity ("hotel load"), the amount of time used

[1] Supported by the Office of Naval Research Grant N00014-98-1-0649 and the National Science Foundation Grant BES-9502477
[2] Supported in part by US Air Force Grant F49620-98-1-0242

to travel between two desired configurations becomes a minimization criteria for our performance requirements. In this paper we deal with the time-optimal problem, although ideally a combination of the energy and the time should be considered. This will be analyzed in future work.

2. TIME OPTIMALITY FOR MECHANICAL SYSTEMS

Let us consider the *Lagrangian*

$$L(q,\dot{q}) = \frac{1}{2}\dot{q}^t M(q)\dot{q} - V(q)$$

defined on the phase space, where the symmetric positive definite $n \times n-$matrix $M(q)$ is called the *inertia matrix* and $V(q)$ is the *potential energy*. We assume M and V to be smooth with respect to q. The equations of motion are given by

$$Q(q(t))u(t) = M(q(t))\ddot{q}(t) + N(q(t),\dot{q}(t)) \quad (1)$$

where $N(q,\dot{q}) = C(q,\dot{q})\dot{q} - \frac{\partial V}{\partial q}^t(q)$ is a vector of dimension n quadratic in $\dot{q}$, $Q(q)$ is a smooth $n \times m$-matrix of rank m and the *control* $u : [0,T] \to \mathcal{U} \subset I\!R^m$ is a measurable bounded function. A system described by equation (1) will be referred to as a *controlled mechanical system*. When $m = n$, the system is fully actuated. It is underactuated if $m < n$. We will assume the domain of control $\mathcal{U}$ is defined by the following constraints: $\alpha_i \le u_i \le \beta_i, i = 1,\cdots,m$. Such a system can be rewritten as a *control system* of dimension $2n$ as follows. Let us introduce new variables $x = (y_1, y_2) \in I\!R^{2n}$ by $y_1 = \psi(q)$, $y_2 = P(q)\dot{q}$ where $q \mapsto \psi(q)$ is a smooth diffeomorphism and $P(q)$ is a smooth $n \times n$-invertible matrix. As we will see with the example on underwater vehicles, it is sometimes convenient to consider such changes of variables, instead of simply $x = (q,\dot{q})^t$, to have simpler formulas. Then, from equation (1) we have

$$\dot{x}(t) = f(x(t)) + G(x(t))u(t) \quad (2)$$

where $f(x)$ is a smooth vector field called the drift and

$$G(x) = \begin{pmatrix} 0 \\ PM^{-1}Q \end{pmatrix}_{2n\times m} \quad (3)$$

where the argument of the matrices is $\psi^{-1}(y_1)$. Let us denote by $(g_i(x))_{i=1,\cdots,m}$ the columns of $G(x)$; they are smooth functions of the n first state variables.

The goal is to determine the *time-optimal* trajectories: given $x_0, x_1 \in I\!R^{2n}$, find an admissible control $u(.)$ such that the corresponding trajectory steers the system (2) from x_0 to x_1 in the *shortest time*.

From the maximum principle, see (Pontryagin *et al.*, 1962), if the control $u : [0,T] \to \mathcal{U}$ and the corresponding trajectory $x(.)$ solution of (2) defined on the same interval are time-optimal, then there exists an absolutely continuous vector $\lambda : [0,T] \to I\!R^{2n}$, $\lambda(t) \neq 0$, such that for $j = 1,\cdots,2n$

$$\dot{x}_j = \frac{\partial H}{\partial \lambda_j}, \qquad \dot{\lambda}_j = -\frac{\partial H}{\partial x_j} \quad (4)$$

holds almost everywhere, where $H(\lambda,x,u) = \lambda^t f(x) + \sum_{i=1}^m \lambda^t g_i(x)u_i$ is the Hamiltonian function, and the following maximum condition is satisfied a.e.:

$$H(\lambda(t),x(t),u(t)) = \max_{v\in\mathcal{U}} H(\lambda(t),x(t),v) = \lambda_0,$$

$\lambda_0 \le 0$. The vector $\lambda(.)$ is called the *adjoint* vector. A triple (x,λ,u) which satisfies the maximum principle, in the sense just stated, is called an *extremal*. When the constant λ_0 is zero, the extremal is called *abnormal*. If there exists a nonempty interval $[t_1,t_2]$ such that $\phi_i(t) = \lambda(t)^t g_i(x(t)) = 0$ is identically zero, the corresponding extremal is called *u_i-singular* on $[t_1,t_2]$. If $\phi_i(t) \neq 0$ for almost all $t \in [t_1,t_2]$, the extremal is called *u_i-nonsingular* on $[t_1,t_2]$. The maximum principle implies that on a u_i-nonsingular extremal the component u_i of the control is bang-bang, which means that it takes its values in $\{\alpha_i,\beta_i\}$ for almost every $t \in [t_1,t_2]$. A u_i-nonsingular extremal is called *u_i-regular* if u_i has a finite number of switches. An extremal is said to be *nonsingular* if it is u_i-nonsingular for all i (resp. *regular*); otherwise, it is called *singular*. It is *totally singular* if it is u_i-singular for all i.

The structure of the time-optimal trajectories is related to the Lie algebra generated by the vector fields $f, g_1,\cdots,g_m$ describing the control system. Hence, our results involve conditions on the Lie brackets of these vector fields. The first result states that under certain Lie bracket conditions, an extremal cannot be totally singular. Let us define $I = \{1,\cdots,m\}$.

Theorem 1. Assume the extremal (x,λ,u) is u_i-singular for all $i \neq k$. If along the extremal there exists a nonempty subset $J \subset I$, $k \notin J$, such that

(1) If $l \in J$, then $[g_j,[f,g_l]] \in Span\{g_1,\cdots g_m\}$ for all $j \in I$;
(2) $Span\{g_j,[f,g_j],ad_f^2 g_l\}_{j\in I; l\in J} = I\!R^{2n}$

then it cannot be u_k-singular. More precisely, the component u_k of the control is bang-bang with a finite number of switches. Moreover, if we have for $j,s \in I, s \neq k$ and $l \in J$, that

$$Span\{g_j,[f,g_s],ad_f^2 g_l\}_{jsl} = I\!R^{2n}$$

along the extremal, then u_k is constant.

We remark that in the fully actuated case ($m = n$), both hypotheses on the Lie brackets in the theorem are automatically satisfied. Indeed, because we are dealing with Lagrangian mechanical system and because $\frac{\partial \psi}{\partial \theta}$ and $P(\theta)$ are invertible, the vector fields $\{g_i, [f, g_i]\}_{i=1,\cdots,n}$ constitute a frame for $\mathbb{R}^{2n}$. As a consequence, for a fully actuated controlled mechanical system, an extremal can never be totally singular. Moreover, the first assumption is true with $J = I \backslash \{k\}$, but this property is relevant in the fully actuated case only for the second part of the theorem. In (Chyba *et al.*, 2000), the theorem is extended and we consider u_i-singular extremals for no more than l controls with $l < m - 1$.

Notice that if the assumptions in Theorem 1 are satisfied, we have local accessibility from any point on the extremal. In particular, for a fully actuated system, this is true for any point $x \in \mathbb{R}^{2n}$.

In (Sontag and Sussmann, 1986), the authors give an algorithm to compute the $n-1$-singular components of the control in the fully actuated situation. The computations in the underactuated case are similar. Let us briefly review the algorithm because it will be useful for the study of underwater vehicles. Assume the extremal to be u_i-singular for $i \neq k$ on the interval $[t_1, t_2]$. It is equivalent to say that the functions $\phi_i = \lambda^t g_i(x)$, $i \neq k$, are identically zero along the extremal on this time interval. Computing the derivative of ϕ_i with respect to time and using the property $[g_j, g_i] = 0$ for all i, j, we obtain that $\lambda^t [f, g_i](x) = 0$, $i \neq k$, for all $t \in [t_1, t_2]$. This is an absolutely continuous function, so we can then compute the second derivative of ϕ_i and we have

$$\lambda^t ad_f^2 g_i(x) + \sum_{j=1}^{m} \lambda^t [g_j, [f, g_i]](x) u_j = 0$$

for almost every t, $i \neq k$. If the matrix $(\lambda^t [g_j, [f, g_i]](x))_{i,j \neq k}$ is invertible along the extremal, the $m-1$ singular components of the control are determined by the equations above. We remark that if assumption 1 of Theorem 1 is satisfied for all $l \neq k$ (which is the case for a fully actuated system) there exist smooth functions $\alpha_{ij}^s(x)$ such that $[g_j, [f, g_i]](x) = \sum_{s=1}^{m} \alpha_{ij}^s(x) g_s(x)$ for $i \neq k$. Assume $\lambda^t g_k(x)$ not vanishing. Then u_k is a constant and using the fact that the switching functions ϕ_i vanish for $i \neq k$, the equations reduce to

$$-\frac{\lambda^t ad_f^2 g_i(x) + \alpha_{ik}^k \lambda^t g_k(x) u_k}{\lambda^t g_k(x)} = \sum_{j \neq k} \alpha_{ij}^k u_j \quad (5)$$

and the remaining $m - 1$ components of the control are completely determined if $\det(\alpha_{ij}^k) \neq 0$. From Theorem 1, the points in the phase space corresponding to a switch of the nonsingular component of the control u_k are contained in the complement S_k^c of the set

$$S_k = \{x;\ Span\{g_i, [f, g_l], ad_f^2 g_j\} = \mathbb{R}^{2n}\}$$

where $i, l \in I; l \neq k; j \in J$.

The following slight modification of Theorem 1 will be useful for our applications in the underactuated case. It provides a gives result in cases when the assumption 2 is not satisfied.

Theorem 2. Assume the extremal (x, λ, u) is u_i-singular for all $i \neq k$. If along the extremal there exist nonempty subsets $J_1, J_2 \subset I$, $k \notin J_1$ and $J_2 \subset J_1$ such that

(1) $l \in J_1, r \in J_2$, we have

$$[g_j, [f, g_l]] \in Span\{g_r, [f, g_r]\}_{r \in I},$$

$$[g_j, ad_f^2 g_v] \in Span\{g_r, [f, g_r], ad_f^2 g_l\}_{r \in I, l \in J_1}$$

for all $j \in I$;

(2) $Span\{g_j, [f, g_j], ad_f^2 g_l, ad_f^3 g_v\}_{jlv} = \mathbb{R}^{2n}$ for $j \in I, l \in J_1, v \in J_2$

then the component u_k of the control is bang-bang with a finite number of switches.

Both situations, fully actuated and underactuated, will be illustrated on the example of underwater vehicles.

3. APPLICATION TO UNDERWATER VEHICLES

The dynamics of underwater vehicles can be described with the equations of motion as follows, see (Leonard, 1997) for more details. The position and orientation of the underwater vehicle are identified with the group of rigid-body motions in $\mathbb{R}^3$: $SE(3) = \{(R, b);\ R \in SO(3), b \in \mathbb{R}^3\}$. If we define $\Omega = (\Omega_1, \Omega_2, \Omega_3)$ and $v = (v_1, v_2, v_3)$ to be the angular and translational velocity of the vehicle in body coordinates, then the kinematic equations are

$$\dot{R} = R\hat{\Omega}, \qquad \hat{\Omega} = \begin{pmatrix} 0 & -\Omega_3 & \Omega_2 \\ \Omega_3 & 0 & -\Omega_1 \\ -\Omega_2 & \Omega_1 & 0 \end{pmatrix} \quad (6)$$

$$\dot{b} = Rv. \quad (7)$$

We begin by assuming the vehicle is submerged in an infinitely large volume of incompressible, irrotational and inviscid fluid at rest at infinity. Let us denote by Π and P the angular and linear components of the impulse (roughly equivalent to momentum) of the body-fluid system with respect to the body-frame. The Kirchhoff equations of motion for a rigid body in such an ideal fluid are given by

$$\dot{\Pi} = \Pi \times \Omega + P \times v + \tau, \quad \dot{P} = P \times \Omega + \mathcal{F} \quad (8)$$

where τ and $\mathcal{F}$ are external torque and force vectors. τ and $\mathcal{F}$ can be used to include gravity, buoyancy and control forces as well as viscous forces such as lift and drag. Notice that Π and P can be computed from the total kinetic energy T of the body fluid system: $T = \frac{1}{2}(\Omega^t J\Omega + 2\Omega^t Dv + v^t Mv)$, where M and J are respectively the body-fluid mass and inertia matrices. In (Leonard, 1997), the Hamiltonian structure of the dynamics of the underwater vehicle is described.

In this paper, we consider a neutrally buoyant, uniformly distributed, ellipsoidal vehicle moving in the vertical inertial plane and we neglect viscous effects. We denote by (x, z) the absolute position of the vehicle where x is the horizontal position and z the vertical position. θ describes its orientation in this plane so that $q = (x, z, \theta)$. Given our assumptions, M and J are diagonal and $D = 0$ so that for our vehicle restricted to the plane $T = \frac{1}{2}(I\Omega^2 + m_1 v_1^2 + m_3 v_3^2)$ where I is the body-fluid moment of inertia in the plane and m_1, m_3 are body-fluid mass terms in the body horizontal and vertical directions, respectively. We assume that $m_1 \neq m_3$, i.e., the planar vehicle is not a circle. We choose the state vector to be $w = (x, z, \theta, v_1, v_3, \Omega)$. Here Ω is the scalar angular rate in the plane. The equations of motion are

$$\begin{aligned} \dot{x} &= \cos\theta v_1 + \sin\theta v_3 \\ \dot{z} &= \cos\theta v_3 - \sin\theta v_1 \\ \dot{\theta} &= \Omega \\ \dot{v}_1 &= -v_3\Omega\frac{m_3}{m_1} \\ \dot{v}_3 &= v_1\Omega\frac{m_1}{m_3} \\ \dot{\Omega} &= v_1 v_3\frac{m_3 - m_1}{I} \end{aligned} \tag{9}$$

With respect to the notations of Section 2 we have $P^{-1}(q) = Q(q)$ where $Q(q)$ is a rotation matrix $\begin{pmatrix} R(\theta) & 0 \\ 0 & 1 \end{pmatrix}$, $R(\theta) = \begin{pmatrix} \cos\theta & \sin\theta \\ -\sin\theta & \cos\theta \end{pmatrix}$. This formulation of the problem leads to simpler computations and to nice geometric interpretation of the results.

3.1 *Fully actuated case*

We first consider the fully actuated case: the control vector is $u = (u_1, u_2, u_3)$ where u_1 is a force in the body 1-axis, u_2 is a force in the body 3-axis and u_3 is a pure torque in the plane. Accordingly, with the drift vector field f given by the planar equations of motion (9), the input vector fields g_i are given by $g_1 = (0, 0, 0, \frac{1}{m_1}, 0, 0)^t$, $g_2 = (0, 0, 0, 0, \frac{1}{m_3}, 0)^t$, $g_3 = (0, 0, 0, 0, 0, \frac{1}{I})^t$. The Lie brackets of f and these vector fields are given by

$$\begin{aligned} [f, g_1] &= \frac{1}{m_1}(-\cos\theta, \sin\theta, 0, 0, -\Omega\frac{m_1}{m_3}, -v_3\alpha)^t, \\ [f, g_2] &= \frac{1}{m_3}(-\sin\theta, -\cos\theta, 0, \Omega\frac{m_3}{m_1}, 0, -v_1\alpha)^t, \\ [f, g_3] &= \frac{1}{I}(0, 0, -1, v_3\frac{m_3}{m_1}, -v_1\frac{m_1}{m_3}, 0)^t \end{aligned}$$

where $\alpha = \frac{m_3 - m_1}{I}$. The assumptions of Theorem 1 are satisfied. We can conclude that there is no totally singular extremal.

Let us study the extremals with 2 components of the control being singular. In order to determine the values of these singular components, we will use the algorithm described in Section 2, in particular equation (5). First, we have to compute the Lie brackets of length 3. It is easy to verify that we have $[g_j, [f, g_j]] = 0$ for $j = 1, 2, 3$, that the following relations are satisfied

$$[g_2, [f, g_1]] = [g_1, [f, g_2]] = -\frac{m_3 - m_1}{m_1 m_3} g_3, \tag{10}$$

$$[g_3, [f, g_1]] = [g_1, [f, g_3]] = -\frac{1}{I} g_2, \tag{11}$$

$$[g_3, [f, g_2]] = [g_2, [f, g_3]] = \frac{1}{I} g_1, \tag{12}$$

and that the Lie brackets of the form $ad_f^2 g_i$ are

$$\begin{aligned} ad_f^2 g_1 &= (\sin\theta\Omega\beta, \cos\theta\Omega\beta, v_3\frac{\alpha}{m_1}, -\frac{\Omega^2}{m_1} \\ &\quad -v_3^2\frac{m_3\alpha}{m_1^2}, 0, 0)^t \\ ad_f^2 g_2 &= (-\cos\theta\Omega\beta, \sin\theta\Omega\beta, v_1\frac{\alpha}{m_3}, 0, \\ &\quad -\frac{\Omega^2}{m_3} + v_1^2\frac{m_1\alpha}{m_3^2}, 0)^t \\ ad_f^2 g_3 &= \frac{1}{I}(\sin\theta v_1\gamma_1 - \cos\theta v_3\gamma_2, \sin\theta v_3\gamma_2 \\ &\quad + \cos\theta v_1\gamma_1, 0, 0, 0, (v_1^2\frac{m_1}{m_3} - v_3^2\frac{m_3}{m_1})\alpha)^t \end{aligned}$$

where $\beta = \frac{1}{m_1} + \frac{1}{m_3}$, $\gamma_1 = \frac{m_1}{m_3} - 1$, $\gamma_2 = \frac{m_3}{m_1} - 1$ and α is as above.

3.1.1. *u_1, u_2-singular extremals* We start with the u_1, u_2-singular extremals. From Theorem 1, the switches of u_3 are located in the set of points $S_3^c = A_1 \cap A_2$ where

$$A_i = \{w;\ \det(g_1, g_2, g_3, [f, g_1], [f, g_2], ad_f^2 g_i) = 0\},$$

$i = 1, 2$. Computing these sets, we find $A_1 = \{w;\ \frac{m_1 - m_3}{I m_1^3 m_3^2} v_3 = 0\}$, $A_2 = \{w;\ \frac{m_1 - m_3}{I m_1^2 m_3^3} v_1 = 0\}$, and

$$S_3^c = \{w;\ v_1 = v_3 = 0\}.$$

By using the Hamiltonian equations (4), we can prove that along a u_1, u_2-singular extremal there is at most one switch and there is one only if $v_1 \equiv v_3 \equiv 0$ along the extremal. This corresponds to a pure rotating motion for the underwater vehicle. Intuitively, this motion is a good candidate for optimality, this will be discussed in (Chyba *et al.*, 2000). The switching time is determined by the initial and final configurations of the phase space.

Let us determine the singular controls u_1 and u_2 when v_1 and v_3 are not both identically zero (it is equivalent to say that the velocities cannot vanish at the same time along the extremal). By Theorem 1, u_3 is constant. Using the algorithm described in Section 2, the equations (10),(11),(12) and the fact that $\lambda^t g_i = 0$, $i = 1, 2$, we have

$$u_1 = \frac{\lambda^t ad_f^2 g_2}{\lambda^t g_3} \frac{m_1 m_3}{m_3 - m_1} = \Omega v_3 (1 + \frac{m_3}{m_1}) + v_1 \frac{\lambda_3}{\lambda_6},$$

$$u_2 = \frac{\lambda^t ad_f^2 g_1}{\lambda^t g_3} \frac{m_1 m_3}{m_3 - m_1} = \Omega v_1 (\frac{m_1}{m_3} - 1) + v_3 \frac{\lambda_3}{\lambda_6}.$$

We remark that these equations are well defined as $\lambda^t g_3 \neq 0$ and the matrix

$$(\alpha_{ij}^3) = \begin{pmatrix} 0 & \dfrac{m_1 - m_3}{m_1 m_3} \\ \dfrac{m_1 - m_3}{m_1 m_3} & 0 \end{pmatrix}$$

is always invertible. If we denote by $\lambda = (\lambda_1, \cdots, \lambda_6)^t$ the adjoint vector, the differential Hamiltonian system corresponding to the u_1, u_2-singular extremals without switches on u_3 can be reduced from 12 to 8 equations. We can prove that

$$\begin{aligned} \dot{v}_1 &= \Omega v_3 + v_1 \frac{\lambda_3}{\lambda_6} \\ \dot{v}_3 &= -\Omega v_1 + v_3 \frac{\lambda_3}{\lambda_6} \\ \dot{\Omega} &= v_1 v_3 \frac{m_3 - m_1}{I} + \frac{u_3}{I} \\ \dot{\lambda}_3 &= \frac{m_1 - m_3}{I} \lambda_6 (v_1^2 + v_3^2) \\ \dot{\lambda}_6 &= \lambda_3 \end{aligned}$$

where u_3 is a constant, $(\dot{x}, \dot{z}, \dot{\theta})$ being given by the equations of motion (9).

Simulations. To perform numerical simulations, we consider an ellipsoidal underwater vehicle with body 1-axis length $L_1 = 0.457$ m and body 3-axis length $L_3 = 0.305$ m. The corresponding mass and inertia terms for the vehicle in water are $m_1 = 13.2$kg, $m_3 = 25.6$kg and $I = 0.12$kg-m^2. In Figure 1 we represent the motion described by the projection of a u_1, u_2-singular extremal in the x, z-plane as well as the motion of the angle θ. Figures 2 and 3 show respectively the corresponding velocities and the values of the singular controls (along this extremal, the function λ_6 is strictly positive and we assume the bound on u_3 is 0.8 N-m). Remark that because the velocities v_1 and v_3 are not identically zero, the underwater vehicle does not move tangentially or perpendicularly to the trajectory. The velocity v_3 vanishes exactly at the point where the trajectory in the x, z-plane has a vertical tangent. At this point the underwater vehicle has made a rotation of $\frac{\pi}{2}$ from its intial configuration $(x, z, \theta) = (0, 0, 0)$.

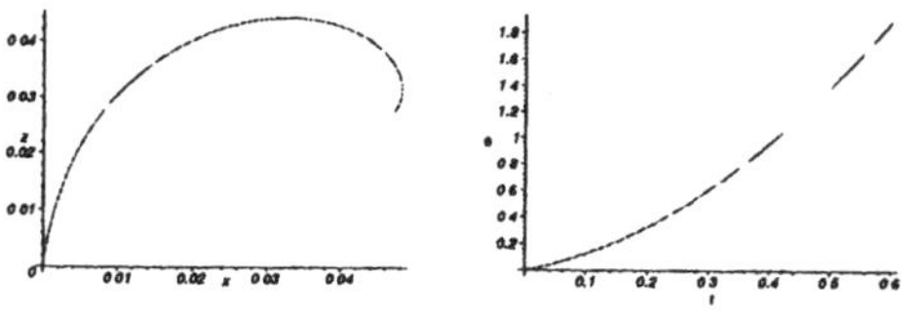

Fig. 1. projection in the x, z-plane; angle

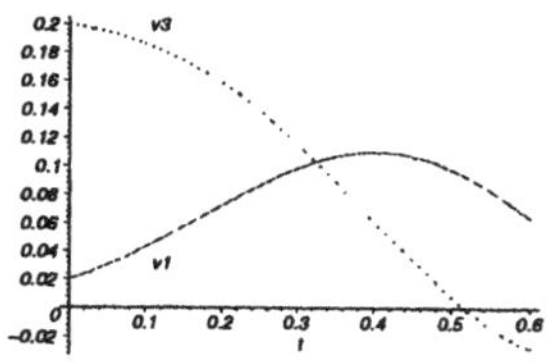

Fig. 2. translational velocities v_1, v_3

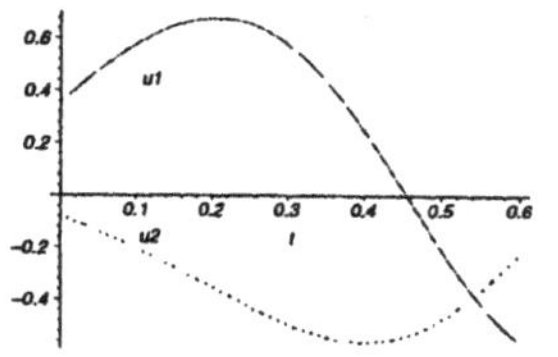

Fig. 3. components u_1, u_2 of the control

3.1.2. u_1, u_3*-singular extremals* In this case, the switches of the control u_2 are located in $S_2^c = B_1 \cap B_3$ where

$$B_i = \{w;\ \det(g_1, g_2, g_3, [f, g_1], [f, g_3], ad_f^2 g_i) = 0\},$$

$i = 1$ or 3. We have $B_1 = \{w;\ \frac{\Omega}{I^2 m_1^2 m_3}(\frac{1}{m_1} + \frac{1}{m_3}) = 0\}$, $B_3 = \{w;\ \frac{v_1}{I^2 m_1^2 m_3}(\frac{m_1}{m_3} - 1) = 0\}$ and $S_2^c = \{w;\ v_1 = \Omega = 0\}$. If Ω and v_1 are identically zero, there is at most one switch and the motion corresponds to a translation in the direction of the vertical axis of the body frame coordinates. If the extremal belongs to the set S_2, then u_2 is constant, $\lambda^t g_2 \neq 0$, and the singular components of the control u_1 and u_3 are given by $u_1 = \frac{\lambda^t ad_f^2 g_3}{\lambda^t g_2} I$ and $u_3 = \frac{\lambda^t ad_f^2 g_1}{\lambda^t g_2} I$.

3.1.3. u_2, u_3*-singular extremals* This situation is completely analogous to the preceding one. The switches of u_1 belongs to $S_1^c = C_2 \cap C_3$, $C_i = \{w;\ \det(g_1, g_2, g_3, [f, g_2], [f, g_3], ad_f^2 g_i) = 0\}$, $i = 2, 3$ with $C_2 = \{w;\ \frac{\Omega}{I^2 m_1 m_3^2}(\frac{1}{m_1} + \frac{1}{m_3}) = 0\}$, $C_3 = \{w;\ \frac{v_3}{I^2 m_1 m_3^2}(\frac{m_3}{m_1} - 1) = 0\}$ and $S_1^c = \{w;\ v_3 = \Omega = 0\}$. When $\Omega \equiv v_3 \equiv 0$, then the motion has at most one switch and is a translation in the direction of the horizontal axis of the body frame coordinates. The singular controls u_2, u_3 for an extremal in the set S_1 are given by $u_2 = -\frac{\lambda^t ad_f^2 g_3}{\lambda^t g_1} I$ and $u_3 = -\frac{\lambda^t ad_f^2 g_2}{\lambda^t g_1} I$.

3.2 *Underactuated case*

Let us consider the case when the control is 2-dimensional $u = (u_2, u_3)$ and $g_2 = (0, 0, 0, -\frac{\sin\theta}{m_1}, \frac{\cos\theta}{m_3}, 0)^t$, $g_3 = (0, 0, 0, 0, 0, \frac{1}{I})^t$. In this case u_2 corresponds to a force in the *inertial* vertical direction and u_3 is again a pure torque in the plane. This choice of control is motivated by a buoyancy driven underwater glider. The Lie brackets involving only the vector fields f and g_3 are given in Section 3.1. We also have

$$[f, g_2] = (\frac{\delta \sin 2\theta}{2}, -\frac{\sin^2\theta}{m_1} - \frac{\cos^2\theta}{m_3}, 0, 0, 0, h)^t$$

where $\delta = (\frac{1}{m_1} - \frac{1}{m_3})$, $\alpha = \frac{m_3 - m_1}{I}$ and $h = \alpha(\frac{v_3 \sin\theta}{m_1} - \frac{v_1 \cos\theta}{m_3})$. The Lie brackets of length 3 are given by

$$[g_2, [f, g_2]] = \frac{\alpha I \sin 2\theta}{m_1 m_3} g_3,$$

$$[g_3, [f, g_2]] = [g_i, [f, g_3]] = 0, \quad i = 2, 3$$

and

$$ad_f^2 g_2 = (\delta \cos 2\theta\Omega, -\delta \sin 2\theta\Omega, -h, \frac{v_3 m_3}{m_1} h, -\frac{v_1 m_1}{m_3} h, \tilde{h})^t$$

where $\tilde{h} = 2\Omega\alpha(\frac{v_1 \sin\theta}{m_3} + \frac{v_3 \cos\theta}{m_1})$. Assumption 1 of Theorem 1 is satisfied with $J = \{2, 3\}$ but the vector fields $\{g_i, [f, g_i], ad_f^2 g_i\}_{i=2,3}$ are linearly dependent everywhere. Therefore we cannot use Theorem 1 to draw conclusions. However, if we assume $\Omega \neq 0$ we have

$$[g_2, ad_f^2 g_3] = h[f, g_3] - \frac{I}{\Omega} ad_f^2 g_2.$$

As $[g_3, ad_f^2 g_3] = 0$, assumption 1 of Theorem 2 is then satisfied. By computing the vector field $ad_f^3 g_3$ we can prove that if $w \notin U$ where $U = \{w;\ \Omega = 0 \text{ and/or } v_1 = v_3 = 0\}$, then assumption 2 of Theorem 2 is verified. As a consequence, if the extremal belongs to the complement of the set U, it cannot be totally singular. More precisely, if one control is singular the other one must be bang-bang with a finite number of switches. More details and the case when the extremal does not belong to U^c everywhere are analysed in (Chyba *et al.*, 2000).

4. CONCLUSION

In this paper, we study the singular extremals of a controlled mechanical system for the time-optimal problem. Characterizing singular extremals is a first step in the study of optimal strategies. Under Lie-algebraic assumptions, we proved that along $m-1$-singular extremals there can only be a finite number of switches. Although one might expect the existence in general of only finitely many switches along optimal motions, many optimal control problems lead to accumulation points of switches, see for instance (Sussmann, 1997), a behavior which is often referred to as the "Fuller phenomenon". Indeed, the maximum principle gives no direct information on the way to connect the pieces of an extremal or a bound on the number of switches, along an optimal trajectory for an optimal control. In (Sussmann, 1979), the author gives conditions for scalar control systems, under which a bang-bang theorem with bounds on the number of switches can be proved. This result can be applied to the underwater vehicle in the underactuated situation with only one control; see (Chyba *et al.*, 2000) for more details.

Optimizing a cost function is clearly interesting by itself. Furthermore, in dealing with the motion planning problem in the presence of obstacles, steering methods based on optimality guarantee the convergence of general motion planning schemes, see (Laumond *et al.*, 1990; Sekhavat and Laumond, 1996).

5. REFERENCES

Chyba, M., N.E. Leonard and E.D. Sontag (2000). Time-optimality for controlled mechanical systems: Application to underwater vehicles. *in preparation.*

Laumond, J.P., M. Taix and P. Jacobs (1990). A motion planner for car-like robots based on a mixed global/local approach. *IEEE Int. Conf. Intelligent Robot and Systems* pp. 765–773.

Leonard, N.E. (1997). Stability of a bottom-heavy underwater vehicle. *Automatica* **33**, 331–346.

Pontryagin, L.S., B. Boltyanski, R. Gamkrelidze and E. Michtchenko (1962). The mathematical theory of optimal processes.

Sekhavat, S. and J.P. Laumond (1996). Topological property for collision-free nonholonomic motion planning: The case of sinusoidal inputs for chained form systems. *Proc. IEEE Int. Conf. on Robotics and Automation* pp. 1692–1697.

Sontag, E.D. (1989). Remarks on the time-optimal control of a class of Hamiltonian systems. *IEEE Conf. on Decision and Control* pp. 317–221.

Sontag, E.D. and H.J. Sussmann (1986). Time-optimal control of manipulators. *Proc. IEEE Int. Conf. on Robotics and Automation* pp. 1692–1697.

Sussmann, H.J. (1979). A bang-bang theorem with bounds on the number of switchings. *SIAM J.Control and Optimization* **17**, 629–651.

Sussmann, H.J. (1997). The Markov-Dubins problem with angular acceleration control. *IEEE Conf. on Decision and Control* pp. 2639–2643.

AFFINE CONNECTION CONTROL SYSTEMS

Andrew D. Lewis *

* *Department of Mathematics & Statistics, Queen's University, Kingston, ON K7L 3N6, Canada*

Abstract: The affine connection formalism provides a useful framework for the investigation of a large class of mechanical systems. Mechanical systems with kinetic energy Lagrangians and possibly with nonholonomic constraints are fit naturally into the formalism, and some results are stated in the areas of controllability and optimal control for affine connection control systems. *Copyright ©2000 IFAC*

Keywords: mechanical systems, connections, controllability, optimal control

1. INTRODUCTION

Control theory for mechanical systems is a topic which has received a certain degree of attention in the past decade. Apart from the fact that many control applications are mechanical in nature, the differential geometric flavour of aspects of both mechanics and nonlinear control theory provides compelling theoretical motivation for this interest.

When one begins to think about studying control theory for mechanical systems, one must in some sense choose something from each of the two subjects—control theory and mechanics—in order to initiate the investigation. In the author's own work, the choice from control theory was nonlinear controllability, and the choice from mechanics was so-called "simple mechanical systems," those whose Lagrangians are kinetic minus potential energy. When this choice is made, the equipment made available by the choice often dictates the nature of the results one obtains. For example, in the author's initial work in the area, the investigation of a certain type of controllability for simple mechanical systems led to the "symmetric product." A readable overview of this work with Richard Murray may be found in a recent SIAM Review paper (Lewis and Murray, 1999). Interestingly, the symmetric product also appears in the somewhat unrelated work of Crouch (1981). The symmetric product is an object which one might consider in terms of affine differential geometry, quite apart from any mechanical or control theoretic context. This is done, along with other related work, in the paper (Lewis, 1998). This differential geometric interpretation of the symmetric product may then be brought *back* to control theory, and provides an interesting interpretation of reachable sets for simple mechanical control systems (Lewis and Murray, 1997*b*).

Recent work of the author has centred on optimal control theory for mechanical systems, again utilising the affine connection framework. Here one can produce a geometric version of the Maximum Principle where the essential ingredient is the so-called "adjoint Jacobi equation" which forms that part of the equation describing the evolution of the adjoint vector which is independent of the cost function, i.e., that part which depends only on the control system. This equation is, as the name suggests, related to the Jacobi equation of geodesic variation. The full development is somewhat lengthy, and here we present an abbreviated form of these results, noting that their full statement has not yet appeared in the literature.

The impression might then be gotten that there is a connection, possibly a deep one, between affine differential geometry and control theory for simple mechanical systems. This impression has been reinforced by other work in this area, for example (Noakes *et*

[1] Partially supported by the Natural Sciences and Engineering Research Council

al., 1989; Crouch and Silva Leite, 1991; Bloch and Crouch, 1995*b*; Lewis, 1997*b*; Bullo *et al.*, 1997; Lewis, 1999*c*; Bullo, 1999*a*; Baillieul, 1999; Bullo, 1999*b*). We will touch on the content of some of these and other papers as they comes up in the sequel.

2. MECHANICAL SYSTEMS AS AFFINE CONNECTION CONTROL SYSTEMS

We begin by motivating a discussion of what we shall in Section 3 refer to as "affine connection control systems." We do this by showing how affine connections naturally arise when discussing mechanical systems with kinetic energy Lagrangians. Thus we have a configuration manifold Q which possesses a Riemannian metric g giving rise to the Lagrangian $L(v_q) = \frac{1}{2}g(v_q, v_q)$. Often, of course, one is interested in including potential forces in the Lagrangian, and it is indeed possible to do this. For example, the issue of potential shaping is touched upon in some recent work (Weibel and Baillieul, 1998; Bullo, 1999*b*; Bloch *et al.*, 1999). The initial work on controllability of Lewis and Murray (1997*a*) also includes potential forces.

Our aim is to show, in as concise a manner as possible, how one makes the step from mechanics to affine differential geometry. To do this we use local coordinates $(q^1, \ldots, q^n)$ for Q and remark that a simple calculation shows that if we take $L = \frac{1}{2}g_{ij}\dot{q}^i\dot{q}^j$ (here we use the summation convention where repeated indices are summed) we obtain the equivalence between the corresponding Euler-Lagrange and the equations

$$\ddot{q}^i + \overset{g}{\Gamma}{}^i_{jk}\dot{q}^j\dot{q}^j = 0, \quad i = 1, \ldots, n, \tag{1}$$

where

$$\overset{g}{\Gamma}{}^i_{jk} = \frac{1}{2}g^{i\ell}\Big(\frac{\partial g_{\ell j}}{\partial q^k} + \frac{\partial g_{\ell k}}{\partial q^j} - \frac{\partial g_{jk}}{\partial q^\ell}\Big).$$

The n^3 functions $\overset{g}{\Gamma}{}^i_{jk}$, $i, j, k = 1, \ldots, n$ are the *Christoffel symbols* for an affine connection $\overset{g}{\nabla}$ called the *Levi-Civita connection*. The equation (1) asserts that the solutions of the Euler-Lagrange equations are exactly *geodesics* for the affine connection $\overset{g}{\nabla}$. A thorough discussion of affine connections may be found in (Kobayashi and Nomizu, 1963), but we shall say a few cursory words on the subject in the next section.

Interestingly, it is also true that one may use the affine connection formalism to describe the motion of a system with a kinetic energy Lagrangian, and with constraints linear in velocity. This idea seems to originate with Synge (1928), and the author was first made aware of it via the paper of Bloch and Crouch (1995*a*) at the CDC in 1995. Briefly the setup is this. One has a distribution D on Q and the system's velocities are constrained to lie in D. To write the equations of motion for such systems, one uses the Lagrange-d'Alembert principle, and doing so yields equations in geodesic form for the affine connection defined by $\nabla_X Y = \overset{g}{\nabla}_X Y + (\overset{g}{\nabla}_X P^\perp)(Y)$ where $P^\perp$ is the orthogonal projection onto the orthogonal complement of D. Details aside, the bottom line is that for mechanical systems, constrained or unconstrained, with kinetic energy Lagrangians, the unforced equations are geodesic equations, and as such are typically written $\nabla_{c'(t)}c'(t) = 0$ for a curve $c\colon I \to Q$ with $I \subset \mathbb{R}$ an interval.

The preceding discussion is of a purely mechanical nature, and has naught to do with control theory. To make the mechanical systems into *control* systems, we add forces to the picture. The idea is to select one input force associated with each direction in which one may apply a force, and take as the control force a linear combination of these forces. We make the assumption that the directions in which one may apply forces vary only with the configuration of the system, and not with, for example, velocity or time. Doing so means that control forces may be modelled as vector fields $\{Y_1, \ldots, Y_m\}$ on Q, and the control system we consider is this one:

$$\nabla_{c'(t)}c'(t) = u^a(t)Y_a(c(t)). \tag{2}$$

This control system forms the basis of discussion for the remainder of the paper.

3. AFFINE CONNECTION CONTROL SYSTEMS

The motivation of Section 2 serves to provide a mechanical backdrop for this section, where we look formally, but briefly, at affine connections and control systems formed by them. We refer to the bibliography, principally (Kobayashi and Nomizu, 1963), for details on the plethora of under-justified assertions in this section.

An *affine connection* on Q assigns to each pair of vector fields X and Y on Q a vector field $\nabla_X Y$ with the assignment satisfying

(1) the map $(X, Y) \mapsto \nabla_X Y$ is $\mathbb{R}$-bilinear,
(2) $\nabla_{fX}Y = f\nabla_X Y$, and
(3) $\nabla_X(fY) = f\nabla_X Y + (\mathscr{L}_X f)Y$

for all vector fields X and Y on Q, all functions f on Q, and where $\mathscr{L}_X$ denotes the Lie derivative with respect to X. The association of this abstract object with the Christoffel symbols of the previous section occurs when we choose local coordinates $(q^1, \ldots, q^n)$. Then we may apply the affine connection to a pair of coordinate vector fields $\frac{\partial}{\partial q^i}$:

$$\nabla_{\frac{\partial}{\partial q^j}}\frac{\partial}{\partial q^k} = \Gamma^i_{jk}\frac{\partial}{\partial q^i},$$

which provides the definition of the Christoffel symbols for an arbitrary affine connection. The vector field $\nabla_X Y$ is called the *covariant derivative* of Y with respect to X, and if we define ∇_X on smooth functions by $\nabla_X f = \mathscr{L}_X f$, then we may extend ∇_X to a derivation on the tensor algebra over Q in the usual manner. That is, it is possible to define the covariant derivative $\nabla_X A$ of an (r,s) tensor field A with respect to X. The *torsion tensor* and the *curvature tensor* for an affine connection are the $(1,2)$ tensor field T on Q and the $(1,3)$ tensor field R on Q defined by

$$T(X,Y) = \nabla_X Y - \nabla_Y X - [X,Y]$$
$$R(X,Y)Z = \nabla_X \nabla_Y Z - \nabla_Y \nabla_X Z - \nabla_{[X,Y]} Z,$$

respectively. The affine connection $\overset{g}{\nabla}$ associated with the Riemannian metric g is then the unique torsion-free affine connection with the property that $\overset{g}{\nabla}_X g = 0$ for every vector field X on Q. Thus far the objects we have discussed are classical to affine differential geometry. However the control considerations of (Lewis and Murray, 1997*a*) led to the symmetric product of two vector fields which we define by $\langle X : Y \rangle = \nabla_X Y + \nabla_Y X$. The geometric meaning of the symmetric product has been provided by the author (Lewis, 1998), and we refer to (Crouch, 1981) for an appearance of the symmetric product in another setting.

A *geodesic* for an affine connection ∇ is a curve $c\colon I \to Q$ from an interval $I \subset \mathbb{R}$ which has the property that $\nabla_{c'(t)} c'(t) = 0$ for $\in I$. If $c_s\colon I \to Q$ is a smooth family of geodesics defined for $s \in]-\varepsilon, \varepsilon[$, and which has the property that $c_0 = c$, we define a *Jacobi field* along c to be any vector field along c of the form

$$\xi(t) = \left.\frac{\mathrm{d}}{\mathrm{d}s}\right|_{s=0} c_s(t).$$

Jacobi fields may be shown to satisfy the *Jacobi equation:*

$$\nabla^2_{c'(t)} \xi(t) + R(\xi(t), c'(t))c'(t) + \nabla_{c'(t)}(T(\xi(t), c'(t))) = 0.$$

Thus the Jacobi equation may be thought of as the equation of "geodesic variation."

Yet another piece of equipment is the *geodesic spray* associated with an affine connection ∇. This is the second-order vector field with the property that the projection of its integral curves to Q are geodesics for ∇. In coordinates we have

$$Z = v^i \frac{\partial}{\partial q^i} - \Gamma^i_{jk} v^j v^k \frac{\partial}{\partial v^i}.$$

If one wishes to treat the system (2) as a first-order control affine nonlinear control system on TQ, the vector field Z is the drift vector field. In a treatment such as this, the control vector fields are the m vector fields on TQ denoted Y_a^{lift}, $a = 1, \ldots, m$ which are the vertical lifts of the control vector fields on Q to TQ. With all this notation the control system (2) is given by

$$\dot{v}(t) = Z(v(t)) + u^a(t) Y_a^{\mathrm{lift}}(v(t))$$

in first-order form on TQ.

Let us now formally define the class of control systems we discuss, and introduce some associated notation. An *affine connection control system* is comprised of a triple $(Q, \nabla, \mathscr{Y})$ where Q is a finite-dimensional manifold, ∇ is an affine connection on Q, and $\mathscr{Y} = \{Y_1, \ldots, Y_m\}$ is a collection of vector fields on Q. To an affine connection control system we obviously associate a control system of the form (2). The inputs $u\colon I \to \mathbb{R}^m$ we consider are measurable functions from an interval $I \subset \mathbb{R}$, and we denote this set of inputs by $\mathscr{U}$. A *controlled trajectory* is then a pair (c,u) where $u\colon I \to \mathbb{R}^m$ is a map from the set of inputs $\mathscr{U}$ and where $c\colon I \to Q$ satisfies (2). A *controlled arc* is a controlled trajectory defined on a compact interval. Let $q_0, q_1 \in Q$ and let $v_{q_0} \in T_{q_0}Q$ and $v_{q_1} \in T_{q_1}Q$. We denote by $\mathrm{Carc}(\Sigma, q_0, q_1)$ (resp. $\mathrm{Carc}(\Sigma, v_{q_0}, v_{q_1})$) the set of controlled arcs (u,c) for which $c(a) = q_0$ (resp. $c'(a) = v_{q_0}$) and $c(b) = q_1$ (resp. $c'(b) = v_{q_1}$) where (u,c) is defined on some interval $[a,b]$. One can also consider controlled arcs defined on a *fixed* interval, but for brevity we do not do so. For $q \in Q$, U a neighbourhood of q, and $T > 0$ define

$$\mathcal{R}_Q^U(q,T) = \{c(T) \mid \exists\, u \in \mathscr{U} \text{ so that } (u,c) \text{ is a controlled trajectory defined on } [0,T] \text{ with } c'(0) = 0_q \text{ and } c(t) \in U\}.$$

Thus $R_Q^U(q,T)$ are those configurations reachable in exactly time T from q starting with zero initial velocity (0_q is the zero vector in T_qQ). Note that we do not restrict the final velocity. We also define

$$\mathcal{R}_Q^U(q, \leq T) = \bigcup_{0 \leq t \leq T} \mathcal{R}_Q^U(q,t).$$

4. CONTROLLABILITY FOR AFFINE CONNECTION CONTROL SYSTEMS

The initial impetus for the investigation of the class of systems we are describing was controllability theory. Here one wishes to exploit the special structure of the system, in conjunction with well-known techniques in nonlinear controllability, to derive useful controllability tests. In this section we suppose that we have an analytic affine connection control system $(Q, \nabla, \mathscr{Y} = \{Y_1, \ldots, Y_m\})$. The controllability tests from nonlinear control theory which we adapt are those from standard accessibility theory (e.g., Sussmann and Jurdjevic, 1972) and the small-time local controllability results of Sussmann (1987). Because affine connection control systems, although they have a state space of TQ, are defined in terms of objects on the configuration manifold Q, one would like to obtain results which are expressed in terms of conditions on Q. Furthermore, it makes a great deal of sense to formulate controllability definitions on the configuration manifold.

Definition 1. Let $\Sigma = (Q, \nabla, \mathscr{Y})$ be an affine connection control system.

(i) Σ is *locally configuration accessible* at q if for each neighbourhood U of q there exists $T > 0$

so that $\mathcal{R}_Q^U(q,t)$ has nonempty interior for each $0 < t \leq T$.

(ii) Σ is *locally configuration controllable* at q if it is locally configuration accessible at q and if for each neighbourhood U of q there exists $T > 0$ so that $q \in \text{int}(\mathcal{R}_Q^U(q,t))$ for each $0 < t \leq T$.

(iii) Σ is *equilibrium controllable* if for each $q_1, q_2 \in Q$ there exists a controlled trajectory (u,c) defined on $[0,T]$ so that $q_1 = c(0)$, $q_2 = c(T)$, $c'(0) = 0_{q_1}$ and 0_{q_2}.

One wishes to study this so-called "configuration controllability" for a couple of reasons. One of the most compelling is that it is possible for a system to be locally configuration accessible (resp. controllable) and *not* be locally accessible (resp. controllable) in state space. If one is interested only in what is happening to configurations anyway, it makes sense to have controllability definitions and tests which reflect this. Also, as we shall see, it is possible to provide simple tests for the configuration controllability definitions we provide.

Let us first look at the accessibility conditions which were first presented in complete form in (Lewis and Murray, 1997*a*). We let $\text{Sym}(\mathscr{Y})$ be the smallest subspace of vector fields on Q which contains $\mathscr{Y}$ and which is closed under symmetric product, and we let $\text{Lie}(\text{Sym}(\mathscr{Y}))$ be the smallest subspace of vector fields on Q which contains $\text{Sym}(\mathscr{Y})$ and which is closed under Lie bracket. We then define

$$\text{Lie}(\text{Sym}(\mathscr{Y}))_q = \{X(q) \mid X \in \text{Lie}(\text{Sym}(\mathscr{Y}))\}.$$

For analytic systems we have the following sharp result for configuration accessibility.

Theorem 2. An analytic affine connection control system $\Sigma = (Q, \nabla, \mathscr{Y})$ is locally configuration accessible at $q \in Q$ if and only if $\text{Lie}(\text{Sym}(\mathscr{Y}))_q = T_qQ$.

For C^∞ systems, the condition $\text{Lie}(\text{Sym}(\mathscr{Y}))_q = T_qQ$ is sufficient but not necessary for local configuration accessibility. This is explored in detail in the original paper of Lewis and Murray—the proof requires delving into detail the bracket computations for an affine connection control system when thought of as a nonlinear control system. In that paper, a good deal of effort is also devoted to deriving the conditions for local configuration accessibility for systems with potential energy; this significantly complicates the statement of the result, so we shall not go into this here. We also mention that the geometric interpretation of the symmetric product (see Lewis, 1998) may be applied to give an interpretation of Theorem 2 (Lewis and Murray, 1997*b*).

The configuration controllability result requires that we look at the behaviour of certain types of symmetric product. A symmetric product from the set $\mathscr{Y} = \{Y_1, \ldots, Y_m\}$ is *bad* if it contains an even number of each of the vector fields Y_a, $a = 1, \ldots, m$, and is *good* if it is not bad. The *degree* of a symmetric product is the total number of vector fields of which it comprised.[2] For example, the symmetric product $\langle Y_a : \langle Y_b : Y_a \rangle\rangle$ is good and of degree 3 and the symmetric product $\langle\langle Y_a : Y_b\rangle : \langle Y_a : Y_b\rangle\rangle$ is bad and of degree 4.

Theorem 3. Let $\Sigma = (Q, \nabla, \mathscr{Y})$ be an affine connection control system.

(i) Σ is locally configuration controllable at $q \in Q$ if every bad symmetric product P can be written at q as a finite $\mathbb{R}$-linear combination of good symmetric products of degree lower than that of P.

(ii) Σ is equilibrium controllable if the conditions of (i) are satisfied at each $q \in Q$.

In practice it suffices to check the condition (i) on a set of symmetric products which form a basis for $\text{Sym}(\mathscr{Y})_q$. These conditions are a watered down version of the general controllability conditions proved by Sussmann (1987). It would be interesting if one were able to provide sharp conditions for analytic systems in terms only of symmetric products (and potentially Lie brackets) of the input vector fields $\mathscr{Y}$. That the control Lie algebra for affine connection control systems is simpler than that for generic nonlinear control systems may be observed by noting that sharp conditions exist for single-input systems (Lewis, 1997*a*). In this case one can show that the system is locally configuration controllable if and only if $\dim(Q) = 1$, i.e., we have controllability only in the uninteresting case when the system is fully actuated.

Unfortunately, space does not permit the presentation of examples which illustrate the range of behaviours relating to configuration accessibility and controllability, and we refer the reader to the references for details concerning some simple physical examples.

5. OPTIMAL CONTROL FOR AFFINE CONNECTION CONTROL SYSTEMS

The structure of mechanical systems in some sense makes the problem of optimal control a natural one. In particular, if one possesses a kinetic energy Riemannian metric, this encourages the definition of some natural cost functions. It turns out to be possible to formulate for affine connection control systems a powerful version of the Maximum Principle. The general buildup is rather substantial, and we refer to (Lewis, 1999*b*) for details. The referenced work relies on a wonderful paper by Sussmann (1997) which provides a general and geometric formulation for the Maximum Principle.

[2] Of course, we are speaking imprecisely here—to do this rigorously requires that one work with free algebraic quantities, and this is explained in detail by Lewis and Murray (1997*a*).

To eliminate a significant part of the generality of (Lewis, 1999*b*) we choose a specific cost function, and the one with which it is the simplest to deal. We let $\Sigma = (Q, \nabla, \mathscr{Y} = \{Y_1, \ldots, Y_m\})$ be a C^∞ affine connection control system and we suppose that we have a Riemannian metric g on Q. We define a cost function $F: \mathbb{R}^m \times Q \to \mathbb{R}$ by $F(u,q) = g(u^a Y_a(q), u^b Y_b(q))$. The objective of the optimal control problem is to minimise

$$J(u,c) = \int_I F(u(t), c(t))\,dt$$

over a class of controlled trajectories (u,c) defined on an interval I.

Let us precisely state the optimal control problems we shall look at.

Definition 4. Let $\Sigma = (Q, \nabla, \mathscr{Y})$ be an affine connection control system, let $q_0, q_1 \in Q$, and let $v_{q_0} \in T_{q_0}Q$ and $v_{q_1} \in T_{q_1}Q$.

(i) A controlled arc (u_*, c_*) is a *solution of* $\mathscr{F}(\Sigma, v_{q_0}, v_{q_1})$ if $J(u_*, c_*) \le J(u,c)$ for every $(u,c) \in \mathrm{Carc}(\Sigma, v_{q_0}, v_{q_1})$.

(ii) A controlled arc $\gamma = (u_*, c_*)$ is a *solution of* $\mathscr{F}(\Sigma, q_0, q_1)$ if $J(u_*, c_*) \le J(u,c)$ for every $(u,c) \in \mathrm{Carc}(\Sigma, q_0, q_1)$.

Lewis (1999*b*) provides a general statement of the Maximum Principle for affine connection control systems which we will here distill to the cost function at hand. We let $P: TQ \to TQ$ be the orthogonal projection onto the distribution spanned by the input vector fields $\mathscr{Y}$, and define a $(2,0)$ tensor field h on Q by

$$h(\alpha, \beta) = g^{-1}(P^*(\alpha), P^*(\beta))$$

for one-forms α and β, where g^{-1} is the $(2,0)$ tensor field associated with g (i.e., that one whose components in coordinates are the inverse of the components of g) and where $P^*: T^*Q \to T^*Q$ is the dual endomorphism of P. We also define some notation for the curvature and torsion tensors. For vector fields X, Y, and Z and a one-form α define R^* and T^* by

$$\langle T^*(\alpha, X); Y \rangle = \langle \alpha; T(Y,X) \rangle,$$
$$\langle R^*(\alpha, X)Y; Z \rangle = \langle \alpha; R(Z,X)Y \rangle$$

where $\langle \cdot ; \cdot \rangle$ denotes the natural pairing of a one-form and a vector field.

The following result describes the normal extremals for the optimal problems defined above.

Theorem 5. Let $\Sigma = (Q, \nabla, \mathscr{Y})$ be a C^∞ affine connection control system. Suppose that (u,c) is a solution of one of the two problems of Definition 4 with u and c defined on $[a,b]$. If (u,c) is a normal extremal then c is of class C^∞ and there exists a C^∞ one-form field λ along c so that c and λ together satisfy the differential equations

$$\begin{aligned} &\nabla_{c'(t)} c'(t) = -h^\sharp(\lambda(t)) \\ &\nabla^2_{c'(t)} \lambda(t) + R^*(\lambda(t), c'(t))c'(t) - \\ &\quad T^*(\nabla_{c'(t)}\lambda(t), c'(t)) = \tfrac{1}{2}\nabla h(\lambda(t), \lambda(t)) - \\ &\quad T^*(\lambda(t), h^\sharp(\lambda(t))). \end{aligned} \tag{3}$$

If $\gamma = (u,c)$ is a solution of $\mathscr{F}(\Sigma, q_0, q_1)$ then we additionally have $\lambda(a) = 0$ and $\lambda(b) = 0$.

The left-hand side of the second of equations (3) is the *adjoint Jacobi equation*, and bears some formal resemblance to the Jacobi equation. The precise nature of this resemblance takes some work to fully elucidate, and we refer to (Lewis, 1999*b*) for the details.

If we suppose $\mathscr{Y}$ contains vector fields which span T_qQ for each $q \in Q$, i.e., that the system is fully actuated, and that ∇ is the Levi-Civita connection associated with the Riemannian metric g used in the definition of the cost function, then, with some straightforward manipulations, we recover the results of Noakes *et al.* (1989) and Crouch and Silva Leite (1991)—namely the necessary condition for minimisers is

$$\overset{g}{\nabla}{}^3_{c'(t)} c'(t) + R(\overset{g}{\nabla}_{c'(t)} c'(t))c'(t) = 0.$$

Of course one can define other natural cost functions, and explore other questions associated with optimal control for affine connection control systems. There is much to be done here, and doubtless some beautiful results await discovery.

6. CLOSING REMARKS

The idea of this paper is to give a flavour of the types of results which one may obtain using the affine connection formalism. That this formalism has an intimate relation to mechanics, as outlined in Section 2, makes the exploration of this affine connection setting a bit more enticing. The emphasis here was on affine connection control systems as a class of systems in and of itself, and dues to space constraints, not much attention has been paid to mechanics per se. However, the reader is invited to look into the references for examples of how the theory may be applied to physical examples. Even some simple examples exhibit surprisingly subtle behaviour.

Also, we have only touched on certain aspects of the author's own work to date. A potentially promising area which has not been discussed is that of whether affine connection control systems may simplify. In particular, the issue of "feedback transformations" for affine connection control systems is one which is unexplored. The paper (Lewis, 1999*c*) gives a very limited type of simplification, and a general setting for feedback transformations for affine connection control systems is described by the author (Lewis, 1999*a*).

Another possible avenue of exploration is that concerning the rôle of symmetry. In mechanics, symmetry plays an important rôle, but how this impinges on control theory, and in particular on the affine connection setting, has not been explored (but see Bloch and Crouch, 1995*b*; Bloch and Crouch, 1998). A case which *has* seen some attention is that when Q is a Lie group, and the problem data is left-invariant. In this case, Bullo *et al.* (1997) provide some explicit trajectory generation algorithms, including an exponential stabilisation algorithm. A different approach for systems with symmetry and nonholonomic constraints is taken by Ostrowski and Burdick (1997).

We hope we have explicated the value of the affine connection formalism in studying control theory for a class of mechanical systems.

7. REFERENCES

Baillieul, John (1999). The geometry of controlled mechanical systems. In: *Mathematical Control Theory*. pp. 322–354. Springer-Verlag. New York-Heidelberg-Berlin.

Bloch, Anthony M. and Peter E. Crouch (1995*a*). Another view of nonholonomic mechanical control systems. In: *Proceedings of the 34th IEEE CDC*. IEEE. New Orleans, LA. pp. 1066–1071.

Bloch, Anthony M. and Peter E. Crouch (1995*b*). Nonholonomic and vakonomic control systems on Riemannian manifolds. *SIAM J. Control Optim.* **33**(1), 126–148.

Bloch, Anthony M. and Peter E. Crouch (1998). Newton's law and integrability of nonholonomic systems. *SIAM J. Control Optim.* **36**(6), 2020–2039.

Bloch, AnthonyM., Naomi Ehrich Leonard and Jerrold E. Marsden (1999). Potential shaping and the method of controlled Lagranigans. In: *Proceedings of the 38th IEEE CDC*. IEEE. Phoenix, AZ. pp. 1652–1658.

Bullo, Francesco (1999*a*). A series describing the evolution of mechanical control systems. In: *Proceedings of the IFAC World Congress*. IFAC. Beijing, China. pp. 479–485.

Bullo, Francesco (1999*b*). Vibrational control of mechanical systems. Submitted to *SIAM Journal on Control and Optimization*.

Bullo, Francesco, Naomi E. Leonard and Andrew D. Lewis (1997). Controllability and motion algorithms for underactuated Lagrangian systems on Lie groups. To appear in *IEEE Transactions on Automatic Control*.

Crouch, Peter E. (1981). Geometric structures in systems theory. *Proc. IEE-D* **128**(5), 242–252.

Crouch, Peter E. and Fátima Silva Leite (1991). Geometry and the dynamic interpolation problem. In: *Proceedings of the ACC*. Boston, MA. pp. 1131–1136.

Kobayashi, Shoshichi and Katsumi Nomizu (1963). *Foundations of Differential Geometry*. Vol. I and II of *Interscience Tracts in Pure and Applied Mathematics*. Interscience Publishers. New York.

Lewis, Andrew D. (1997*a*). Local configuration controllability for a class of mechanical systems with a single input. In: *Proceedings of the ECC*. Brussels, Belgium.

Lewis, Andrew D. (1997*b*). Simple mechanical control systems with constraints. To appear in *IEEE Transactions on Automatic Control*.

Lewis, Andrew D. (1998). Affine connections and distributions with applications to nonholonomic mechanics. *Rep. Math. Phys.* **42**(1/2), 135–164.

Lewis, Andrew D. (1999*a*). The category of affine connection control systems. in preparation.

Lewis, Andrew D. (1999*b*). The geometry of optimal control for affine connection control systems. in preparation.

Lewis, Andrew D. (1999*c*). When is a mechanical control system kinematic?. In: *Proceedings of the 38th IEEE CDC*. IEEE. Phoenix, AZ. pp. 1162–1167.

Lewis, Andrew D. and Richard M. Murray (1997*a*). Controllability of simple mechanical control systems. *SIAM J. Control Optim.* **35**(3), 766–790.

Lewis, Andrew D. and Richard M. Murray (1997*b*). Decompositions of control systems on manifolds with an affine connection. *Systems Control Lett.* **31**(4), 199–205.

Lewis, Andrew D. and Richard M. Murray (1999). Controllability of simple mechanical control systems. *SIAM Rev.* **41**(3), 555–574.

Noakes, Lyle, Greg Heinzinger and Brad Paden (1989). Cubic splines on curved spaces. *IMA J. Math. Control Inform.* **6**(4), 465–473.

Ostrowski, James Patrick and Joel Wakeman Burdick (1997). Controllability tests for mechanical systems with constraints and symmetries. *J. Appl. Math. and Comp. Sci.* **7**(2), 101–127.

Sussmann, Héctor J. (1987). A general theorem on local controllability. *SIAM J. Control Optim.* **25**(1), 158–194.

Sussmann, Héctor J. (1997). An introduction to the coordinate-free maximum principle. In: *Geometry of Feedback and Optimal Control* (B. Jakubczyk and W. Respondek, Eds.). pp. 463–557. Dekker Marcel Dekker. New York.

Sussmann, Héctor J. and Velimir Jurdjevic (1972). Controllability of nonlinear systems. *J. Differential Equations* **12**, 95–116.

Synge, John Lighton (1928). Geodesics in nonholonomic geometry. *Math. Ann.* **99**, 738–751.

Weibel, Steven P. and John Baillieul (1998). Averaging and energy methods for robust open-loop control of mechanical systems. In:. *Essays on Mathematical Robotics, (Minneapolis, MN, 1993)*. Vol. 104 of *The IMA Volumes in Mathematics and its Applications*. pp. 203–269. Springer-Verlag. New York-Heidelberg-Berlin.

CONTROL ALGORITHMS USING AFFINE CONNECTIONS ON PRINCIPAL FIBER BUNDLES

Hong Zhang **James P. Ostrowski**

General Robotics, Automation, Sensing, and Perception Lab
University of Pennsylvania, 3401 Walnut Street,
Philadelphia, PA 19104-6228
E-mail: {hozhang, jpo}@grip.cis.upenn.edu

Abstract: In this paper, we develop tools for studying the control of underactuated mechanical systems that evolve on a configuration space with a principal fiber bundle structure. We utilize the affine connection to describe the equations of motion and study the controllability of such systems. The invariance of the system with respect to the action of a Lie group leads to a reduced form of the affine connection. This connection, and its corresponding symmetric product, are then used to develop constructive, cyclic control inputs to steer the mechanical system. We explore the development of the equations, based on the original formulation by Bullo, and its application to the steering of a blimp-like vehicle with directional thrusters.

Keywords: Nonlinear Control, Reduction, Mechanical Systems

1. INTRODUCTION

We are motivated in this paper by a class of problems from robotics in which the dynamics evolve on a product bundle between a Lie group and a general "shape" manifold. This leads us to consider mechanical systems on principal fiber bundles, in which the motion of the system is generated through a complex interaction of thrusts/forces and internal changes in the shape or configuration of the robot. There is an extensive literature studying such systems, including kinematic versions (Kelly and Murray 1995), dynamic systems purely on Lie groups (Bullo *et al.* 1998), and dynamic systems with nonholonomic constraints (Ostrowski 1995).

In studying the controllability of a mechanical system, classical tools from nonlinear control theory (Sastry 1999) suggest that one compute the closure of the Lie bracket of all the control inputs. When the control inputs enter in as second order inputs, such as forces or torques instead of velocity, this procedure requires the system to be transformed into a kinematic one, with the associated cost of increasing the rank of state space and losing much of the intrinsic structure of the mechanical system. However, work by Lewis and Murray (Lewis and Murray 1995) has shown that a proper geometric mechanism for examining controllability of mechanical systems with dynamic inputs is through the use of the affine connection and symmetric product. They showed that if the closure of the Lie brackets and symmetric products of all the control inputs is full rank, the system will be configuration controllable. Bullo, Leonard, and Lewis later applied these results to underactuated Lagrangian systems evolving on a Lie group (Bullo *et al.* 1998). Further, they took advantage of the special Lie group structure to derive an algorithm for generating the control inputs that lead to motion along the directions generated through the operation of the symmetric product.

Lagrangian reduction provides a powerful tool in analyzing mechanical systems that have components evolving on a Lie group. Generally, the Lie group describes the position and orientation of the system, while the remaining variables constitute an internal shape space. Some examples of the shape variables that result are the thruster angle of a blimp (Zhang and Ostrowski 1999), the leg angle of a robot leg (Lewis and Murray 1995), or the wheel direction angles of a snakeboard (Ostrowski 1998). Through a local trivialization, we can use the internal symmetries of the system to decouple the dynamics into two parts, vertical and horizontal, and connect them with a mechanical connection (or constraints) (Bloch *et al.* 1996). Likewise, we can apply the same technique to the computation of covariant derivatives by finding the vertical and horizontal parts, and then use the Lie bracket and symmetric product to take advantage of the geometric structure of the system.

In this paper, we review the some of the tools used in Lagrangian reduction and the development of the mechanical connection, and then derive the explicit expression for the covariant derivative of vectors on a principal fiber bundle (or its local trivialization). We present the computations in terms of local forms of the quantities that arise, including the mechanical connection and the locked inertia tensor, since these allow for a reduced and compact representation, and are the form which would generally be used for computations. We also derive explicit cyclic control inputs for driving such a system in the case that the metric is constant, i.e., the Lagrangian function is independent of the configuration variables. We apply these results to the blimp example to show that it is straightforward to generate explicit open-loop controls for such systems based on the affine connection.

In Section 2 we give some background on fiber bundles and the method of reduction for Lagrangian systems. In Section 3 we present the reduced version of the Levi-Civita affine connection for principal fiber bundles. Section 4 provides a sketch of how the general notions of controllability for a mechanical system reduce for a principal fiber bundle. Finally, in Section 5, we use a perturbation expansion to derive control algorithms based on small-amplitude cyclic inputs to generate directions associated with the symmetric product.

2. REDUCTION IN LAGRANGIAN SYSTEMS

Traditionally, we can describe a dynamic system (including robotic systems and many other examples) using Lagrange's equation:

$$\frac{d}{dt}\left(\frac{\partial L}{\partial \dot{q}}\right) - \frac{\partial L}{\partial q} = \tau, \tag{1}$$

where $L(q, \dot{q}, t) = T - V$ is the Lagrangian of the mechanical system, T and V are the kinetic energy and potential energy of the system, $q \in Q$ are the generalized coordinates in the configuration space, and τ is the function of external forces (including torques). For a mechanical control system, we can further divide the external forces into the control inputs $C(q, \dot{q}, t, u)$ and any additional uncontrolled forces F_{unc}. For simplicity, we will assume in this paper that $V = 0$ and the system has no other external forces so that $F_{unc} = 0$. The case that $V \neq 0$ can be treated in the general setting of affine connections, although it requires significant additional work. We will term a simple mechanical system as one for which we can expand the expression of the Lagrangian to be

$$L(q, \dot{q}) = \frac{1}{2}\dot{q}^T \mathbb{G} \dot{q}, \tag{2}$$

where $\mathbb{G}$ is the metric of the system.

For a mobile robot such as a blimp or a satellite with either a thruster or a rotor, examination of the configuration space shows that it can be divided into two subspaces. One is the ***pose space***, G, which describes position and orientation. This can generally be considered to be a Lie group (with group elements $g \in G$). The other one is the ***base (or shape) space***, M, which describes the internal shape of the robot or any other variables, and it can be any manifold (with element denoted by $r \in M$). Thus, we can decompose the original configuration space into the structure $Q = G \times M$.

Mathematically, this structure is called a ***trivial principal fiber bundle***, where Q is said to be composed of ***fibers***, G, over the ***base*** manifold M. It is a special case of a principal fiber bundle, where M is just the quotient space $M = Q \;/\; G$. However, for a general system, we can always choose a local trivialization to give the simple product bundle structure (locally) as $Q = G \times M$. Therefore, we can rephrase the Lagrangian with the divided variables as:

$$L(q, \dot{q}) = L(r, g, \dot{r}, \dot{g}) = \left(\dot{g}^T \; \dot{r}^T\right) \mathbb{G} \begin{pmatrix} \dot{g} \\ \dot{r} \end{pmatrix}.$$

For a mechanical system, the motion along the fibers corresponds to the translation and rotation of the system, while the motion on the base space represents the internal changes in shape. These loosely correspond to the motion in the vertical and horizontal spaces, respectively. Strictly speaking, if we construct the projection map as $\pi : Q \to Q/G$, then the vertical space of the bundle at the point q is the kernel $\ker(T_q\pi)$, or the tangent space to the group orbit through q. The choice of horizontal space is then generally derived through either the metric, as shown below, or through additional constraints.

The process of reduction requires that the Lagrangian be independent of the group variables. A Lagrangian L is said to be ***G-invariant*** under the group action if L is invariant under the induced action of G on TQ. For many robotic locomotion systems, including even free-flying or neutrally buoyant robots, the system is G-invariant. In such cases, the dependence of L on g appears only through the velocity $\xi = g^{-1}\dot{g}$ where $\xi \in \mathfrak{g}$, is an element of the Lie algebra of G. This induces a function, which we call the ***reduced Lagrangian***, l, as

$$\begin{aligned} L(q, \dot{q}) &= l(g^{-1}g, r, g^{-1}\dot{g}, \dot{r}) = l(r, \xi, \dot{r}) \\ &= \left(\xi^T \; \dot{r}^T\right) \hat{\mathbb{G}}(r) \begin{pmatrix} \xi \\ \dot{r} \end{pmatrix}, \end{aligned} \tag{3}$$

with a reduced metric (Ostrowski 1995):

$$\hat{\mathbb{G}}(r) = \begin{pmatrix} I(r) & I(r)\,A(r) \\ A(r)^T I(r) & m(r) \end{pmatrix}. \tag{4}$$

In the reduced metric, there are two important quantities, namely the local forms of the ***locked inertia tensor***, I, and the ***mechanical connection***, A.

In a mechanical system, the locked inertia tensor is a map, $I(r) : \mathfrak{g} \times \mathfrak{g} \to \mathbb{R}$, that defines a inner product on $\mathfrak{g}$. This fact is important, since we can use I below to define a connection restricted to the vertical subbundle. In a G-invariant mechanical system, $I(r)$ has the interpretation of the inertia of the system with respect to the body-fixed frame when all the shape variables (such as the joints) are fixed.

On the other hand, the ***mechanical connection***, A, is a Lie algebra-valued one-form on M. It can be considered as a map $A(r) : T_rM \to \mathfrak{g}$ for $r \in M$. In a robotic system, A relates internal shape changes, $\dot{r}$, to changes in pose, $\dot{g}$. It should be noted that both I and A are functions of base variables r only, and for this reason are referred to as "local" quantities.

The mechanical connection leads to a natural way to define the horizontal subspace of the system. A vector, $(\dot{g}, \dot{r}) \in T_gG \times T_rM$ is said to be ***horizontal*** if it satisfies $g^{-1}\dot{g} + A(r)\dot{r} = 0$. This can be extended pointwise to define a horizontal subbundle that is complementary to the vertical subbundle.

An alternative (and more intrinsically "motivated") derivation of these equations can be derived by viewing things according to this splitting. Note that a general vector, $(\dot{g}, \dot{r})$ can be split into horizontal and vertical components as $(-gA(r)\dot{r}, \dot{r})$ and $(\dot{g} + gA(r)\dot{r}, 0)$. If write the reduced metric in terms of $\dot{r}$ and the ***locked body angular velocity***, $\Omega = \xi + A(r)\dot{r}$, (it is induced from the body momentum via the locked inertia tensor) then the reduced metric is block diagonalized to

$$\begin{pmatrix} I(r) & 0 \\ 0 & m(r) - A(r)I(r)A(r) \end{pmatrix} = \begin{pmatrix} I(r) & 0 \\ 0 & \Delta(r) \end{pmatrix}.$$

The term Δ will also be used below, in order to define a reduced affine connection on T_rM.

We also note that there are certain properties based on taking derivatives of the connection and the locked inertia tensor that play an important role in the analysis here. The ***covariant exterior derivative*** (or ***local curvature form***) of the local form of the mechanical connection, A, is

$$B(v, w) = DA(v, w) = dA(v, w) - [Av,\ Aw] \qquad (5)$$

where $v, w \in T_rM$, or, in local coordinates,

$$B^b(v, w) = \left(\frac{\partial A^b_\alpha}{\partial r^\beta} - \frac{\partial A^b_\beta}{\partial r^\alpha} - E^b_{\alpha c} A^c_\beta \right) v^\alpha w^\beta,$$

where $E^b_{\alpha c} = C^b_{ac} A^a_\alpha$ depends on the structure constants, C^b_{ac} of the Lie algebra, and appears frequently in the derivation.

We also define two additional derivate operators. For a map $\kappa(r) : \mathfrak{g} \to \mathbb{R}$, we define $D\kappa(r) : \mathfrak{g} \times T_rM \to \mathbb{R}$ to satisfy

$$D\kappa(r)(\xi, v) = \left(\frac{\partial \kappa_a}{\partial r^\alpha} - \kappa_b E^b_{\alpha a} \right) \xi^a v^\alpha,$$

for $\xi \in \mathfrak{g}$ and $v \in T_rM$. Similarly, for $I(r) : \mathfrak{g} \times \mathfrak{g} \to \mathbb{R}$, we define $DI : \mathfrak{g} \times \mathfrak{g} \times T_rM \to \mathbb{R}$ via

$$DI(r)(\xi, \eta, v) = \left(\frac{\partial I_\alpha}{\partial r^\alpha} - I_{ac}E^c_{\alpha b} - I_{bc}E^c_{\alpha a} \right) \xi^a \eta^b v^\alpha.$$

3. THE AFFINE CONNECTION

A useful mechanism for representing the equations of motion for a mechanical system is through the ***affine connection***. The affine connection, denoted ∇, represents a differentiation operator (***covariant derivative***) of a vector field along another vector field. It plays a central role as it defines the equations of motion for mechanical systems in terms of geodesics, as shown below.

There are a variety of ways to interpret the covariant derivative and its associated quantities (see (Spivak 1979, Abraham and Marsden 1978) for details and references). For mechanical systems, we will focus on a special case that can be derived from a kinetic energy metric. This is the unique, torsion-free affine connection known as the ***Levi-Civita connection***, which can be defined as

$$\nabla_X Y = \left(\frac{\partial Y^i}{\partial q^j} X^j + \Gamma^i_{jk} X^j Y^k\right) \frac{\partial}{\partial q^i}, \qquad (6)$$

where the ***Christoffel symbols***, Γ, are given by

$$\Gamma^i_{jk} = \frac{1}{2} \mathbb{G}^{il} \left(\frac{\partial \mathbb{G}_{lj}}{\partial q^k} + \frac{\partial \mathbb{G}_{lk}}{\partial q^j} - \frac{\partial \mathbb{G}_{jk}}{\partial q^l} \right), \qquad (7)$$

with $\mathbb{G}_{ij}$ the elements of the metric $\mathbb{G}$, and $\mathbb{G}^{il}$ the elements of $\mathbb{G}^{-1}$.

The affine connection in our setting reduces in several respects as shown in the following proposition, which we present without proof due to restricted space.

Proposition 1. Given G-invariant vector fields, $X = (g\xi, v)$ and $Y = (g\eta, w)$ on Q, with $\xi, \eta \in \mathfrak{g}$ and $v, w \in T_rM$, then the covariant derivative of Y along X will be

$$\nabla_X Y = \begin{pmatrix} g\overset{I}{\nabla}_\xi \eta \\ \overset{\Delta}{\nabla}_v w \end{pmatrix} - \frac{1}{2} \begin{pmatrix} g \\ 0 \end{pmatrix} \begin{pmatrix} I & 0 \\ 0 & \Delta \end{pmatrix}^{-1} \begin{pmatrix} \mathbb{L} \\ \mathbb{S} \end{pmatrix}, \qquad (8)$$

where

$$\begin{aligned} \mathbb{L} = {} & -D(I\Omega)(\cdot, w) - D(I\Psi)(\cdot, v) \\ & + I_{\lambda\pi} A^\pi_\alpha \left(\bar{\nabla}_v w\right)^\alpha - 2 I_{\lambda\pi} A^\pi_\alpha \frac{\partial w^\alpha}{\partial r^\beta} v^\beta \} \\ \mathbb{S} = {} & I\Omega B(w, \cdot) + I\Psi B(v, \cdot) - DI(\Omega, \Psi). \end{aligned}$$

and $\overset{I}{\nabla}_\xi \eta$ and $\overset{\Delta}{\nabla}_v w$ are the covariant derivative using metrics of I and $\Delta = m - A^T I\, A$, respectively, $\Omega = \xi + A(r) \cdot v$, $\Psi = \eta + A(r) \cdot w$ and $D(I\Omega)$ represents the derivative defined above on $I\Omega : \mathfrak{g} \to \mathbb{R}$.

4. CONTROLLABILITY

Recall the definition of the Lie bracket:

$$\begin{aligned} [X, Y] &= \nabla_X Y - \nabla_Y X \\ &= \frac{\partial Y^i}{\partial q^j} X^j - \frac{\partial X^i}{\partial q^j} Y^j. \end{aligned} \qquad (9)$$

It is interesting to note that this represents the asymmetric part of the covariant derivative for a torsion-free affine connection. If we write the Lie bracket explicitly in our fiber bundle coordinates, we have

$$[X, Y] = \begin{pmatrix} g[\xi, \eta] + g\left(\frac{\partial \eta}{\partial r} v - \frac{\partial \xi}{\partial r} w\right) \\ [v, w] \end{pmatrix}. \qquad (10)$$

An equally important quantity that arises when studying the controllability of mechanical systems is the

symmetric product, $\langle \cdot : \cdot \rangle$, which is essentially the symmetric part of the affine connection:

$$\langle X : Y \rangle = \nabla_X Y + \nabla_Y X. \tag{11}$$

Likewise, we can write down the terms of the symmetric product in our case explicitly as:

$$< X : Y >= \begin{pmatrix} g(< \xi \,\tilde{:}\, \eta > - I^{-1}\mathbb{L}) \\ < v \,\tilde{:}\, w > - \Delta^{-1}\mathbb{S} \end{pmatrix} \tag{12}$$

where $< \xi \,\tilde{:}\, \eta >= \tilde{\nabla}_\xi \eta + \tilde{\nabla}_\eta \xi$ and $< v \tilde{:} w >= \tilde{\nabla}_v w + \tilde{\nabla}_w v$. Notice that the terms $\mathbb{L}$ and $\mathbb{S}$ play a similar role as the additional derivative terms found in Equation 10, as they are the cross-coupling terms that arise when dealing with the principal bundle.

Since the main focus of this paper is on constructing controls using cyclic inputs, we remark here only briefly on the natural extension of the controllability analysis developed by Lewis, Murray, and Bullo (Lewis and Murray 1997, Bullo *et al.* 1998) to the setting of principal fiber bundles. Given a set of vector fields $\mathcal{X} = \{X_1, X_2, \ldots, X_n\}$, we denote all the iterative symmetric products of the elements in $\mathcal{X}$ as $Sym(\mathcal{X})$. Then, we can define good and bad symmetric products (c.f., (Lewis and Murray 1997)) by denoting an element of $Sym\mathcal{X}$ as ***bad*** if it contains an even number of each element of $\mathcal{X}$ and ***good*** if it is not bad.

Theorem 2. Controllability [Lewis and Murray (Lewis and Murray 1997)]: Consider the mechanical control system on the configuration manifold Q whose Lagrangian is the kinetic energy with respect to a Riemannian metric $\mathbb{G}$ and whose input vector fields are $Y = \{Y_1, ..., Y_n\}$. Then

- the system is locally configuration accessible at q if the distribution defined by $\overline{Lie(\overline{Sym}(Y))}$ has maximal rank at q, and
- the system is equilibrium controllable if it is locally configuration accessible and if every bad symmetric product is a linear combination of good symmetric products of lower degree.

Given $Y = \{Y_1, ..., Y_n\}$, we denote the reduced vector fields by $B = \{B_1, ..., B_n\}$, such that correspondence $Y = gB$ or

$$Y_i = gB_i, \quad i = 1, ..., n.$$

Proposition 3. Consider the control system given in theorem 2 on trivial principal fiber bundle with input vector fields $B = \{B_1, ..., B_m\}$ (in the body frame). Then

- the system is locally configuration accessible at q if the distribution defined by $\overline{Lie(\overline{Sym}(B))}$ has maximal rank at q, and
- the system is equilibrium controllable if it is locally configuration accessible and if every bad symmetric product is a linear combination of good symmetric products of lower degree.

We also note that it is likely that one could define a similar notion of fiber configuration controllability for mechanical systems, as was done for kinematic systems in (Kelly and Murray 1995).

Example 1. Consider a rigid body moving in $SE(2)$ with a thruster to adjust its pose. See Fig. 1. We remark that the original motivation for this problem is actually our blimp system (Zhang and Ostrowski 1999) restricted to a vertical plane. The control inputs are the thruster force u and its direction relative to the body axis X^b. Note that this distinguishes this example in an important sense from the usual rigid body problem with a single thruster, in that one must control the heading of the thruster indirectly through a pointing angle, instead of directly using, for example, two independent, orthogonally-mounted thrusters. The acting point of the thruster is located assumed to be along the body's long axis, at a distance h from the center of the mass. The control inputs are the force and the orientation of the thruster.

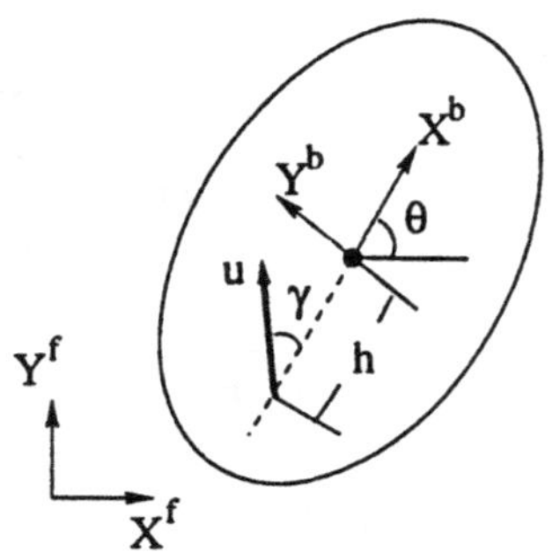

Fig. 1. A planar blimp with rotating thruster input.

For simplicity, we assume the thruster is massless but comes with a rotational inertia, J_2. The Lagrangian of the system is then

$$L = \frac{1}{2}m(\dot{x}^2 + \dot{y}^2) + \frac{1}{2}J_1\dot{\theta}^2 + \frac{1}{2}J_2(\dot{\gamma} + \dot{\theta})^2, \tag{13}$$

while the dynamics of the system in body frame is

$$\nabla_{\begin{pmatrix} \xi \\ \dot{r} \end{pmatrix}} \begin{pmatrix} \xi \\ \dot{r} \end{pmatrix} = Bu_1 + B^r u_2 \tag{14}$$

with $B = \frac{1}{m}\begin{pmatrix} \cos\gamma \\ \sin\gamma \\ -hm/(J_1 + J_2)\sin\gamma \\ 0 \end{pmatrix}$ and $B^r = (0, 0, 0, 1)^T$ representing the forcing in the group and shape directions, respectively.

From the Lagrangian, we find the metric of the system to be

$$\mathbb{G} = \begin{pmatrix} m & 0 & 0 & 0 \\ 0 & m & 0 & 0 \\ 0 & 0 & J_1 + J_2 & J_2 \\ 0 & 0 & J_2 & J_2 \end{pmatrix},$$

which is constant and G-invariant. Thus, in order to check the controllability, we can just compute $\overline{Lie(\overline{Sym}(B))}$ instead of $\overline{Lie(\overline{Sym}(Y))}$. With some simple computations we get

$$\langle B:B\rangle = \frac{2h\sin\gamma}{m(J_1+J_2)}\begin{pmatrix}\sin\gamma\\ \cos\gamma\\ 0\\ 0\end{pmatrix},$$

$$\langle B:\langle B:B\rangle\rangle = k_1(r)B$$

$$[B,\langle B:B\rangle] = k_2(r)B,$$

$$\langle\langle B:B\rangle:\langle B:B\rangle\rangle = 0,$$

$$B_r = \langle B:B^r\rangle = \frac{\partial B}{\partial r} = \frac{1}{m}\begin{pmatrix}-\sin\gamma\\ \cos\gamma\\ -\dfrac{hm}{J_1+J_2}\cos\gamma\\ 0\end{pmatrix}$$

$$\langle B:\langle B:B^r\rangle\rangle = \frac{-h}{m^2(J_1+J_2)}\begin{pmatrix}-\sin 2\gamma\\ \cos 2\gamma\\ 0\\ 0\end{pmatrix}$$

$$\langle\langle B:B^r\rangle:B^r\rangle = \frac{\partial^2 B}{\partial r^2} = -B$$

Hence, $\overline{Lie}(\overline{Sym}(B)) = \{B, \langle B:B^r\rangle, \langle B:\langle B:B^r\rangle\rangle\}$ has full rank, which means the system is locally configuration accessible. Furthermore, since all of the three elements of the defined $\overline{Lie}(\overline{Sym}(B))$ are good symmetric products and have full rank, all the other iterative symmetric products of the inputs, including the "bad" ones, will be a linear combination of these three elements (as we showed this for $\langle B:B\rangle$ and $\langle\langle B:B\rangle:\langle B:B\rangle\rangle$). Therefore, the system is also equilibrium controllable at any point.

5. CONTROL ALGORITHM

From the analysis above, we see that we can simplify the analysis of the controllability by restricting the computations to a reduced space. We can use these same computations to build constructive control inputs to our system, given some additional assumptions. First, we assume that the shape space is fully actuated, so that the unactuated degrees of freedom all lie in the directions tangent to the group orbit. In addition, we assume that all symmetric products and Lie brackets do not generate additional terms in the shape direction. In other words, all elements of $\overline{Lie}(\overline{Sym}(B))$, except those found in B, are of the form $(\xi, 0)^T$ for $\xi \in \mathfrak{g}$. Finally, we assume that the metric is a constant, so that all of the shape dependence is found in the inputs, $B(r)$.

We then follow along the lines of Bullo (Bullo 1999) in first using a series solution to the second-order dynamics to yield a first-order time varying solution to our equations. This will provide a basis for a perturbation expansion in terms of the shape inputs that leads to a simplified representation of the explicit dependence of the equations on the symmetric product.

Let $\bar{X}(T) = \int_0^T X(t)dt$ and small-amplitude inputs of the form $Y = \varepsilon B$. For $q \in Q$, a first-order expansion of the evolution of q can be written as (Bullo 1999)

$$\dot{q} = \varepsilon\overline{Y} - \frac{\varepsilon^2}{2}\overline{\langle\overline{Y}:\overline{Y}\rangle} + \frac{\varepsilon^3}{2}\overline{\langle\overline{\langle\overline{Y}:\overline{Y}\rangle}:\overline{Y}\rangle} + O(\varepsilon^4).$$

This general solution can also be interpreted on the principal fiber bundle, where $\dot{q} = (g\xi, \dot{r})^T$. The expansion then reduces to

$$\begin{pmatrix}\xi\\ \dot{r}\end{pmatrix} = \varepsilon\overline{B} - \frac{\varepsilon^2}{2}\overline{\langle\overline{B}:\overline{B}\rangle} + \frac{\varepsilon^3}{2}\overline{\langle\overline{\langle\overline{B}:\overline{B}\rangle}:\overline{B}\rangle} + O(\varepsilon^4),$$

where we have again used $Y = gB$ to denote the inputs pulled back via the group action.

Given the assumptions stated above, we find that we can restrict our attention to the fiber variables, since $\dot{r} = \varepsilon\overline{B^r}$. In this case, we will assume that $r(t,\varepsilon) = r_0 + \varepsilon r_1(t)$, with $\dot{r} = \varepsilon\dot{r}_1$, and $\ddot{r} = \varepsilon\ddot{r}_1 = u_j$. Assume $\dot{r}(0) = 0$ then we have $\dot{r} = \bar{u}_j$. Also, let the control input for the group variables be εu_i.

If we assume a solution for the group velocities to be

$$\xi(t,\varepsilon) = \varepsilon\xi^1(t) + \varepsilon^2\xi^2(t) + \varepsilon^3\xi^3(t) + O(\varepsilon^4), \tag{15}$$

with $\xi(t) = 0$, then a straightforward (though messy) perturbation analysis that involves Taylor expanding the input vectors, $B(r)$, about r_0, yields

$$\xi^1 = B_0\overline{u_i}$$

$$\xi^2 = -\frac{1}{2}(\langle B:B\rangle_0\overline{\overline{u_i}\,\overline{u_i}} + \langle B:B^r\rangle_0\overline{\overline{u_i}\,\dot{r}_1}) + \left.\frac{\partial B}{\partial r}\right|_{r_0}\overline{\overline{u_i}\,r_1}$$

$$\xi^3 = \frac{1}{2}\Bigg(\langle\langle B:B\rangle:B\rangle_0\overline{\overline{u_i}\,\overline{\overline{u_i}\,\overline{u_i}}} - \left.\frac{\partial\langle B:B\rangle}{\partial r}\right|_{r_0}\overline{\overline{\overline{u_i}\,\overline{u_i}}\,\dot{r}_1}$$
$$+ \langle\langle B:B\rangle:B^r\rangle|_{r_0}\overline{\overline{u_i}\,\overline{\dot{r}_1 u_i}} + \left.\frac{\partial\langle B:B^r\rangle}{\partial r}\right|_{r_0}\overline{r_1\dot{r}_1\,\overline{u_i}}$$
$$+ \frac{\partial^2 B}{\partial r^2}|_{r_0}\overline{r_1^2\,u_i}\Bigg),$$

where B_0 is B evaluated at $r = r_0$.

The computations for higher orders of symmetric products yield increasingly cumbersome terms at each order, so this type of treatment is likely to be practical only for systems in which the desired directions of motion are generated with low order symmetric products and Lie brackets.

We note that the approximations given here only include control of the velocity terms in the Lie group directions. A full solution using the Magnus single exponential representation could be performed to lead to solutions for $g(t)$ as is done in (Leonard and Krishnaprasad 1994, Radford and Burdick n.d., Ostrowski 1999). This would lead to additional terms involving the Lie bracket. Also, in developing control inputs, our analysis in the example below will be greatly simplified by the fact that all bad symmetric products vanish at r_0. If this were not the case, we would need to introduce a scaled secondary input (i.e., $\varepsilon u_i \to \varepsilon u_i^1 + \varepsilon^2 u_i^2$). This could then be used to offset any contributions by bad symmetric products, as done in (Bullo *et al.* 1998).

Example 1 (cont'd). Consider the example of the planar blimp again, which we wish to control using small-amplitude, sinusoidal inputs for both u and γ. We utilize both the inputs, B and B^r, as before. If we assume $\gamma_0 = 0$, we can evaluate the following terms at $r_0 = 0$: $b_1 = B_0 = (1/m, 0, 0, 0)^T$, $b_2 = \langle B:B^r\rangle_0 = \left((0, \frac{1}{m}, \frac{-h}{J_1+J_2}, 0\right)^T$, and $b_3 = \langle B:\langle B:B^r\rangle\rangle_0 = \frac{-h}{m^2(J_1+J_2)}(0,1,0,0)^T$, while all other products are just the scaling of one of these three vectors. Since there is only one control input on the Lie algebra, and the bad symmetric product $\langle B : B \rangle_0 = 0$, we do not need to worry about the effect of the bad products. Therefore, the approximate solution of the dynamic equation becomes

$$\begin{aligned}\xi(t) &= \varepsilon B_0 \overline{u} - \frac{\varepsilon^2}{2}\{\langle B : B\rangle_0 \overline{\overline{u}\, u} + 2B_{r0}\overline{\overline{\gamma_1} u}\} \\ &+ \frac{\varepsilon^3}{2}\langle\langle B : B\rangle : B\rangle_0 \overline{\overline{\overline{u}\, u}\, u} + \frac{\partial \langle B : B\rangle}{\partial r}|_{\gamma_0} \overline{\overline{\overline{u}\, u}\, \gamma_1} \\ &+ 2\langle B : B_r\rangle_0\, \overline{u\, \overline{\gamma_1 u}} + 2\frac{\partial^2 B}{\partial r^2}|_{\gamma_0} \overline{\overline{\overline{u}\, \gamma_1}\, \gamma_1}\} \\ &= \alpha_1 b_1 + \alpha_2 b_2 + \alpha_3 b_3\end{aligned}$$

where α_1, α_2 and α_3 correspond to the functions of control inputs. Any choice of γ_1 and u which can single out one direction can be used to steer the blimp. We choose to use sinusoidal inputs.

(1) Set $\gamma_1 = 0$, and u as a constant, say $u = \frac{1}{2\pi}$, then $\xi(2\pi) \approx \varepsilon b_1$.
(2) Set $u = \gamma_1 = \frac{1}{\sqrt{\pi}} \sin t$, then $\xi(2\pi) \approx \varepsilon^2 b_2$.
(3) Set $\gamma_1 = \cos t$, and $u^1 = \frac{1}{\sqrt{\pi}} \sin t$, then $\xi(2\pi) \approx \frac{1}{2}\varepsilon^3 b_3$

Assume $m = 2kg$, $J_1 = 1kg \cdot m^2$, $J_2 = 0.2kg \cdot m^2$, and $h = 0.5m$, then we have the following simulation result based on the above sets of inputs with $\varepsilon = 0.4$.

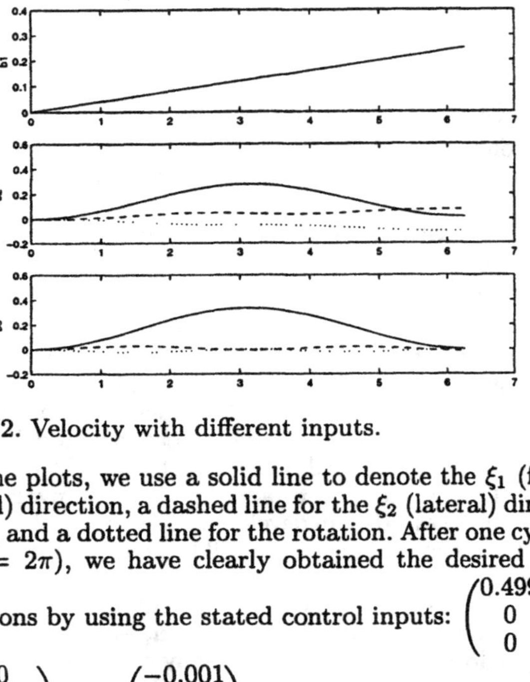

Fig. 2. Velocity with different inputs.

In the plots, we use a solid line to denote the ξ_1 (forward) direction, a dashed line for the ξ_2 (lateral) direction, and a dotted line for the rotation. After one cycle ($T = 2\pi$), we have clearly obtained the desired directions by using the stated control inputs: $\begin{pmatrix}0.499\\0\\0\end{pmatrix}$, $\begin{pmatrix}0\\0.46\\-0.42\end{pmatrix}$, and $\begin{pmatrix}-0.001\\-0.104\\0\end{pmatrix}$, where the theoretical directions are $\begin{pmatrix}0.5\\0\\0\end{pmatrix}$, $\begin{pmatrix}0\\0.5\\-0.42\end{pmatrix}$, and $\begin{pmatrix}0\\-0.104\\0\end{pmatrix}$. In order to get faster speeds, we can increase ε to a larger value, although the error will be comparably greater.

6. ACKNOWLEDGMENT

The authors would like to express sincere gratitude to Francesco Bullo and Andrew Lewis, for sharing their notes and ideas with us. The derivations in this paper are clearly based on the foundation assembled by Bullo, while Lewis' notes have served as a useful check on the correctness of the equations. We also would like to acknowledge the support of NSF grants MIP94-20397 and IRI-9711834, and ARO grants P-34150-MA-AAS, DAAH04-96-1-0007, and DURIP DAAG55-97-1-0064.

7. REFERENCES

Abraham, Ralph and Jerrold E. Marsden (1978). *Foundations of Mechanics*. second ed.. Addison-Wesley. Reading, MA.

Bloch, Anthony M., P.S. Krishnaprasad, Jerrold E. Marsden and Richard M. Murry (1996). Nonholonomic mechanical system with symmetry. *Archive for Rational Mechanics and Analysis* **136**, 21–99.

Bullo, Francesco (1999). A series describing the evolution of mechanical control systems. In: *IFAC World Congress*. pp. 479–485.

Bullo, Francesco, Naomi Ehrich Leonard and Andrew D. Lewis (1998). Controllability and motion algorithms for underactuated Lagrangian systems on Lie groups. Submitted to IEEE *Transactions on Automatic Control*.

Kelly, Scott D. and Richard M. Murray (1995). Geometric phases and locomotion. *J. Robotic Systems* **12(6)**, 417–431.

Leonard, Naomi Elrich and P. S. Krisnaprasad (1994). Motion control of drift-free, left-invariant systems on Lie groups part I: Averaging and controllability. Technical Report TR 94-8. University of Maryland.

Lewis, Andrew D. and Richard M. Murray (1995). Configuration controllability for a class of mechanical systems. In: *IEEE Conference on Decision and Control*. Vol. 1-5.

Lewis, Andrew D. and Richard M. Murray (1997). Configuration controllability of simple mechanical control systems. *SIAM Journal on control and optimization* **35(3)**, 766–790.

Ostrowski, James P. (1995). The Mechanics and Control of Undulatory Robotic Locomotion. PhD thesis. California Institute of Technology. Pasadena, CA. Available electronically at http://www.cis.upenn.edu/~jpo/papers.html.

Ostrowski, James P. (1998). Computing reduced equations for robotic systems with constraints and symmetries. *To appear: IEEE Transactions of Robotics and Automation.*

Ostrowski, James P. (1999). Steering for a class of dynamic nonholonomic systems. To appear in IEEE *Transactions on Automatic Control*.

Radford, Jim and Joel Burdick (1998). Local motion planning for nonholonomic control systems evolving on principal bundles. Submitted to *Conf. Mathematical Theory of Networks and Systems.*

Sastry, Shankar (1999). *Nonlinear Systems*. Interdisciplinary applied mathematics-System and control. Springer-Verlag. New York.

Spivak, Michael (1979). *A Comprehensive Introduction to Differential Geometry*. Vol. 1-5. 2nd ed.. Publish or Perish Inc.

Zhang, Hong and James P. Ostrowski (1999). Visual servoing with dynamics: Control of an unmanned blimp. In: *International Conference on Robotics and Automation*. pp. 618–623.

KINEMATIC ASYMMETRIES AND THE CONTROL OF LAGRANGIAN SYSTEMS WITH OSCILLATORY INPUTS

J. Baillieul [*,1]

[*] *Aerospace/Mechanical Engineering*
Boston University
Boston, MA 02215
johnb@bu.edu

Abstract: Some curvature-like geometric invariants of controlled Lagrangian systems are defined and shown to be of interest in determining the systems' response to high-frequency oscillatory inputs. The theory is illustrated by several examples of controlled rotating systems. For a rotating two d.o.f. pendulum with unequal cross-sectional moments of inertia, it is shown that a small-amplitude high-frequency oscillatory component of the rotational velocity can stabilize the neutral (hanging) equilibrium at average rates of rotation higher than the critical rate at which the neutral equilibrium loses stability for constant-velocity rotations. This and other examples illustrate the ways in which geometric and mechanical asymmetries affect a system's response to oscillatory forcing. *Copyright ©2000 IFAC*

Keywords: averaged potential, input connection, kinematic asymmetry

1. INTRODUCTION

This paper presents new results on the geometric control theory of ***super-articulated*** (or ***under-actuated***) mechanical systems with oscillatory inputs. Specifically we present a detailed analysis of rotating mechanisms with dynamics prescribed by a Lagrangian $L(r, q, \dot{r}, \dot{q})$ together with equations of motion:

$$\frac{d}{dt}\frac{\partial L}{\partial \dot{r}} - \frac{\partial L}{\partial r} = u, \tag{1}$$

$$\frac{d}{dt}\frac{\partial L}{\partial \dot{q}} - \frac{\partial L}{\partial q} = 0. \tag{2}$$

[1] Support from the Army Research Office under the ODDR&E MURI97 Program Grant No. DAAG55-97-1-0114 to the Center for Dynamics and Control of Smart Structures (through Harvard University) is gratefully acknowledged.

In each of the examples, the controlled coordinate r will represent angular displacement with respect to a spatially fixed axis.

The principal tools in this study are the geometric theory of averaging and energy methods developed in earlier work ([1]-[8]) whereby stable responses of (2) to oscillatory motions of the r variable are associated with local minima of an energy-like quantity called the ***averaged potential***. Our previous work has employed a variety of both classical and geometric averaging methods. In [3], for instance, we introduced Rayleigh-type dissipation to justify the use of results from the classical theory of averaging. Recent work by Bullo ([10]) follows the same approach while also using ideas from the modern theory of geometric mechanics. Rayleigh dissipation is not an essential part of the stability story, however, and in [4] it was shown that under certain symmetry conditions, a more geometric approach, based on ideas from Floquet theory, could be used to prove the stability of motions around local minima of the averaged potential. This analysis was pursued in depth in [5], and it was shown in [6] how the results could be applied to identify a rich class of stable responses in rapidly forced multi-link (heavy) kinematic chains. Recently, it was shown ([8]) how the theory depended characteristic length parameters of links, and in particular it was shown that stable motions could be produced in chains whose link-lengths were a millimeter or less. Related but independent research by Fraser ([9]) has shown that high-frequency longitudinal forcing can be used to effectively stiffen flexible beams.

The rotating mechanism examples treated below extend the class of applications of oscillation-controlled Lagrangian systems in interesting ways and allow us to make connections between geometric invariants (such as curvatures) associated with the models and mechanical properties of the of the underlying physical systems. In making these connections, we obtain some surprising (but experimentally verified) results on the way in which kinematic and inertial asymmetries can affect a mechanism's response to oscillatory forcing. Specifically, we show that for a two degree-of-freedom (Hooke-joint) rotating pendulum with certain inertial asymmetries, we can use small-amplitude, high-frequency oscillations of the velocity of rotation to extend the range of rotation rates for which the vertical equilibrium is stable.

The starting point in this analysis is to recall certain curvature-like quantities associated with the Lagrangian dynamics (1)-(2) which had been shown to be useful in determining how the unactuated degrees of freedom (the q-variables) are influenced by oscillatory motions of the directly controlled degrees of freedom (the r-variables). In [7], it was shown that certain symmetry conditions were necessary for there to be a coordinate system in which the q-variables did not depend of accelerations of the r-variables. These symmetry conditions are similar to the conditions used in the stability analysis of [4] and also similar to the matching conditions which play a role in the method of controlled Lagrangians studied in [11]. Section 2 introduces some curvature-like invariants that arise in the theory of controlled Lagrangian systems. The quantities are defined in terms of the coefficients in a certain *reduced* Lagrangian wherein we regard the variables $r, \dot{r}, \ddot{r}$ (jointly) as the control inputs—rather than u. Section 3 reviews recent work on averaging oscillation-controlled Lagrangian systems and describes how this work is related to the invariants of Section 2. Section 4 treats two rotating system examples in detail. In both cases, the controlled coordinate r enters the Lagrangian model as a cyclic variable, so that in the reduced Lagrangian, the pair $(\dot{r}, \ddot{r})$ is regarded as the input. Depending on the value of the invariants of Section 2, it may or may not be possible to eliminate the dependence on the acceleration $\ddot{r}$. It will be shown that for the Hooke-joint pendulum, it is not possible to eliminate this dependence.

In the present paper, we present a detailed analysis of the geometric conditions of [7] in terms of the physical characteristics of several rotating mechanical systems.

2. VELOCITY- AND ACCELERATION-CONTROLLED LAGRANGIAN SYSTEMS

In contrast to the recent work of Bullo ([10]), we shall study the dynamics of (2) assuming that the variable $r(t)$ rather than $u(t)$ is the control input (whose influence on the evolution of $q(t)$ is also directly felt through first and second derivative terms $\dot{r}(t)$ and $\ddot{r}(t)$). We assume the Lagrangian has a simple structure consisting of kinetic minus potential energy terms—i.e. $L = \frac{1}{2}\dot{y}^T M \dot{y} - V(y)$, and the inertia matrix is partitioned conformably with (r, q), $M = \begin{pmatrix} \mathcal{N} & \mathcal{A} \\ \mathcal{A}^T & \mathcal{M} \end{pmatrix}$. The point of view which we adopt is that the details of the dependence on the input u may be ignored, and that the triple $(r, \dot{r}, \ddot{r})$ may be thought of as the input which drives the system (2). We may thus understand the dynamics of this system by applying the Euler-Lagrange operator $\frac{d}{dt}\frac{\partial}{\partial \dot{q}} - \frac{\partial}{\partial q}$ to the *reduced Lagrangian*

$$\mathcal{L}(q, \dot{q}; t) = \frac{1}{2}\dot{q}^T \mathcal{M} \dot{q} + \dot{r}^T \mathcal{A} \dot{q} - \mathcal{V}(q, t), \quad (3)$$

where $\mathcal{V}_a(q, t) = V(r, q) - \frac{1}{2}\dot{r}(t)^T \mathcal{N}(r(t), q)\dot{r}(t)$ is the *amended potential.* The t-dependence of the reduced Lagrangian also involves the inertia terms $\mathcal{M}(r(t), q)$ and $\mathcal{A}(r(t), q)$. The *reduced* equations of motion may be written componentwise as

$$\sum_{j=1}^{n} m_{kj}\ddot{q}_j + \sum_{\ell=1}^{m} a_{\ell k}\dot{v}_\ell + \sum_{i,j=1}^{n} \Gamma_{ijk}\dot{q}_i\dot{q}_j$$

$$+\sum_{j=1}^{n}\sum_{\ell=1}^{m}\hat{\Gamma}_{\ell jk}v_\ell\dot{q}_j = F(t), \quad (k=1,\ldots,n), \quad (4)$$

where $v_\ell = \dot{r}$,

$$\Gamma_{ijk} = \frac{1}{2}\Big(\frac{\partial m_{ki}}{\partial q_j} + \frac{\partial m_{kj}}{\partial q_i} - \frac{\partial m_{ij}}{\partial q_k}\Big),$$

$$\hat{\Gamma}_{\ell jk} = \frac{\partial m_{kj}}{\partial r_\ell} + \frac{\partial a_{\ell k}}{\partial q_j} - \frac{\partial a_{\ell j}}{\partial q_k},$$

and a_{ij} and m_{ij} are the ij-th entries in the $m \times n$ and $n \times n$ matrices $\mathcal{A}(r,q)$ and $\mathcal{M}(r,q)$ respectively. $F(t)$ is a vector of *generalized forces* $F_i(t) = \frac{\partial \mathcal{V}}{\partial q_i} - \sum_{k,\ell=1}^{m}\frac{\partial a_{\ell i}}{\partial r_k}v_\ell v_k$. These may be thought of as coming from a velocity-dependent potential if

$$\frac{\partial^2 a_{\ell i}}{\partial q_j \partial r_k} = \frac{\partial^2 a_{\ell j}}{\partial q_i \partial r_k}$$

for all $k,\ell = 1,\ldots,m$ and $i,j = 1\ldots,n$. In classical mechanics, the quantities Γ_{ijk} defined in this way in terms of the inertia tensor $\mathcal{M}$ are called *Christoffel symbols of the first kind.* To be consistent with this nomenclature, we call the $\hat{\Gamma}_{\ell jk}$ *input symbols of the first kind.*

In the next section, we outline a theory of averaging for systems of the form (4) forced by high-frequency oscillatory inputs $(r,\dot{r},\ddot{r})$. As pointed out in [7], the form of the reduced equations of motion (4) is invariant under a change of coordinates and even invariant under an input-dependent change of coordinates. We call such a system a *velocity-controlled Lagrangian system* if there is a choice of coordinates such that $a_{ij} \equiv 0$ for all i,j. If there is no choice of coordinates in which all the a_{ij}'s vanish, the system (4) is said to be *acceleration-controlled.* In Section 4, we shall prove that the pendulum mechanism depicted in Fig. 1 is acceleration-controlled. A simple example of a velocity-controlled system is a rotating simple pendulum such as the mechanism of Fig. 1 with ψ locked at $\psi = 0$ (or equivalently with ϕ locked at $\phi = 0$). For this mechanism and the given coordinate variables, there will be no terms in the Lagrangian coupling the velocities $v = \dot{\theta}$ and $\dot{\phi}$. The Euler-Lagrange equations for ϕ thus contain no terms involving the acceleration $\dot{v} = \ddot{\theta}$.

It is important to distinguish between these acceleration- and velocity-controlled systems since for high-frequency inputs, the influence of acceleration will clearly be dominant. A slightly more subtle reason to make the distinction is that for acceleration-controlled systems, as Example 1 of Section 4 shows, velocity and acceleration terms may have opposing effects on stability.

In [7], it was pointed out that curvature-like quantities defined in terms of the Γ_{ijk}'s and $\hat{\Gamma}_{\ell jk}$'s are useful in determining whether a system is velocity- or acceleration-controlled. In recalling the basic elements of this theory it will be useful to define *Christoffel symbols of the second kind:* $\Gamma^{\sigma}_{ij} = m^{\sigma k}\Gamma_{ijk}$, and *input symbols of the second kind:* $\hat{\Gamma}^{\sigma}_{\ell j} = m^{\sigma k}\hat{\Gamma}_{\ell jk}$. The Christoffel symbols of either the first or second kind define the Levi-Civita connection associated with the inertia tensor $\mathcal{M}$, and the similarly, the input symbols (of either kind) define the *input connection.* The question of whether coordinates can be found in terms of which the coupling terms $\mathcal{A}$ in (3) vanish is addressed by the following:

Theorem 1. (Baillieul, [7]) Consider the Lagrangian control system (3). Let $U \times V \subset \mathbb{R}^n \times \mathbb{R}^m$, and suppose that $F : U \times V \to \mathbb{R}^n$ is an input-dependent change of coordinates $\bar{q} = F(q,r)$ such that for each x the metric tensor $\mathcal{M}$ expressed in $\bar{q}-$coordinates has δ_{ij} as its $ij-$th entry. Suppose, moreover, that in the $\bar{q}-$coordinates all cross coupling terms $\bar{\mathcal{A}}(\bar{q},r)$ vanish. Then the system (defined by (3)) is flat and has flat inputs, which is to say both the hatted and unhatted Riemann symbols of the second kind,

$$R_j{}^{\gamma}{}_{\alpha\beta} = \frac{\partial\Gamma^{\gamma}_{\alpha j}}{\partial q_\beta} - \frac{\partial\Gamma^{\gamma}_{\beta j}}{\partial q_\alpha} + \sum_{k=1}^{n}(\Gamma^{k}_{\alpha j}\Gamma^{\gamma}_{\beta k} - \Gamma^{k}_{\beta j}\Gamma^{\gamma}_{\alpha k})$$
$$j,\gamma,\alpha,\beta = 1,\ldots,n,$$

and

$$\hat{R}_j{}^{\gamma}{}_{\ell i} = \frac{\partial\hat{\Gamma}^{\gamma}_{\ell j}}{\partial r_i} - \frac{\partial\hat{\Gamma}^{\gamma}_{ij}}{\partial r_\ell} + \sum_{k=1}^{n}(\hat{\Gamma}^{k}_{\ell j}\hat{\Gamma}^{\gamma}_{ik} - \hat{\Gamma}^{k}_{ij}\hat{\Gamma}^{\gamma}_{\ell k})$$
$$i,\ell = 1,\ldots,m; \quad j,\gamma = 1,\ldots,n,$$

vanish.□

The Riemann symbols $R_j{}^{\gamma}{}_{\alpha\beta}$, $\hat{R}_j{}^{\gamma}{}_{\ell i}$ determine the curvatures of the Levi-Civita connection and the input connection respectively. It is a classical result that there exists a choice of coordinates with respect to which the inertia tensor $\mathcal{M}$ is represented by the identity matrix if and only if all Riemann symbols $R_j{}^{\gamma}{}_{\alpha\beta} = 0$. The above theorem is a partial extension to the case of controlled

Lagrangian systems. The reader is referred to [7] for further details.

3. STABILITY OF RAPIDLY FORCED SUPER-ARTICULATED MECHANICAL SYSTEMS

We take (3) as the starting point of our discussion. The dynamics (4) are the explicit rendering of the Euler-Lagrange equations

$$\frac{d}{dt}\frac{\partial \mathcal{L}}{\partial \dot q} - \frac{\partial \mathcal{L}}{\partial q} = 0. \qquad (5)$$

By means of the Legendre transformation to (q,p)-coordinates, where

$$\begin{aligned} p &= \frac{\partial \mathcal{L}}{\partial \dot q} \\ &= \mathcal{M}(q)\dot q + \mathcal{A}^T(q)\dot r \end{aligned} \qquad (6)$$

we obtain an associated Hamiltonian system

$$\mathcal{H}(q,p;r,v) = \frac{1}{2}(p - \mathcal{A}^T v)^T \mathcal{M}^{-1}(p - \mathcal{M}^T v) + \mathcal{V}, \qquad (7)$$

which depends on t through the input variables $(r,\dot r)$. We apply simple averaging to (7) as discussed in [7], to obtain the *averaged Hamiltonian*

$$\begin{aligned}\overline{\mathcal{H}}(q,p) = \frac{1}{2}p^T\overline{\mathcal{M}^{-1}}p - \overline{v^T\mathcal{A}\mathcal{M}^{-1}}p \\ + \frac{1}{2}\overline{v^T\mathcal{A}\mathcal{M}^{-1}\mathcal{A}^T v} + \overline{\mathcal{V}},\end{aligned} \qquad (8)$$

where the overbars indicate that simple averages over one period (T) have been taken. As in [7], $\overline{\mathcal{H}}(q,p)$ may be re-written as the sum of an *averaged kinetic energy* term and an *averaged potential*:

$$\begin{aligned}\mathcal{V}_A(q) = \frac{1}{2}\overline{v^T\mathcal{A}\mathcal{M}^{-1}\mathcal{A}^T v} \\ -\frac{1}{2}\overline{v^T\mathcal{A}\mathcal{M}^{-1}}\left(\overline{\mathcal{M}^{-1}}\right)^{-1}\overline{\mathcal{M}^{-1}\mathcal{A}^T v} + \overline{\mathcal{V}}.\end{aligned} \qquad (9)$$

There is a growing body of literature on the role of the averaged potential in determining the stability of motions of Lagrangian systems subject to high-frequency oscillatory forcing. The following theorem was proved in [4].

Theorem 2. (Baillieul, [4]) Consider the Lagrangian control system (3), where we assume $v(\cdot)$ is an $\mathbb{R}^m$-valued piecewise continuous periodic function of period $T > 0$ such that $\bar v = \frac{1}{T}\int_0^T v(s)\,ds = 0$. Let $\mathcal{V}_A(q)$ be the averaged potential (9), and let q_0 be a critical point of $\mathcal{V}_A$. (I.e., all partial derivatives of $\mathcal{V}_A$ vanish when evaluated at q_0.) Then if $\mathcal{A}(q_0) = 0$, the following holds:

(1) The averaged potential $\mathcal{V}_A(q)$ agrees up to terms of order 2 with the averaged potential associated with the *linear Lagrangian system*

$$\mathcal{M}_0\,\ddot q + \sum_{\ell=1}^m\left(\dot v_\ell \mathcal{A}_1^{\ell\,T} q + v_\ell(\mathcal{A}_1^{\ell\,T} - \mathcal{A}_1^\ell)\dot q\right) + K\,q = 0. \qquad (10)$$

where $\mathcal{M}_0 = \mathcal{M}(q_0)$, $K = \frac{\partial^2 \mathcal{V}_A}{\partial q^2}(q_0)$, and where we define the $n \times n$ matrices $\mathcal{A}_1^\ell$ as follows. If we write

$$v^T\mathcal{A}(q)\dot q = \sum_{\ell=1}^m v_\ell A^\ell(q)\dot q,$$

then we define $\mathcal{A}_1^\ell = \frac{\partial A^\ell}{\partial q}(q_0)$.

(2) Suppose $w(\cdot)$ is an $\mathbb{R}^m$-valued piecewise continuous periodic function of period $T > 0$ such that $\bar w = \frac{1}{T}\int_0^T w(s)\,ds = 0$. Consider the linear Lagrangian system (10) with input $v(t) = w(\omega t)$. The averaged potential for this system is given by

$$\hat{\mathcal{V}}_A(q) = \frac{1}{2}q^T\Big(K + \sum_{i,j=1}^m \sigma_{ij}\mathcal{A}_1^i\mathcal{M}_0^{-1}\mathcal{A}_1^{j\,T}\Big)q, \qquad (11)$$

where $\sigma_{ij} = (1/T)\int_0^T w_i(s)w_j(s)\,ds$. If the matrix $\frac{\partial^2\mathcal{V}_A}{\partial q^2}(q_0) = \frac{\partial^2\hat{\mathcal{V}}_A}{\partial q^2}(0)$ is positive definite, and if $\mathcal{A}_1^{\ell\,T} = \mathcal{A}_1^\ell$ for $\ell = 1,\ldots,m$, the origin $(q,\dot q) = (0,0)$ of the phase space is stable in the sense of Lyapunov provided ω is sufficiently large.□

Remark 1. (Reduction and averaging) Theorem 2 illustrates the importance of feature of approaching the linearized analysis of controlled Lagrangian systems with care. In particular, equation (10) is obtained from equation (4) by linearization with respect to $(q,\dot q)$ about $(q_0,0)$ but with *full nonlinear dependence on* $(r,\dot r,\ddot r)$ *being maintained.* In other words, because $\dot r$ and $\ddot r$ are assumed to have large magnitude, the operations of linearization and reduction (from L to $\mathcal{L}$) do not commute, and r is not treated as a configuration variable in the reduced Lagrangian (3).

Under the (essential) assumption that $\mathcal{A}(q_0) = 0$, the theorem states that the operations of linearization (w.r.t. q and $\dot{q}$) and averaging of the reduced model $\mathcal{L}$ do commute. When $\mathcal{A}(q_0) \neq 0$, it is much less straightforward to use linearized models in this type of analysis, and we refer to [12] for the details of this case.

Remark 2. The assumption that $A_1^\ell = A_1^{\ell T}$ is explicitly used in proving Theorem 2, but as noted in [4], alternative approaches (e.g. introducing dissipation) do not require it. If the inertia tensor $\mathcal{M}$ does not depend on r, the symmetry $A_1^\ell = A_1^{\ell T}$ is equivalent to the vanishing of the input connection at the point q_0. It is also one of the matching conditions studied in [11]. In the next section, we examine several interesting examples in which this symmetry condition does not hold.

4. CONTROLLED ROTATIONS OF SYSTEMS WITH KINEMATIC AND DYNAMIC ASYMMETRIES

The essential features of the asymmetry condition $A_1^\ell \neq A_1^{\ell T}$ are best illustrated by considering the case $m = \dim v = 1$. We consider forced rotations of various mechanisms about a prescribed axis. Note that because $m = 1$, the input connections are always flat (i.e. $\hat{R}_j{}^\gamma{}_{\ell i} = 0$ for all possible choices of the indices $\ell = i = 1;\ j, \gamma = 1, 2$).

Example 1. Consider the dynamics of the two-degree-of-freedom rotating pendulum mechanism depicted in Fig. 1. Let θ denote the configuration variable defining rotation about the vertical axis, and $v = \dot{\theta}$. Let ψ and ϕ be the Hooke joint angles as depicted. Letting

$$y = \begin{pmatrix} \theta \\ \psi \\ \phi \end{pmatrix},$$

we have the Lagrangian $L = \frac{1}{2}\dot{y}^T M(\psi, \phi)\dot{y} - V(\psi, \phi)$, where $M = \begin{pmatrix} \mathcal{N} & \mathcal{A} \\ \mathcal{A}^T & \mathcal{M} \end{pmatrix}$ with

$$\mathcal{N}(\psi, \phi) = m[(\ell^2 + \frac{b^2}{2})s_1^2 + (\ell^2 + \frac{a^2}{2})c_1^2 s_2^2 + \frac{a^2 + b^2}{2}c_1^2 c_2^2]$$

$$\mathcal{A}(\psi, \phi) = \left(m(\ell^2 - \frac{b^2}{2})c_1 s_2 c_2,\ \ -m(\ell^2 + \frac{b^2}{2})s_1\right),$$

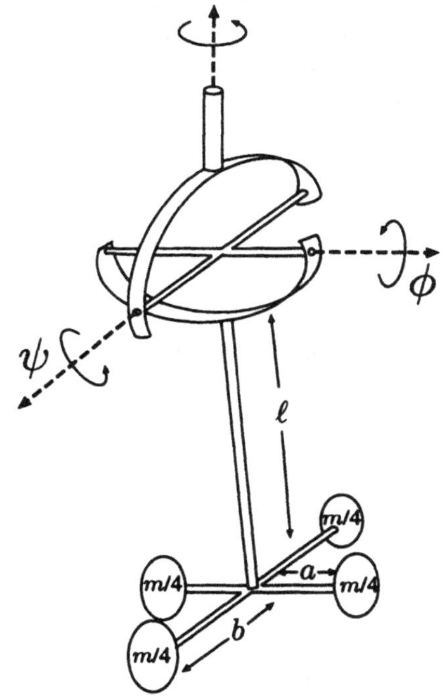

Fig. 1. A rotating pendulum suspended by a universal or Hooke joint. The masses are distributed at the bottom of a slender shaft so that the pendulum has unequal moments of inertia computed about the point of suspension.

$$\mathcal{M}(\phi) = \begin{pmatrix} m[(\ell^2 + \frac{a^2}{2})c_2^2 + \frac{a^2 + b^2}{2}s_2^2] & 0 \\ 0 & m(\ell^2 + \frac{b^2}{2}) \end{pmatrix}$$

and

$$V(\phi, \psi) = mg\ell \cos\phi \cos\psi.$$

We have adopted the shorthand notation $c_1 = \cos\psi$, $s_1 = \sin\psi$, $c_2 = \cos\phi$, and $s_2 = \sin\phi$.

From this, we may write the *reduced Lagrangian*

$$\begin{aligned} \mathcal{L}(\psi, \phi, \dot{\psi}, \dot{\phi}) &= \frac{1}{2}(\dot{\psi}, \dot{\phi})\mathcal{M}(\phi)(\dot{\psi}, \dot{\phi})^T \\ &+ v\mathcal{A}(\psi, \phi)(\dot{\psi}, \dot{\phi})^T - \mathcal{V}_a(\psi, \phi; v). \end{aligned} \quad (12)$$

The averaged potential, introduced in Section 3, specializes to

$$\begin{aligned} \mathcal{V}_A(q) &= \frac{1}{2}\mathcal{A}(q)^T \mathcal{M}(q)^{-1}\mathcal{A}(q)\,(\overline{v^2} - \bar{v}^2) \\ &- \overline{v^2}\mathcal{N}(\psi, \phi) + V(\psi, \phi), \end{aligned}$$

where again, the overbars denote time averages over one period of the velocity input $v(t)$. To be more explicit, we take a particular periodic forcing function $v(t) = \omega + \alpha \sin \beta t$. Then the form of the averaged potential highlights the importance of our distinction between *velocity-* and *acceleration-controlled* Lagrangian systems. The input-dependent parameters are

$$\overline{v^2} = \omega^2 + \frac{\alpha^2}{2}$$

and

$$\overline{v^2} - \bar{v}^2 = \frac{\alpha^2}{2}.$$

In the case of a velocity-controlled system, α and ω always appear together in the sum $\omega^2 + \frac{\alpha^2}{2}$, and adding an oscillatory component changes the averaged potential in the same manner as increasing the average (dc) component $\bar{v} = \omega$ of the rotational velocity. For acceleration-controlled systems, on the other hand, adding an oscillatory component to the forcing can have an effect on $\mathcal{V}_A$ which is quite distinct from the effect of increasing ω.

It is somewhat remarkable that, depending on the relative magnitude of the moments of inertia in the rotating pendulum of Fig. 1, the addition of a small-amplitude high-frequency oscillation can have a stabilizing effect. The analysis is carried out as follows. The averaged potential is written explicitly as

$$\mathcal{V}_A(\psi,\phi) = \mathcal{V}_g(\psi,\phi) + \mathcal{V}_\omega(\psi,\phi) + \mathcal{V}_\alpha(\psi,\phi), \quad (13)$$

where $\mathcal{V}_g(\psi,\phi) = -mg\ell c_1 c_2$ is the term due to gravity, $\mathcal{V}_\omega(\psi,\phi) = -\left(I_x s_1^2 + I_y c_1^2 s_2^2 + I_z c_1^2 c_2^2\right)\frac{\omega^2}{2}$ is the term due to the dc component of rotational velocity, and $\mathcal{V}_\alpha(\psi,\phi) = -\frac{I_y I_z c_1^2}{I_y c_2^2 + I_z s_2^2}\frac{\alpha^2}{4}$ is the term due to the sinusoidal signal superimposed on the dc velocity ω The inertias are $I_x = m(\ell^2 + \frac{b^2}{2})$, $I_y = m(\ell^2 + \frac{a^2}{2})$, and $I_z = m\frac{a^2+b^2}{2}$.

Since $a_{11}(\psi,\phi) = m(\ell^2 - \frac{b^2}{2})\cos\psi\sin\phi\cos\phi$ and $a_{12}(\psi,\phi) = -m(\ell^2 + \frac{b^2}{2})\sin\psi$, we have

$$\frac{\partial a_{11}}{\partial\phi} = m(\ell^2 - \frac{b^2}{2})\cos\psi(\cos^2\phi - \sin^2\phi)$$

and

$$\frac{\partial a_{12}}{\partial\psi} = -m(\ell^2 + \frac{b^2}{2})\cos\psi.$$

Even restricted to the operating point of interest $(\psi,\phi) = (0,0)$, these two quantities are generally unequal. This is reflected in the non-symmetry of the matrix coefficient of the acceleration term $\dot{v}$ in the linearization. (See Fig. 3 below.) The rotating Hooke-joint pendulum thus provides a simple example of a controlled mechanism for which the input connection does not vanish (although the input connection is flat as mentioned above). While the asymmetry also violates the hypothesis used to prove the statement regarding Lyapunov stability in Theorem 2, it is possible to use other means to associate stable motions of the physical system to local minima of the averaged potential.

To pursue the geometric analysis of this example, we examine the curvature of the Levi-Civita connections associated with the 3×3 (unreduced) inertia matrix $M = \begin{pmatrix} \mathcal{N} & \mathcal{A} \\ \mathcal{A}^T & \mathcal{M} \end{pmatrix}$ and the 2×2 (reduced) inertia matrix $\mathcal{M}$. It is a straightforward calculation (which is tedious, if you don't use computer algebra) to show that none of the six independent Riemann symbols of the first kind associated with the 3×3 inertia tensor $M(\psi,\phi)$ is zero. We can compare this with the results obtained by evaluating corresponding Riemann symbols and input symbols for the reduced Lagrangian system defined by $\mathcal{M}(\psi,\phi)$ and $\mathcal{A}((\psi,\phi)$. For the 2×2 tensor $\mathcal{M}(\psi,\phi)$, the curvature tensor is defined by the single Riemann symbol

$$R_{1212} = -m\,(l^2 - \frac{a^2}{2})\Big(\cos 2\phi + \frac{(l^2 - \frac{a^2}{2})\sin^2 2\phi}{4(l^2\cos^2\phi + \frac{a^2}{2}[1+\sin^2\phi])}\Big).$$

Thus, although the input connection is flat, the necessary conditions of Theorem 1 are not satisfied, indicating that this is an example of a *bona fide* acceleration-controlled system. This will be confirmed by showing that the acceleration component of our input $v(t) = \omega + \alpha\sin\beta t$ affects the averaged potential differently from the effect of the average of the rotational forcing ω.

To examine the effect of these sinusoidal variations, we write down the Hessian of the averaged potential evaluated at $(\psi,\phi) = (0,0)$:

$$m\begin{pmatrix} g\ell - \omega^2(\ell^2 - \frac{a^2}{2}) + \frac{\alpha^2}{4}(a^2+b^2) & 0 \\ 0 & g\ell - \omega^2(\ell^2 - \frac{b^2}{2}) - \frac{\alpha^2}{4}(a^2+b^2)\left(\frac{2\ell^2 - b^2}{2\ell^2 + a^2}\right) \end{pmatrix}.$$

Assume that the pendulum length is greater than either of the cross-sectional dimensions (a or b) (so that the z-axis moment of inertia is the smallest).

We find that oscillatory forcing will have a stabilizing effect precisely when $b > a$ (or equivalently when $I_x > I_y$). Referring to Figure 1 and the Hessian, we see that when there is no oscillatory component of rotational velocity (i.e. $\alpha = 0$), and $I_x > I_y$, the joint ψ, hinged in the upper yoke, "loses stability" first as the constant rotational velocity ω is increased through the critical value $\omega_{cr} = \sqrt{mg\ell/(I_x - I_z)} = \sqrt{mg\ell/(\ell^2 - a^2/2)}$. Careful inspection of this Hessian shows that adding a small amplitude oscillatory component to $v(\cdot)$ adds a positive component to the eigenvalue $\frac{\partial^2 \mathcal{V}_A}{\partial \psi^2}(0,0)$ and a negative component to the eigenvalue $\frac{\partial^2 \mathcal{V}_A}{\partial \phi^2}(0,0)$. It thus makes the "$\psi$-direction" slightly more stable while rendering the "ϕ-direction" slightly less stable. This is apparently due to the slight asymmetry in the way in which the ψ and ϕ variables "feel" the rotational acceleration $\ddot{\theta} = \dot{v}$. Note that in Figure 1, ψ is the degree-of-freedom associated with the "upper" hinge in the yoke, and ϕ is coordinate of the "lower" hinge displacement. The linearized dynamics displayed in Fig. 3 show the asymmetry in the way $\dot{v}$ enters the equations of motion. It is because of this asymmetry that high frequency oscillations tend to stabilize the neutral configuration $(\psi, \phi) = (0,0)$ when $b > a$. It is remarkable that because of the slight asymmetry in the possible motions of ψ and ϕ, no stabilizing effect is obtained from oscillatory rotations in the case $a > b$.

A Parametric Analysis of the Averaged Potential of the Rotating Hooke-joint Pendulum

$I_y > I_x > I_z$ (i.e. $a > b$)	Oscillatory forcing destabilizes
$I_x > I_y > I_z$ (i.e. $b > a$)	Oscillatory forcing stabilizes super – critical rotations with $\frac{mg\ell}{I_x - I_z} < \omega^2 < \frac{mg\ell}{I_y - I_z}$

Example 2. To understand the richness in the dynamics of Example 1, it is interesting to also investigate a rotating mechanism in which the two pendulum degrees of freedom are replaced by linear spring degrees of freedom. Once again we let θ denote the configuration variable defining the controlled rotation of a platform-supported mechanism about a fixed axis. Again, we wish to investigate the role kinematic (and dynamic) asymmetries play in determining the dynamic effects of adding an oscillatory component to the controlled velocity of rotation. We shall assume that a spring with spring constant k_1 connects the point mass m to the upper yoke, while a spring with constant k_2 connects this yoke to another yoke which is rigidly connected to the rotating platform. We assume the upper yoke has mass M, while the platform has moment of inertia I about its axis of rotation. There is no loss of generality in assuming all the mass m of the load in the upper yoke is concentrated at a point. The configuration variables for this system are x, denoting the displacement of the point mass with respect to the center of the upper yoke, and y denoting the displacement of the upper yoke with respect to the center of the platform. Letting $z = (x,y)^T$, we have Lagrangian $L = \frac{1}{2}\dot{z}^T M(x,y)\dot{z} - V(x,y)$, where $M = \begin{pmatrix} \mathcal{N} & \mathcal{A} \\ \mathcal{A}^T & \mathcal{M} \end{pmatrix}$ and

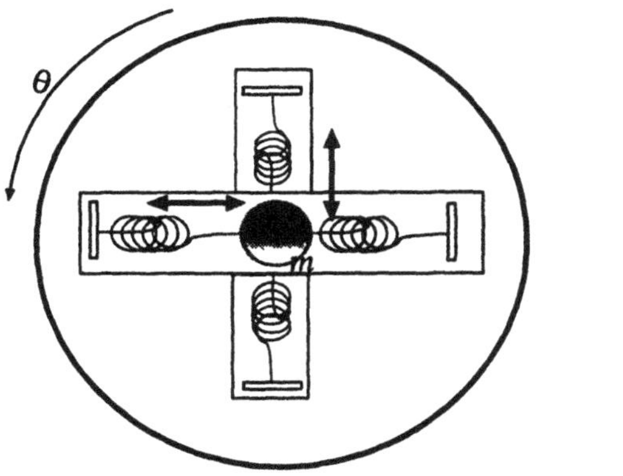

Fig. 2. One linearization of the rotating pendulum mechanism is physically realized as this rotating spring-yoke system.

$$\mathcal{N}(x,y) = I + My^2 + m(x^2 + y^2)$$

$$\mathcal{A}(x,y) = (-my,\ mx),$$

$$\mathcal{M} = \begin{pmatrix} m & 0 \\ 0 & m + M \end{pmatrix},$$

and

$$V(x,y) = \frac{1}{2}(k_1 x^2 + k_2 y^2).$$

Again, we construct the averaged potential using the general formula of Section 3:

Comparison of linearized rotating Hooke-joint dynamics and linearized rotating spring-yoke dynamics

Example 1 Linearization of reduced dynamics of the Hooke-joint mechanism about $(\psi,\phi)=(0,0)$	**Example 2** Linear reduced rotating spring-yoke mechanism dynamics
$\begin{pmatrix} m(\ell^2+\frac{a^2}{2}) & 0 \\ 0 & m(\ell^2+\frac{b^2}{2}) \end{pmatrix}\begin{pmatrix}\ddot{\psi}\\ \ddot{\phi}\end{pmatrix} + \dot{v}\begin{pmatrix} 0 & m(\ell^2-\frac{b^2}{2}) \\ -m(\ell^2+\frac{b^2}{2}) & 0 \end{pmatrix}\begin{pmatrix}\psi\\ \phi\end{pmatrix} + v\begin{pmatrix} 0 & 2m\ell^2 \\ -2m\ell^2 & 0 \end{pmatrix}\begin{pmatrix}\dot{\psi}\\ \dot{\phi}\end{pmatrix} + \begin{pmatrix} mg\ell - mv^2(\ell^2-\frac{a^2}{2}) & 0 \\ 0 & mg\ell - mv^2(\ell^2-\frac{b^2}{2}) \end{pmatrix}\begin{pmatrix}\psi\\ \phi\end{pmatrix} = \begin{pmatrix}0\\0\end{pmatrix}.$	$\begin{pmatrix} m & 0 \\ 0 & m+M \end{pmatrix}\begin{pmatrix}\ddot{x}\\ \ddot{y}\end{pmatrix} + \dot{v}\begin{pmatrix} 0 & -m \\ m & 0 \end{pmatrix}\begin{pmatrix}x\\ y\end{pmatrix} + v\begin{pmatrix} 0 & -2m \\ 2m & 0 \end{pmatrix}\begin{pmatrix}\dot{x}\\ \dot{y}\end{pmatrix} + \begin{pmatrix} k_1-mv^2 & 0 \\ 0 & k_2-(m+M)v^2 \end{pmatrix} = \begin{pmatrix}0\\0\end{pmatrix}$

Fig. 3. The asymmetry in the roles of the two hinges comprising the Hooke-joint in the pendulum of Example 1 give rise to the possibility of using high-frequency oscillation to stabilize the vertical configuration in high-velocity (super-critical) rotations. Despite kinematic asymmetries in the rotating spring-yoke mechanism of Example 2, no such oscillation-induced stabilization is possible.

$$\begin{aligned}\mathcal{V}_A(x,y) &= \frac{1}{2}\Big(k_1 - \frac{mM}{m+M}\overline{v^2} - \frac{m^2}{m+M}\bar{v}^2\Big)x^2 \\ &\quad +\frac{1}{2}\Big(k_2 - M\overline{v^2} - m\bar{v}^2\Big)y^2 \\ &= \frac{1}{2}\Big(k_1 - m(\omega^2 + \frac{M}{m+M}\frac{\alpha^2}{2})\Big)x^2 \\ &\quad + \frac{1}{2}\Big(k_2 - (m+M)(\omega^2 + \frac{M}{m+M}\frac{\alpha^2}{2})\Big)y^2.\end{aligned}$$

The simple form of the averaged potential in this case shows that there is a significant difference between the two mechanisms in terms of the effect of the oscillatory component of rotational velocity. For all choices of mechanism parameters in the spring-yoke mechanism, adding an oscillatory component of rotational forcing tends to destabilize the neutral equilibrium.

Looking at the various connections defined by inertia tensors, we find again that all six independent Riemann symbols of the first kind associated with the 3×3 inertia tensor $M(x,y)$ are nonzero. This means there is no change of coordinates $(\theta,x,y) \mapsto (\tilde{\theta},\tilde{x},\tilde{y})$ which transforms M into the identity. For the 2×2 (reduced) inertia tensor $\mathcal{M}(x,y)$, however, the Riemann symbol $R_{1212} = 0$. Hence the necessary conditions of Theorem 2 *are* satisfied.

While Example 1 illustrates the use of Theorem 1 in proving that the rotating Hooke-joint pendulum is acceleration controlled, Example 2 shows that the necessary conditions are not sufficient. Recall that we have shown that for *velocity-controlled* systems, terms in the averaged potential which involve accelerations will appear jointly with velocities, involving quantities of the form $\overline{v^2} = \omega^2 + \frac{\alpha^2}{2}$. In the present example, terms which involve accelerations also involve velocity and enter as components of the quantity $\omega^2 + \frac{M}{m+M}\frac{\alpha^2}{2}$. Although increasing the dc-component ω has the same effect on the averaged potential as increasing the magnitude α of the oscillatory forcing, it is only in the limiting case $m \to 0$ that the system tends to become *velocity-controlled.* In this limiting case, the mechanism is equivalent to a rotating single degree-of-freedom mass-spring system, which may be shown to be a true velocity-controlled system.

5. CONCLUSION

This short note has examined the open-loop oscillatory control of rotating systems in which there are certain mechanical asymmetries. For a broad class of controlled Lagrangian systems, we distinguish between *acceleration-controlled* systems and *velocity-controlled* systems. In the control of rotating systems, this distinction is important, since we have shown that for acceleration-controlled systems, superimposing an oscillatory component on a forced constant velocity rotation may have a stabilizing effect. Such stabilization is observed in controlling a rotating two degree-of-freedom pendulum with unequal cross-sectional moments of inertia. By superimposing a high-frequency sinusoidal signal on a constant rate of rotation, we have shown that it is possible to stably rotate the system in its neutral, hanging configuration at average rates higher than the critical (dc) rotational velocity at which this configuration would lose stability. We have shown in general that when certain curvature-like quantities associated with the system inertia are not zero, there is no coordinate change which transforms an acceleration-controlled system into a velocity-controlled system. While the pendulum example was shown to be acceleration-controlled, we have also presented a two d.o.f. example (of a rotating spring-yoke mechanism) which we showed was velocity-controlled in a certain limiting sense.

6. REFERENCES

[1] J. BAILLIEUL, 1990. "The Behavior of Superarticulated Mechanisms Subject to Periodic Forcing," in *Analysis of Controlled Dynamical Systems*, Proceedings of a Conference held in Lyon, France 3-6 Juillet, 1990, Gauthier, Bride, Bonnard, Kupka, Eds., Birkhauser.

[2] J. BAILLIEUL & M. LEVI, 1991. "Constrained Relative Motions in Rotational Mechanics," *Archive for Rational Mechanics and Analysis*, No. 115, pp. 101-135.

[3] J. BAILLIEUL, 1993. "Stable Average Motions of Mechanical Systems Subject to Periodic Forcing," *Dynamics and Control of Mechanical Systems: The falling cat and related problems*, Fields Institute Communications, **1**, Michael Enos, Ed., American Mathematical Society, Providence, pp. 1-23.

[4] J. BAILLIEUL, 1995. "Energy Methods for Stability of Bilinear Systems with Oscillatory Inputs," *Int'l J. of Robust and Nonlinear Control*, Special Issue on the Control of Mechanical Systems," H. Nijmeijer & A.J. van der Schaft, Guest Eds., Vol. 5, pp. 205-381.

[5] S. WEIBEL, T. KAPER, & J. BAILLIEUL, 1997. "Global Dynamics of a Rapidly Forced Cart and Pendulum," *Nonlinear Dynamics*, **13**: 131-170, July, 1997.

[6] S. WEIBEL & J. BAILLIEUL, 1998. "Open-loop Stabilization of an n–Pendulum," *Int. J. of Control*, vol. 71, no. 5, pp. 931-957.

[7] J. BAILLIEUL, 1998. "The Geometry of Controlled Mechanical Systems," in *Mathematical Control Theory*, J. Baillieul & J.C. Willems, Eds., Springer-Verlag, New York, 1998, pp. 322-354.

[8] J. BAILLIEUL, 1999. "A control design which respects characteristic length scales in smart systems and smart structures," *Proceedings of SPIE's 6-th Annual Int'l Symposium on Smart Structures and Smart Materials*, March 1-4, Newport Beach, CA, Volume 3667, pp. 202-210.

[9] A.R. CHAMPNEYS & W.B. FRASER, 1999. "The 'Indian rope trick' for a parametrically excited flexible rod; linearized analysis," Preprint, School of Mathematics and Statistics, Univ. of Sydney, NSW 2006, Australia.

[10] F. BULLO, 1999. "Vibrational Control of Mechanical Systems," University of Illinois, CSL Preprint.

[11] A.M. BLOCH, N.E. LEONARD, & J.E. MARSDEN, 1999. "Potential Shaping and the Method of controlled Lagrangians," *Proc. of the 38th IEEE Conf. on Dec. & Control*, Phoenix, AZ, Dec. 8, pp. 1652-1657.

[12] S. WEIBEL & J. BAILLIEUL, 1998. "Averaging and Energy Methods for Robust Open-loop Control of Mechanical Systems," in *Essays on Mathematical Robotics* Volume 104, IMA Volumes in Mathematics and its Applications, Edited by J. Baillieul, S.S. Sastry, and H.J. Sussmann, Springer-Verlag, New York, pp. 203-269.

NEW EXAMPLES IN NONINTERACTING CONTROL FOR HAMILTONIAN SYSTEMS

Alessandro Astolfi * **Laura Menini** **

* *Dep. of Electrical and Electronic Engineering, Imperial College*
Exhibition Road, London SW7 2BT, England
E-mail: a.astolfi@ic.ac.uk
** *Dip. di Informatica, Sistemi e Produzione*
Università di Roma Tor Vergata
Via di Tor Vergata 110, 00133 Roma, Italy.
E-mail: menini@disp.uniroma2.it

Abstract: In this paper, the problem of noninteracting control with stability is studied for generalized Hamiltonian systems with reference to two relevant case studies: the control of the angular velocities of an underactuated rigid body and the control of a third order food chain. For the rigid body, under suitable conditions, noninteraction and asymptotic stability are jointly achievable by means of a static state-feedback control law; whereas, for the food chain, noninteraction and asymptotic stability of a "critical" equilibrium are jointly achievable by means of dynamic state-feedback control laws. *Copyright ©2000 IFAC*

1. INTRODUCTION

The problem of asymptotic stabilization of Hamiltonian and generalized Hamiltonian systems has been widely studied in the control literature, see *e.g.* (van der Schaft, 1986; Ortega *et al.*, 1998; Woolsey and Leonard, 1999; Bloch *et al.*, 1999; Bloch, 1999). This is because, due to their special structure, it is possible to obtain general results, valid for a large set of physically motivated systems.

On the other hand, the problem of noninteracting control for nonlinear systems has received considerable attention (Isidori and Grizzle, 1988; Wagner, 1989; Zhan *et al.*, 1991; Battilotti, 1994; Isidori, 1995). By means of geometric control theory, necessary and sufficient conditions for the existence of either static or dynamic state-feedback control laws, which allow to obtain stable noninteractive closed-loop systems, have been proposed, and systematic design procedures have been given. The essential results of the theory are related with the "so called" P^* and Δ_{mix} dynamics, contained in the zero dynamics Δ^* as depicted in the illustrative diagram in Fig. 1, see (Isidori, 1995, Chapter 7) for details. As a matter of fact, under suitable regularity and stabilizability assumptions, the problem of noninteracting control with stability is solvable by means of a regular static state-feedback if and only if the P^* dynamics are asymptotically stable, whereas it is solvable by means of dynamic state-feedback if and only if the Δ_{mix} dynamics are asymptotically stable.

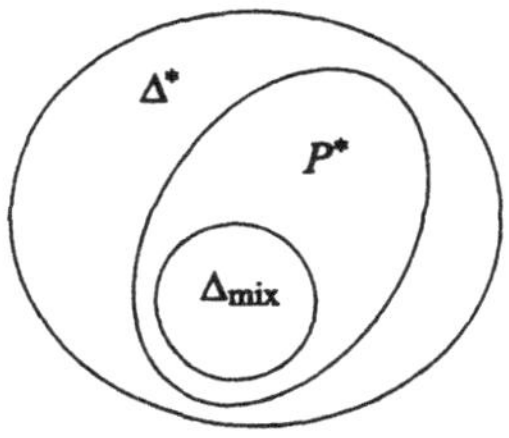

Fig. 1. Inclusion properties relative to the dynamics concerned with the problem of noninteraction with stability.

The problem of noninteracting control with stability for Hamiltonian systems characterized by a non-degenerate Poisson bracket has been dealt with in (Huijberts and van der Schaft, 1990; Astolfi and Menini, 1999; Astolfi and Menini, 2000), with particular attention to the so-called "simple" Hamiltonian systems. Therein, easy to check conditions for the solvability of the problem by means of either static or dynamic state-feedback control laws have been given. To the best of the authors' knowledge, the problem of noninteraction with stability for generalized Hamiltonian systems has not been considered yet, *i.e.*, no general results exploiting the special structure of such systems are available in the literature. In this paper, two significant examples are studied in detail: an underactuated rigid body (Euler's equations) and a three dimensional food-chain. In both cases it is assumed that the state of the system is available for feedback.

2. GENERALIZED HAMILTONIAN SYSTEMS

In this paper we consider systems described by:

$$\dot{x} = (J(x) - R(x))\frac{\partial H}{\partial x}(x) + b(x)u \quad \text{(1a)}$$

$$y = h(x), \quad \text{(1b)}$$

where $x \in \mathbb{R}^n$, $u \in \mathbb{R}^m$, $y \in \mathbb{R}^q$, $J(x) = -J^T(x)$, and $R(x) = R^T(x) \geq 0$, which are also called port-controlled Hamiltonian systems, see (Maschke and van der Schaft, 1992; van der Schaft, 1996; Ortega *et al.*, 1999). Many physical systems can be described by equations (1), with the matrix $J(x)$ describing the internal structure (the interconnection) of the system and the matrix $R(x)$ describing the natural damping.

Observe that the dimension of the state-space of system (1) is in general not even. This has strong influence on the structural properties of the system, as many interesting properties can be proved only if the Poisson structure is non-degenerate, *i.e.*, if the matrix $J(x)$ has full rank. Since the rank of any skew-symmetric matrix is even, any Poisson structure defined on an odd dimensional manifold is degenerate. Generalized Hamiltonian systems with degenerate Poisson structure are also called *Poisson systems.*

Example 1. (Rigid body). Consider a rigid body with principal moments of inertia $I_i > 0$, $i = 1, 2, 3$, and let ω_1, ω_2 and ω_3 denote the angular velocities with respect to the principal axes. Assume that two actuators (*e.g.* gas jet actuators) provide control torques u_1 and u_2, which are the control inputs of the system. Assume also that the two outputs to be decoupled are linear combinations of the ω_i's. In particular, let the system be described by:

$$I_1\dot{\omega}_1 = (I_2 - I_3)\,\omega_2\,\omega_3 + u_1, \quad \text{(2a)}$$

$$I_2\dot{\omega}_2 = (I_3 - I_1)\,\omega_3\,\omega_1 + u_2, \quad \text{(2b)}$$

$$I_3\dot{\omega}_3 = (I_1 - I_2)\,\omega_1\,\omega_2 + b_{3,1}\,u_1 + b_{3,2}\,u_2, \quad \text{(2c)}$$

$$y_1 = \omega_1 + B\,\omega_2 + C\,\omega_3, \quad \text{(2d)}$$

$$y_2 = F\,\omega_1 + \omega_2 + H\,\omega_3. \quad \text{(2e)}$$

Setting $x = [\omega_1 \quad \omega_2 \quad \omega_3]^T$, such a system can be written as a generalized Hamiltonian system, with

$$H(x) = \frac{1}{2}\left(I_1\,\omega_1^2 + I_2\,\omega_2^2 + I_3\,\omega_3^2\right),$$

$$J(x) = \begin{bmatrix} 0 & -\dfrac{I_3\,\omega_3}{I_1\,I_2} & \dfrac{I_2\,\omega_2}{I_1\,I_3} \\ \dfrac{I_3\,\omega_3}{I_1\,I_2} & 0 & -\dfrac{I_1\,\omega_1}{I_2\,I_3} \\ -\dfrac{I_2\,\omega_2}{I_1\,I_3} & \dfrac{I_1\,\omega_1}{I_2\,I_3} & 0 \end{bmatrix},$$

and $R(x) = 0$. Notice that $J(x)$ has rank 2 for every $x \neq 0$.

Example 2. (Food chain). Consider the normalized third order preys-predators system

$$\dot{x}_1 = \phi(x_1, x_2) - x_1 + u_1, \quad \text{(3a)}$$

$$\dot{x}_2 = -\phi(x_1, x_2) - x_2 + \phi(x_2, x_3), \quad \text{(3b)}$$

$$\dot{x}_3 = -\phi(x_2, x_3) - x_3 + u_2, \quad \text{(3c)}$$

$$y_1 = x_1, \quad \text{(3d)}$$

$$y_2 = x_3, \quad \text{(3e)}$$

with Lotka-Volterra predation mechanism, *i.e.*, with $\phi(x_i, x_j) = x_i\,x_j$. Such a system is *positive*, *i.e.*, $x_i(0) \geq 0$, $i = 1, 2, 3$ and $u_i(\tau) \geq 0$, $i = 1, 2$, for all $\tau \in [0, t)$, imply $x_i(t) \geq 0$, for all $t > 0$. Moreover, it can be written in the form (1) with $H(x) = x_1 + x_2 + x_3$, $R(x) = \operatorname{diag}(x_1, x_2, x_3)$, and $J(x) = \begin{bmatrix} 0 & x_1x_2 & 0 \\ -x_1x_2 & 0 & x_2x_3 \\ 0 & -x_2x_3 & 0 \end{bmatrix}$.

3. NONINTERACTING CONTROL WITH STABILITY FOR THE RIGID BODY

Consider the system in Example 1 and let $A_1 = (I_2 - I_3)/I_1$, $A_2 = (I_3 - I_1)/I_2$, $A_3 = (I_1 - I_2)/I_3$, $\alpha = b_{3,1}/I_3$, $\beta = b_{3,2}/I_3$, and $K = [A_1 \quad A_2 \quad A_3 \quad \alpha \quad \beta \quad B \quad C \quad F \quad H]^T$. Assume $A_i \neq 0$, $i = 1, 2, 3$.

The problem of stabilization of the zero equilibrium of the rigid body angular velocity has been widely studied and solved, see *e.g.*, (Brockett, 1983; Outbib and Sallet, 1992; Astolfi, 1999) and the references therein. On the other hand, the problem of noninteracting control is trivially solvable. Goal of this section is to investigate under which conditions both problems can be jointly solved. To be more general, we also investigate the possibility of regulating the two outputs to desired values Y_1^* and Y_2^*, obtaining, at the same time, both noninteraction and stability for the overall system.

To this end, we make the simplifying assumption [1] $FB \neq 1$ and we assume $FB + BH\alpha - H\beta - 1 - C\alpha + C\beta F \neq 0$ so that the system has vector relative degree (1, 1). After a preliminary feedback, called standard noninteracting feedback, system (2) can be rewritten as

$$\dot{y}_1 = v_1, \tag{4a}$$

$$\dot{y}_2 = v_2, \tag{4b}$$

$$\dot{\omega}_3 = f_3(y_1, y_2, \omega_3) + g_{1,3}v_1 + g_{2,3}v_2, \tag{4c}$$

where v_1 and v_2 are the new control inputs, and $g_{1,3}$ and $g_{2,3}$ are two real numbers, depending on the vector K. Letting $f(y_1, y_2, \omega_3) = [0 \;\; 0 \;\; f_3(y_1, y_2, \omega_3)]^T$, $g_1 = [1 \;\; 0 \;\; g_{1,3}]^T$ and $g_2 = [0 \;\; 1 \;\; g_{2,3}]^T$, we can state the following result.

Theorem 1. For system (4), $P^* \equiv \Delta_{\text{mix}} = \emptyset$ if and only if either $L_{g_1}f = 0$ or $L_{g_2}f = 0$.

Proof. The necessity is based on straightforward but tedious computations, which show that, for all possible values of the vector K, if $L_{g_i}f \neq 0$, $i = 1, 2$, then $\dim(\Delta_{\text{mix}}) = \dim(P^*) = 1$. To show that $L_{g_1}f = 0$ implies $P^* = \emptyset$ (the case of $L_{g_2}f = 0$ can be treated similarly), notice that, in such a case, $P_2^* = \text{span}\{g_1\}$, and $P_1^* = \text{span}\{g_2, [0 \;\; 0 \;\; 1]^T\}$.

Corollary 1. If $L_{g_1}f = 0$ and $L_{g_2}f \neq 0$ (or $L_{g_2}f = 0$ and $L_{g_1}f \neq 0$), then the problem of noninteracting control with asymptotic stability is locally solvable for system (4) by means of static state-feedback.

The condition in Corollary 1 can be easily tested in view of the following lemma.

Lemma 1. $L_{g_1}f = 0$ if and only if either

(i) $A_1 A_3 > 0$, $\beta = 0$,
$\alpha = \pm\sqrt{A_3/A_1}$, $H = -F/\alpha$,

or

(ii) $A_2 A_3 > 0$, $\alpha = 0$,
$\beta = \pm\sqrt{A_3/A_2}$, $H = -1/\beta$.

$L_{g_2}f = 0$ if and only if either

(iii) $A_2 A_3 > 0$, $\alpha = 0$,
$\beta = \pm\sqrt{A_3/A_2}$, $C = -B/\beta$,

or

(iv) $A_1 A_3 > 0$, $\beta = 0$,
$\alpha = \pm\sqrt{A_3/A_1}$, $C = -1/\alpha$.

If none of conditions (i)–(iv) holds, then the Δ_{mix} dynamics (coinciding with the P^* dynamics) are described by the equation:

$$\dot{\omega}_3 = Q_0(K, Y_1^*, Y_2^*) + Q_1(K, Y_1^*, Y_2^*)\,\omega_3 + Q_2(K, Y_1^*, Y_2^*)\,\omega_3^2. \tag{5}$$

Therefore, we now study the properties of equation (5). Let $\Delta = Q_1^2 - 4\,Q_2\,Q_0$ (for simplicity we drop the arguments), then, one of the following conditions occurs.

a) System (5) has no equilibria, *i.e.*, either $\Delta < 0$ or $Q_1 = Q_2 = 0$ and $Q_0 \neq 0$.

b) System (5) has 1 equilibrium, *i.e.*, $Q_2 = 0$ and $\Delta > 0$.

c) System (5) has 2 coinciding equilibria, *i.e.*, $\Delta = 0$ and $Q_2 \neq 0$.

d) System (5) has 2 distinct equilibria, *i.e.*, $\Delta > 0$ and $Q_2 \neq 0$.

e) System (5) has an infinite number of equilibria, *i.e.*, $Q_2 = Q_1 = Q_0 = 0$.

Remark 1. If $Y_1^* = 0$ and $Y_2^* = 0$ then only cases c) and e) can occur.

Theorem 2. If $L_{g_1}f \neq 0$ and $L_{g_2}f \neq 0$, then the problem of noninteracting control with asymptotic stability is globally solvable if and only if case b) occurs, with $Q_1 < 0$, and it is locally solvable if case d) occurs. The problem of noninteracting control with simple stability is locally solvable in case e).

3.1 *A numerical example.*

If condition (i) of Lemma 1 holds and condition (iv) does not [2], then, in view of Corollary 1, the problem of noninteracting control with stability can be solved by means of static state-feedback. From equations (4), we have that $P_1^* = \text{span}\{g_2, [f, g_2]\}$ and $P_2^* = \text{span}\{g_1\}$. Then, the change of coordinates $z = -g_{1,3}\,y_1 + \omega_3$, transforms system (4) into the following form:

$$\dot{y}_1 = v_1, \tag{6a}$$

$$\dot{y}_2 = v_2, \tag{6b}$$

$$\dot{z} = f_z(z, y_2) + b\,v_2, \tag{6c}$$

where b is real constant. The stabilization of subsystem (6a) is trivial, whereas, in order to deal with subsystem (6b)-(6c), a further change of coordinates $\tilde{z} = z - b\,y_2$ is convenient. As a matter of fact, the origin $y_2 = 0$, $\tilde{z} = 0$ of subsystem

[1] This assumption can be easily removed.

[2] Note that (i) and (iii) cannot occurr simultaneously because $A_1 A_3 > 0 \Rightarrow A_2 A_3 < 0$.

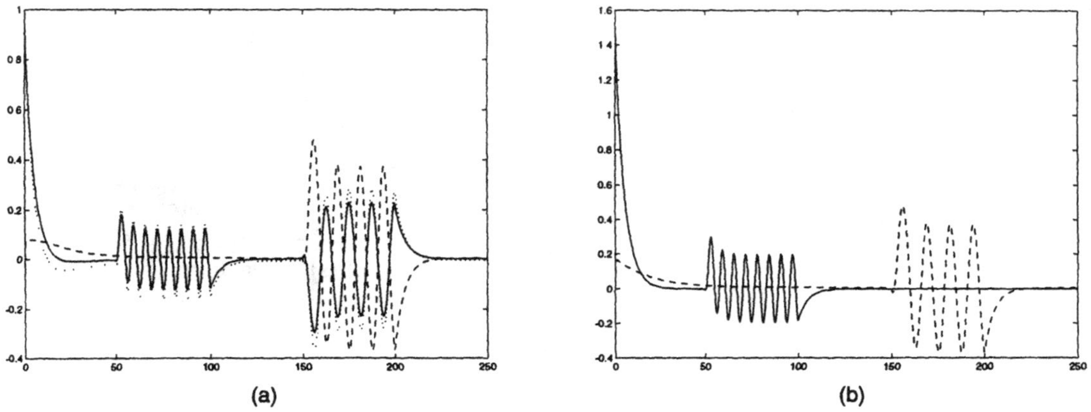

Fig. 2. The rigid body example. (a): $\omega_1(t)$ (solid), $\omega_2(t)$ (dashed) and $\omega_3(t)$ (dotted); (b) $y_1(t)$ (solid) and $y_2(t)$ (dashed).

$$\dot{y}_2 = v_2, \tag{7a}$$
$$\dot{\tilde{z}} = a_2\,\tilde{z}^2 + a_1\,\tilde{z}\,y_2, \tag{7b}$$

being $a_1 \neq 0$, can be easily stabilized by means of a static feedback control law from y_2 and $\tilde{z}$, whose design can be performed by means of standard backstepping techniques (Krstić *et al.*, 1995). The controller proposed here, in order to obtain global asymptotic stability of the origin and noninteraction, is of the form:

$$\begin{aligned} v_1 &= -k_1\,y_1 + r_1, \\ v_2 &= (\gamma_1 + 2\,\gamma_2\,\tilde{z})\left(\delta_3\,\tilde{z}^3 + \frac{\delta_3}{\gamma_2}\,\tilde{z}\,y_2\right) - \\ &\quad k_2\left(y_2 - \left(\gamma_1\,\tilde{z} + \gamma_2\,\tilde{z}^2\right)\right) - a_1\,\tilde{z}^2 + r_2, \end{aligned}$$

where γ_1, γ_2 and δ_3 are suitable constants, and r_1 and r_2 new inputs with respect to which noninteraction is guaranteed. The results of a simulation with the proposed controller, with $A_1 = 1$, $A_2 = -1$, $A_3 = 1.5$ $\alpha = \sqrt{A_3/A_1}$, $\beta = 0$, $B = 1$, $C = 0.5$, $F = 0.5$, $H = -F/\alpha$, $a_1 \approx -1.23$, $a_2 \approx -1.61$, $\gamma_1 \approx -1.32$, $\gamma_2 \approx 612$, $\delta_3 = -1.5$, $k_1 = 0.2$, $k_2 = 0.2$, $\omega_1(0) = 1$, $\omega_2(0) = 0$, $\omega_3(0) = 1$, are reported in Fig. 2. In order to emphasize the property of noninteraction, $r_1(t)$ has been set to zero for $t \notin [50, 100]$, and equal to a sinusoidal signal in such an interval, whereas the input $r_2(t)$ has been set to zero for $t \notin [150, 200]$, and equal to a sinusoidal signal in such an interval.

As for the possibility of regulating the two outputs y_1 and y_2 to constant reference values Y_1^* and Y_2^*, it is clear from the above equations that there are no restrictions on the admissible values for Y_1^*. Furthermore, from equations (7), it is clear that, in general, for every $Y_2^* \neq 0$, only local asymptotic stability of $y_2 = Y_2^*$, $\tilde{z} = \tilde{Z}^*$, being $\tilde{Z}^*$ a real constant depending on K, can be achieved.

4. NONINTERACTING CONTROL WITH STABILITY FOR THE FOOD-CHAIN

The goal of this section is to regulate, if possible, the two outputs of system (3) to desired constant values $Y_1^* > 0$ and $Y_2^* > 0$, obtaining, at the same time, both noninteraction and stability for the overall system. The stabilization problem has been solved in (Ortega *et al.*, 1999) by means of both state and output feedback control laws, by using the concepts of energy-shaping and damping injection. Notice that, after the static state-feedback

$$\begin{aligned} u_1 &= x_1 - x_1\,x_2 + v_1, \\ u_2 &= x_3 + x_2\,x_3 + v_2, \end{aligned}$$

system (3) can be written as

$$\dot{y}_1 = v_1, \tag{8a}$$
$$\dot{y}_2 = v_2, \tag{8b}$$
$$\dot{x}_2 = -x_2\,(1 + y_1 - y_2). \tag{8c}$$

From equation (8c), it is obvious that, if $1 + Y_1^* > Y_2^*$, then the problem is easily solvable (with $\lim_{t\to+\infty} x_2(t) = 0$), whereas, if $1 + Y_1^* < Y_2^*$, then the problem is not solvable at all. Hence, only the special case $1 + Y_1^* = Y_2^*$ needs to be investigated. By means of easy computations, it can be seen that $\Delta_{\text{mix}} = \emptyset$, whereas, at any point of the state space where $x_2 \neq 0$, $P^* = \text{span}\{\partial/\partial x_2\}$. Hence, the problem of noninteraction with asymptotic stability around an equilibrium $x_1 = Y_1^*$, $x_2 = x_2^* > 0$, $x_3 = Y_2^* = 1 + Y_1^*$, is not solvable by means of static state-feedback, but it is solvable by means of dynamic state-feedback. The controller proposed here is based on a dynamic extension as described in (Isidori, 1995, Chapter 7), but has lower dimension than the controller which would result from applying the design technique reported there. Let μ_1 and μ_2 be the two state variables of the controller, and let:

$$\dot{\mu}_1 = -\mu_1\,(1 + y_1) + w_1, \tag{9a}$$

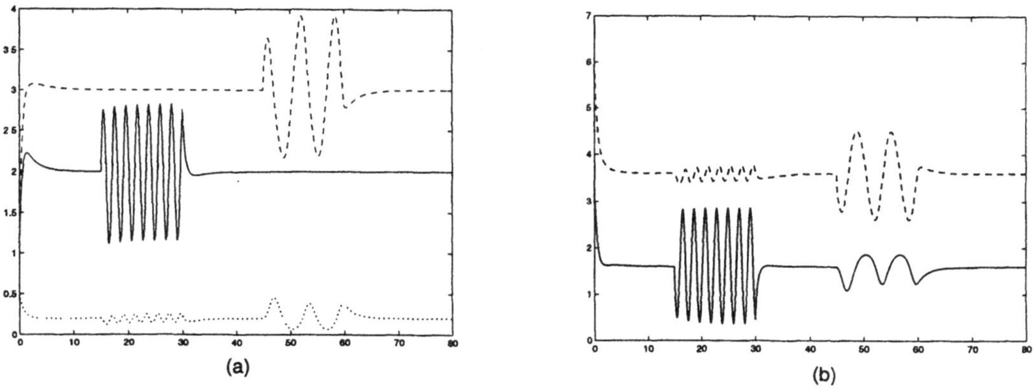

Fig. 3. The food chain example. (a): $x_1(t)$ (solid), $x_2(t)$ (dashed) and $x_3(t)$ (dotted); (b) $u_1(t)$ (solid) and $u_2(t)$ (dashed)

$$\dot{\mu}_2 = -\mu_2 (1 - y_2) + w_2, \tag{9b}$$

where the two fictitious inputs w_1 and w_2 will be specified later as functions of the state of the extended system. After the change of coordinates $\xi_3 = \mu_1 \mu_2 / x_2$, which is valid as long as $x_2 \neq 0$, the extended system (8)-(9), has the following form:

$$\dot{y}_1 = v_1, \tag{10a}$$
$$\dot{\mu}_1 = -\mu_1 (1 + y_1) + w_1, \tag{10b}$$
$$\dot{y}_1 = v_1, \tag{10c}$$
$$\dot{\mu}_2 = -\mu_2 (1 - y_2) + w_2, \tag{10d}$$
$$\dot{\xi}_3 = -\xi_3 \left(1 - \frac{w_1}{\mu_1} - \frac{w_2}{\mu_2}\right). \tag{10e}$$

From equation (10e), it is clear that the choice

$$w_1 = (Y_1^* + 1)\,\mu_1, \tag{11a}$$
$$w_2 = -(Y_2^* - 1)\,\mu_2, \tag{11b}$$

with $1 + Y_1^* = Y_2^*$, implies that the variable ξ_3 remains constant for all $t \geq 0$. Hence, assuming that the two state variables μ_1 and μ_2 are regulated to suitable desired values $\mu_{1,\infty}$ and $\mu_{2,\infty}$, respectively, one has

$$\lim_{t \to +\infty} x_2(t) = \frac{\mu_{1,\infty}\, \mu_{2,\infty}\, x_2(0)}{\mu_1(0)\, \mu_2(0)}.$$

Therefore, not only it is possible to regulate y_1 and y_2, but also the *non-actuated* state value x_2 to a desired positive value X_2^*, by choosing $\mu_1(0) = \mu_{1,\infty}\sqrt{x_2(0)/X_2^*}$, $\mu_2(0) = \mu_{2,\infty}\sqrt{x_2(0)/X_2^*}$. Now, in order to obtain asymptotic stability of the equilibrium $y_1 = Y_1^*$, $\mu_1 = \mu_{1,\infty}$, with domain of attraction $\mu_1 > 0$, for the subsystem (10a)-(10b)-(11a), one can use standard backstepping techniques (Krstić *et al.*, 1995), yielding the control law

$$v_1 = -(\mu_1 + k_1)(y_1 - Y_1^* - \mu_1 + \mu_{1,\infty}) + r_1,$$

where $k_1 > 0$ and r_1 is the new input, with respect to which noninteraction has to be guaranteed. Analogously, the choice

$$v_2 = -(\mu_2 + k_2)(y_2 - Y_2^* + \mu_2 - \mu_{2,\infty}) + r_2,$$

with $k_2 > 0$, guarantees asymptotic stability of the equilibrium $y_2 = Y_2^*$, $\mu_2 = \mu_{2,\infty}$, with domain of attraction $\mu_2 > 0$, for the subsystem (10c)-(10d)-(11b).

Remark 2. It can be shown that, if $X_2^* < 1$ and the initial conditions $x_1(0)$ and $x_3(0)$ are in a suitable neighborhood of the desired values Y_1^* and Y_2^*, then $u_1(t) \geq 0$ and $u_2(t) \geq 0$, for all $t \geq 0$.

Remark 3. It is stressed that the special case $1 + Y_1^* = Y_2^*$ is the only one in which it is possible to obtain a steady state response in which also $X_2^* > 0$.

4.1 *Simulation results*

The results of a simulation with the proposed controller, with $Y_1^* = 2$, $Y_2^* = 3$, $X_2^* = 0.2$, $\mu_{1,\infty} = \mu_{2,\infty} = 0.3$ $x_1(0) = x_3(0) = 1$, $x_2(0) = 0.5$, $k_1 = k_2 = 2$, are reported in Fig. 3. In order to emphasize the property of noninteraction, the input $r_1(t)$ has been set to zero for $t \notin [15,\ 30]$, and equal to a sinusoidal function inside such an interval, whereas the input $r_2(t)$ has been set to zero for $t \notin [45,\ 60]$, and equal to a sinusoidal function inside such an interval.

4.2 *Extensions*

It is easy to see that the property $\Delta_{\text{mix}} = \emptyset$ still holds for more general predation mechanisms than the above mentioned $\phi(x_i, x_j)$. In particular, equation (3b) (which is the only relevant one, since the two inputs can be used to obtain equations (8a) and (8b)) can be substituted by

$$\dot{x}_2 = -x_2\, \theta_1(x_1) - x_2 + x_2\, \theta_3(x_3),$$

where the $\theta_i(\cdot)$, $i = 1, 2$ are arbitrary functions. This fact suggest that it might be possible to extend the results presented above also to this case.

Another interesting extension is suggested by the fact that $\Delta_{\text{mix}} = \emptyset$ also for higher order food chains, provided that a special input and output structure is selected. For example, limiting the attention to systems characterized by the Lotka-Volterra predation mechanism $\phi(x_i, x_j)$, if the dynamics of the system are:

$$\begin{aligned}
\dot{x}_1 &= \phi(x_1, x_2) - x_1 + u_1, \\
&\vdots \\
\dot{x}_i &= -\phi(x_{i-1}, x_i) - x_i + \phi(x_i, x_{i+1}), \\
&\vdots \\
\dot{x}_n &= -\phi(x_{n-1}, x_n) - x_n + u_2, \\
y_1 &= x_r, \quad r \in \{1, 2, \ldots, n-2\} \\
y_2 &= x_{r+2},
\end{aligned}$$

after the change of coordinates $z_1 = y_1$, $z_2 = \dot{y}_1$, ..., $z_r = y_1^{(r-1)}$, $z_{r+1} = y_2$, $z_{r+2} = \dot{y}_2$, ..., $z_{n-1} = y_2^{(n-r-2)}$, $z_n = x_{r+1}$, and a suitable static state-feedback, the system becomes:

$$\begin{aligned}
\dot{z}_1 &= z_2, \\
&\vdots \\
\dot{z}_r &= v_1, \\
\dot{z}_{r+1} &= z_{r+2}, \\
&\vdots \\
\dot{z}_{n-1} &= v_2, \\
\dot{z}_n &= -z_n\,(z_1 + 1 - z_{r+1}), \\
y_1 &= z_1, \\
y_2 &= z_{r+1}.
\end{aligned}$$

It is easy to see that also in this case $\Delta_{\text{mix}} = \emptyset$, whereas $P^* = \text{span}\{\partial/\partial z_n\}$.

5. REFERENCES

Astolfi, A. (1999). Output feedback stabilization of the angular velocity of a rigid body. *Syst. & Contr. Lett.*

Astolfi, A. and L. Menini (1999). Further results on decoupling with stability for Hamiltonian systems. In: *Stability and Stabilization* (D. Aeyels, A. van der Schaft and F. Lamnabhi-Lagarrigue, Eds.). Springer Verlag.

Astolfi, A. and L. Menini (2000). Noninteracting control with stability for Hamiltonian systems. *IEEE Trans. Automatic Control.* To appear.

Battilotti, S. (1994). *Noninteracting control with stability for nonlinear systems.* Springer Verlag.

Bloch, A. M. (1999). Asymptotic stability in energy preserving systems. In: *Proceedings of the 38-th Conference on Decision and Control.* Phoenix, AZ.

Bloch, A. M., N. E. Leonard and J. E. Marsden (1999). Potential shaping and the method of controlled Lagrangians. In: *Proceedings of the 38-th Conference on Decision and Control.* Phoenix, AZ.

Brockett, R. W. (1983). Asymptotic stability and feedback stabilization. *Differential geometry control theory* pp. 181–191.

Huijberts, H. J. C. and A. J. van der Schaft (1990). Input-output decoupling with stability for Hamiltonian systems. *Math. Control Signal Systems* pp. 125–138.

Isidori, A. (1995). *Nonlinear control systems.* Springer Verlag. Third edition.

Isidori, A. and J. W. Grizzle (1988). Fixed modes and nonlinear noninteractive control with stability. *IEEE Trans. Aut. Contr.* **AC-33**, 907–914.

Krstić, M., I. Kanellakopoulos and P. Kokotović (1995). *Nonlinear and Adaptive Control Design.* John Wiley & sons. New York.

Maschke, B. M. and A. J. van der Schaft (1992). Port controlled Hamiltonian systems: modeling origins and system theoretic properties. In: *Proc. 2nd IFAC Symp. on Nonlinear Control Systems design (NOLCOS'92).* Bordeaux. pp. 282–288.

Ortega, R., A. Astolfi, G. Bastin and H. Ro driguez Cortes (1999). Output feedback control of food-chain systems. In: *New trends in nonlinear observer design* (H. Nijmeijer and T. Fossen, Eds.). Springer-Verlag.

Ortega, R., A. Loria, P. J. Nicklasson and H. Sira-Ramirez (1998). *Passivity-Based Control of Euler-Lagrange Systems.* Springer Verlag. Berlin.

Outbib, R. and G. Sallet (1992). Stabilizability of the angular velocity of a rigid body revisited. *Syst. & Contr. Lett.* **18**, 93–98.

van der Schaft, A. J. (1986). Stabilization of Hamiltonian systems. *Nonl. An. Th. Meth. Appl.* **10**, 1021 – 1035.

van der Schaft, A. J. (1996). *L_2-Gain and Passivity Techniques in Nonlinear Control.* Springer-Verlag. Berlin.

Wagner, K. G. (1989). On nonlinear noninteraction with stability. In: *Proceedings of the 28-th Conference on Decision and Control.* Tampa, FL.

Woolsey, C. A. and N. E. Leonard (1999). Global asymptotic stabilization of an underwater vehicle using internal rotors. In: *Proceedings of the 38-th Conference on Decision and Control.* Phoenix, AZ.

Zhan, W., A. Isidori and T. J. Tarn (1991). A canonical dynamic extension algorithm for noninteraction with stability for affine nonlinear systems. *Syst. & Contr. Lett.* **17**, 177–184.

ROBUST OUTPUT-FEEDBACK TRACKER DESIGN FOR NONHOLONOMIC SYSTEMS

Zhong-Ping Jiang *,[1]

* *Department of Electrical Engineering, Polytechnic University,*
Six Metrotech Center, Brooklyn, NY 11201.
E-mail: zjiang@control.poly.edu

Abstract: This paper presents new results for the global output-feedback tracking of nonholonomic systems in chained form. The tracking controllers are obtained on the basis of a recursive technique and a full exploitation of the system structure. Sufficient conditions on the reference paths are given to attain *global exponential* tracking. When disturbances occur in the dynamic extension of a nonholonomic chained system, it is shown how to modify the controller design procedure to yield robust tracking control laws. The proposed method is demonstrated and discussed by means of a benchmark nonholonomic knife-edge mechanical system. *Copyright © 2000 IFAC*

Keywords: Nonholonomic systems; global exponential tracking; output feedback.

1. INTRODUCTION

In recent years, many researchers have been attracted to the open-loop control and feedback control of nonlinear mechanical systems with nonholonomic constraints. This has been known as a particularly challenging class of control systems for nonlinear control, because there is no C^1, or even continuous, state-dependent stabilizing control law (Brockett, 1983). In other words, Brockett's well-known necessary condition for feedback asymptotic stabilization implies that traditional linear methods and existing nonlinear control design schemes are not applicable to solve the asymptotic stabilization of nonholonomic control systems. New ideas and fundamentally nonlinear techniques are necessary and, indeed, have been introduced by a number of researchers (see Kolmanovsky and McClamroch (1995) for an excellent introduction to this field).

In this paper, we focus on the *tracking* problem for *a popular class* of nonholonomic control systems, known as *chained systems* due to Murray–Sastry (1991). In contrast to the stabilization issue, the *tracking* problem turns out to be easier to handle, although it is even practically more meaningful. Some earlier constructive results on this topic are mainly based on the application of linearization and dynamic feedback linearization – see, for instance, Kanayama et al. (1990); Walsh et al. (1994); Fierro and Lewis (1995); Kolmanovsky and McClamroch (1995); Fliess et al. (1995). As it is well understood within the control community, the major drawbacks of these approaches are the smallness of region of feasibility of a linearization-based control law and the singularity issue pertaining to the application of a dynamic linearizing feedback based on differentially flat systems theory. Nevertheless, some exceptions are Samson and Ait-Abderrahim (1991); Escobar et al. (1998); Jiang and Nijmeijer (1997) where global tracking solutions were obtained for particular low-dimensional nonholonomic systems transformable into the simplest chained form, by means of Lyapunov's direct method and the energy-shaping approach. Our present work takes an approach different from Samson and Ait-Abderrahim (1991); Escobar et al. (1998) and is a natural continuation of our recent papers Jiang and Pomet (1996); Jiang and Nijmeijer (1997, 1999b,a), that proposed for the first time a recursive application of

[1] Partially supported by a start-up grant from Polytechnic University.

integrator backstepping to nonholonomic control problems (see Kokotović (1992) for a neat introduction to the backstepping technique). While Jiang and Nijmeijer (1997, 1999b) focus on the state-feedback tracking problem, we propose new solutions to the problem of output-feedback tracking, that improve our earlier results in Jiang and Nijmeijer (1999a). As several authors noticed, the class of chained form systems appears to be the largest class of nonholonomic control systems for which systematic and constructive control methods are available in the past literature Kolmanovsky and McClamroch (1995).

Recall that a chained system is represented by

$$\begin{aligned} \dot{x}_1 &= u_1 \\ \dot{x}_2 &= u_2 \\ \dot{x}_i &= x_{i-1}u_1, \quad 3 \le i \le n \end{aligned} \tag{1}$$

where $u = (u_1, u_2)$ is the control and $x = (x_1, \ldots, x_n)$ is the state. Here, we view the *flat* outputs $y := (x_1, x_n)$ (see Fliess et al. (1995) for the definition of flatness) as the output of (1).

The tracking problem we want to solve in this paper is stated as follows. Given a vector-valued reference trajectory $x_d(t) = (x_{1d}(t), \ldots, x_{nd}(t))$ which is generated by the *replica* chained form

$$\begin{aligned} \dot{x}_{1d} &= u_{1d} \\ \dot{x}_{2d} &= u_{2d} \\ \dot{x}_{id} &= x_{(i-1)d}u_{1d}, \quad 3 \le i \le n \end{aligned} \tag{2}$$

with $u_d = (u_{1d}, u_{2d})$ as the reference control, the goal is to design a C^0 time-varying output-feedback law of the type

$$\dot{\chi} = \nu(t, \chi, y) \,, \quad u = \mu(t, \chi, y) \tag{3}$$

so that the following properties hold:

(1) All the closed-loop trajectories $x_e(t) := x(t) - x_d(t)$ and $\chi(t)$ are bounded over $[0, \infty)$.
(2) Under additional conditions, $x_e(t)$ converges to zero at an *exponential* rate as $t \to \infty$.

The tracking problem for a dynamic extension of (1) with nonlinear disturbances will be stated and solved in Section 3.

2. GLOBAL OUTPUT FEEDBACK TRACKING

In this subsection, we discuss how to solve the problem of global output-feedback exponential tracking for nonholonomic systems in chained form (1). An extension to the dynamic model description follows readily as a direct application of backstepping. For simplicity, we assume that $u_{1d}(t) \ge 0$ for all $t \ge 0$. (This usually means that a virtual object moves forward along the reference path – see, e.g., Kanayama et al. (1990); Jiang and Nijmeijer (1997).) The case when $u_{1d}(t) \le 0$ for all $t \ge 0$ can be treated analogously. See, for instance, Jiang and Nijmeijer (1999a) for discussions on the general case. For simplicity, the proofs are omitted. As we shall see in Section 3, the robustification of our Lyapunov design in front of some disturbances is a rather straighforward matter.

2.1 *Design of a reduced-order observer*

The goal of this section is to design an observer to estimate the unmeasured states $(x_2, \ldots, x_{n-1})$ using the information of output $y = (x_1, x_n)$. Owing to the triangular structure in the chained form (1), a (global) reduced-order time-varying estimator can be easily obtained, under

Assumption 1. $x_{id}(t)\,(2 \le i \le n-1)$, $u_d(t)$ and $u_{1d}^{(i)}(t)$ for all $1 \le i \le n-2$ are bounded over $[0, \infty)$. In addition, there exist a nonnegative integer $\ell \ge n-3$ and a constant $\sigma_1 > 0$ such that

$$\liminf_{t \to \infty} \frac{1}{t} \int_{t_0}^{t_0+t} |u_{1d}(\tau)|^{2\ell+2} d\tau \; > \; \sigma_1, \; \forall t_0 \ge 0 \tag{4}$$

Indeed, for constant design parameters k_i $(1 \le i \le n-2)$, introduce the new (unmeasured) variables

$$\begin{aligned} \xi_1 &= x_{n-1} - k_1 x_n \\ \xi_2 &= x_{n-2} - k_2 x_n \\ &\vdots \\ \xi_{n-2} &= x_2 - k_{n-2} x_n \end{aligned} \tag{5}$$

which can be reconstructed through the time-varying observer that depends on the output x_n and the input u

$$\begin{aligned} \dot{\hat{\xi}}_1 &= \hat{\xi}_2 u_1 + k_2 x_n u_1 - k_1(\hat{\xi}_1 + k_1 x_n) u_1 \\ &\vdots \\ \dot{\hat{\xi}}_{n-3} &= \hat{\xi}_{n-2} u_1 + k_{n-2} x_n u_1 - k_{n-3}(\hat{\xi}_1 + k_1 x_n) u_1 \\ \dot{\hat{\xi}}_{n-2} &= u_2 - k_{n-2}(\hat{\xi}_1 + k_1 x_n) u_1 \end{aligned} \tag{6}$$

For notational simplicity, let us denote $\tilde{\xi} = (\tilde{\xi}_1, \ldots, \tilde{\xi}_{n-2})^T$ as the observation error, with $\tilde{\xi}_i = \xi_i - \hat{\xi}_i$ for all $1 \le i \le n-2$. It is easily seen that the $\tilde{\xi}$-dynamics satisfy

$$\dot{\tilde{\xi}} = u_1 \begin{bmatrix} -k_1 & & \\ \vdots & & I_{n-3} \\ -k_{n-2} & \cdots & 0 \end{bmatrix} \tilde{\xi} := u_1 A_o \tilde{\xi} \tag{7}$$

It is shown that, under sufficient conditions on u_1, the observation error $\tilde{\xi}(t)$ goes to zero at an exponential rate as $t \to \infty$. To this end, let us pick a continuous, nonnegative, bounded function $u_{1d}(t)$ (which will be understood as the reference input signal in subsec. 2.2) and rewrite (7) as

$$\dot{\tilde{\xi}} = u_{1d}(t)A_o\tilde{\xi} + (u_1 - u_{1d}(t))A_o\tilde{\xi} \quad (8)$$

Obviously, the constant matrix A_o can be made asymptotically stable with all eigenvalues assignable arbitrarily in the open left-half complex plane provided that the k_i's are appropriately chosen and the complex conjugate eigenvalues are assigned in pair. Once A_o is rendered asymptotically stable, let $P_o = P_o^T > 0$ be the unique solution to the Lyapunov matrix equation

$$A_o^T P_o + P_o A_o = -I_{n-2} \quad (9)$$

Under the conditions of the following proposition, the observation error $\tilde{\xi}(t)$ globally exponentially converges to zero and, therefore, the unmeasured states $(x_2, \ldots, x_{n-1})$ of system (1) are asymptotically recovered via the reduced-order observer (6).

Proposition 1. If $u_{1d}(t) \geq 0$ and $u_1(t)$ are continuous, bounded functions on $[0, \infty)$, and if there exist a finite time instant $t_0 \geq 0$ and $\gamma > \lambda_{\max}(P_o)/\lambda_{\min}(P_o)$ such that

$$\liminf_{t\to\infty} \frac{1}{t}\int_{t_0}^{t_0+t} u_{1d}(\tau)d\tau > \gamma \sup_{\tau \geq t_0} |u_1(\tau) - u_{1d}(\tau)|$$

then, for any $\tilde{\xi}(0) \in \mathbb{R}^{n-2}$, the solution $\tilde{\xi}(t)$ of (8) exponentially converges to 0 as $t \to \infty$.

In the next subsection, we show how to fulfill the conditions of Proposition 1 by carefully selecting the tracking control law u_1.

2.2 *Output-feedback design procedure*

Introduce the following new variables

$$\begin{aligned} \zeta_1 &= x_n - x_{nd} \\ \zeta_n &= x_1 - x_{1d} \quad (10)\\ \zeta_i &= \hat{\xi}_{i-1} - (x_{(n-i+1)d} - k_{i-1}x_{nd}),\ 1 < i < n \end{aligned}$$

Letting $k_0 = 0$ and, for each $1 \leq i \leq n-2$, denoting $\Xi_i = \zeta_{i+1} + k_i\zeta_1 - k_{i-1}(\zeta_2 + k_1\zeta_1)$, the ζ-dynamics satisfy ODEs

$$\begin{aligned} \dot{\zeta}_1 &= \Xi_1 u_{1d} + \tilde{\xi}_1 u_1 \\ &\quad + (u_1 - u_{1d})(\Xi_1 + x_{(n-1)d}), \\ \dot{\zeta}_i &= \Xi_i u_{1d} + (u_1 - u_{1d})(\Xi_i + x_{(n-i)d} \\ &\quad - k_{i-1}x_{(n-1)d}), \quad 1 < i < n-1 \\ \dot{\zeta}_{n-1} &= u_2 - k_{n-2}(\hat{\xi}_1 + k_1 x_n)u_1 \\ &\quad - (u_{2d} - k_{n-2}x_{(n-1)d}u_{1d}), \\ \dot{\zeta}_n &= u_1 - u_{1d} \quad (11) \end{aligned}$$

Our tracking design procedure will be developed on the interconnected system (11) and (8), using the information of *partial*-state $\zeta = (\zeta_1, \ldots, \zeta_n)^T$. Notice that the state component $\tilde{\xi}$ is unavailable for feedback design.

Observe that the $(\zeta_1, \ldots, \zeta_{n-1})$-subsystem with input u_2 satisfies a lower-triangularity condition and that its nominal part, when ignoring the terms relating to $\tilde{\xi}_1$ and $u_1 - u_{1d}$, is a linear time-varying system. Guided by this important observation and inspired by the earlier backstepping designs of Jiang and Nijmeijer (1997, 1999b), we will first design a backstepping-based tracking control law u_2 and then focus on the design of input u_1 aiming at the fulfillment of the conditions of Proposition 1. To this end, we start with an introduction of a change of coordinates that brings the $(\zeta_1, \ldots, \zeta_{n-1})$-subsystem of (11) into a new system. Such a coordinates transformation can be easily found via backstepping:

$$\begin{aligned} \tilde{\zeta}_1 &= \zeta_1, \quad \tilde{\zeta}_n = \zeta_n \\ \tilde{\zeta}_{i+1} &= \zeta_{i+1} - \tilde{\alpha}_i(\zeta_1, \ldots, \zeta_i, \phi_1, \ldots, \phi_i) \quad (12) \end{aligned}$$

where, for each $1 \leq i \leq n-2$,

$$\begin{aligned} \tilde{\alpha}_i &= -\tilde{c}_i\tilde{\zeta}_i\phi_1(u_{1d}) - \tilde{\zeta}_{i-1} - k_i\zeta_1 + k_{i-1}(\zeta_2 + k_1\zeta_1) \\ &\quad + \sum_{j=1}^{i-1}\frac{\partial\tilde{\alpha}_{i-1}}{\partial\zeta_j}\Xi_j + \sum_{j=1}^{i-1}\frac{\partial\tilde{\alpha}_{i-1}}{\partial\phi_j}\phi_{j+1} \quad (13) \end{aligned}$$

with $\tilde{c}_i > 0$, $\tilde{\zeta}_0 = 0$ and the ϕ_i's defined as $\phi_1 = u_{1d}^{2\ell+2}$ and $\phi_i = \dot{\phi}_{i-1}/u_{1d}$ for $1 < i \leq n-2$.

In new $\tilde{\zeta}$-coordinates, system (11) is rewritten as

$$\begin{aligned} \dot{\tilde{\zeta}}_i &= -u_{1d}\tilde{\zeta}_{i-1} - \tilde{c}_i u_{1d}^{2\ell+2}\tilde{\zeta}_i + u_{1d}\tilde{\zeta}_{i+1} \\ &\quad - \frac{\partial\tilde{\alpha}_{i-1}}{\partial\zeta_1}u_1\tilde{\xi}_1 + (u_1 - u_{1d})\varphi_i, \quad \forall 1 \leq i < n-1 \\ \dot{\tilde{\zeta}}_{n-1} &= u_2 + \varphi_n - \frac{\partial\tilde{\alpha}_{n-2}}{\partial\zeta_1}u_1\tilde{\xi}_1 + (u_1 - u_{1d})\varphi_{n-1} \\ \dot{\tilde{\zeta}}_n &= u_1 - u_{1d} \quad (14) \end{aligned}$$

where φ_n is a known function and every φ_i, $1 \leq i \leq n-1$, is a known function depending on $(\zeta_1, \ldots, \zeta_{i+1}, x_d)$.

To complete the design of the input u_2, consider the function

$$\tilde{V} = \frac{1}{2}\tilde{\zeta}_1^2 + \cdots + \frac{1}{2}\tilde{\zeta}_{n-1}^2 \quad (15)$$

When u_2 is selected as

$$u_2 = -\tilde{c}_{n-1}u_{1d}^{2\ell+2}\tilde{\zeta}_{n-1} - \tilde{\zeta}_{n-2}u_{1d} - \varphi_n \quad (16)$$

with $\tilde{c}_{n-1} > 0$, the time derivative of $\tilde{V}$ along the solutions of (14) satisfies

$$\dot{\tilde{V}} = -\sum_{i=1}^{n-1} \tilde{c}_i u_{1d}^{2\ell+2}\tilde{\zeta}_i^2 + \left(\tilde{\zeta}_1 + \sum_{i=1}^{n-2} \frac{\partial \tilde{\alpha}_i}{\partial \zeta_1}\tilde{\zeta}_{i+1}\right) u_1 \tilde{\xi}_1$$
$$+ \sum_{i=1}^{n-1} \tilde{\zeta}_i \varphi_i (u_1 - u_{1d}) \qquad (17)$$

It is shown that this tracking control law (16) together with

$$u_1 = u_{1d} - \tilde{c}_n \tilde{\zeta}_n, \quad \tilde{c}_n > 0 \qquad (18)$$

solves the output-feedback tracking problem.

Theorem 2. If Assumption 1 holds with $u_{1d}(t) \geq 0$ for all $t \geq 0$, then, all signals $(x_e(t), \tilde{\xi}(t), \zeta(t))$ associated with the resulting closed-loop system (1), (6), (16) and (18) are globally bounded on $[0, \infty)$. Moreover, for any initial instant $t_0 \geq 0$, any initial tracking error $x_e(t_0) \in \mathbb{R}^n$ and any $\tilde{\xi}(t_0) \in \mathbb{R}^{n-1}$, there exists a constant $\tilde{\delta} = \tilde{\delta}(t_0, x_e(t_0), \tilde{\xi}(t_0)) > 0$ such that

$$|x_e(t)| \leq \tilde{\delta} \exp(-\tilde{\sigma}(t - t_0)), \quad \forall t \geq t_0 \qquad (19)$$

with $\tilde{\sigma} > 0$ independent of the initial conditions $(t_0, x_e(t_0), \tilde{\xi}(t_0))$.

3. TRACKING OF A DYNAMIC MODEL

Naturally, the knowledge of more state variables, in addition to $y = (x_1, x_n)$, allows us to reduce the observer order and to improve the performance. Without going into a very general circumstance, we briefly discuss the commonly encountered situation where the whole states x of the kinematic model (1) are measured and the unmeasured states and disturbances only appear in the integrators appended to (1). More specifically, we consider a simplified dynamic system which is composed of (1) and two disturbed integrators:

$$\begin{aligned} \dot{u}_1 &= v_1 \\ \dot{u}_2 &= v_2 + \kappa_1(u_1)\kappa_2(x) \end{aligned} \qquad (20)$$

where κ_1 and κ_2 are known C^1 functions. We assume that the x-state is measured and the (u_1, u_2)-state is not measured. Owing to this assumption, the design of v_2 needs to take into account the presence of the *nonlinear* disturbance driven by u_1. As already said, it deserves our efforts to identify a more general class of disturbances and develop a robustification tool for the tracking problem in the future. This section is only devoted to the demonstration of how the tracking methodology presented in the last section can be extended to the case where a nonholonomic system is transformed into a perturbed dynamic model with incomplete state information. Such an example is given in Section 4.

In order to provide an estimate $\hat{u}_1$ for the unmeasured state u_1, we consider the linear double integrator: $\dot{x}_1 = u_1$ and $\dot{u}_1 = v_1$. According to linear systems theory, a reduced-order observer is of the form

$$\dot{\hat{\tilde{u}}}_1 = -k_1\hat{\tilde{u}}_1 - k_1^2 x_1 + v_1, \quad k_1 > 0 \qquad (21)$$

which gives an estimate $\hat{\tilde{u}}_1$ of the variable $\tilde{u}_1 = u_1 - k_1 x_1$. Thus, $\hat{u}_1 = \hat{\tilde{u}}_1 + k_1 x_1$ is an estimate of u_1. Let us denote the observation error e_1^u as

$$e_1^u = u_1 - \hat{u}_1 = \tilde{u}_1 - \hat{\tilde{u}}_1 \qquad (22)$$

Then, it holds that $\dot{e}_1^u = -k_1 e_1^u$. Similarly, we can derive a reduced-order observer for the unmeasured state u_2 as ($k_2 > 0$)

$$\dot{\hat{\tilde{u}}}_2 = -k_2\hat{\tilde{u}}_2 - k_2^2 x_2 + v_2 + \kappa_1(\hat{u}_1)\kappa_2(x) \qquad (23)$$

which yields an estimate $\hat{\tilde{u}}_2$ of the variable $\tilde{u}_2 = u_2 - k_2 x_2$. As a result, $\hat{u}_2 = \hat{\tilde{u}}_2 + k_2 x_2$ is an estimate of u_2. Then, the observation error $e_2^u := u_2 - \hat{u}_2 = \tilde{u}_2 - \hat{\tilde{u}}_2$ satisfies

$$\dot{e}_2^u = -k_2 e_2^u + (\kappa_1(u_1) - \kappa_1(\hat{u}_1))\kappa_2(x) \qquad (24)$$

Introduce the new variables $\eta_i = x_{n-i+1} - x_{(n-i+1)d}$ for each $1 \leq i \leq n-1$, $\eta_n = \hat{\tilde{u}}_2 + k_2 x_2 - u_{2d}$, $\eta_{n+1} = x_1 - x_{1d}$ and $\eta_{n+2} = \hat{\tilde{u}}_1 + k_1 x_1 - u_{1d}$. By direct computation, we have

$$\begin{aligned} \dot{\eta}_i &= \eta_{i+1}u_{1d} - (\eta_{i+1} + x_{(n-i)d})(\eta_{n+2} + e_1^u) \\ \dot{\eta}_{n-1} &= \eta_n + e_2^u, \quad 1 \leq i < n-1 \\ \dot{\eta}_n &= v_2 + \kappa_1(\hat{u}_1)\kappa_2(x) - \dot{u}_{2d} + k_2 e_2^u \\ \dot{\eta}_{n+1} &= \eta_{n+2} + e_1^u \\ \dot{\eta}_{n+2} &= v_1 - \dot{u}_{1d} + k_1 e_1^u \end{aligned} \qquad (25)$$

Notice that the state η of system (25) is available for feedback design. As previously done, we have converted the tracking problem for the *perturbed* dynamic model (1) and (20) into a stabilization problem for the transformed system (25).

Before proceeding with the design of (v_1, v_2), we need the following assumption on the functions κ_1 and κ_2.

Assumption 2. κ_1 is a polynomial function with degree p and κ_2 satisfies the inequality

$$|\kappa_2(x)| \leq \kappa_2^\star |(x_2, \ldots, x_{n-1})| \qquad (26)$$

for a positive constant $\kappa_2^\star$ and all $x \in \mathbb{R}^n$.

Next, we focus on the design of tracking controllers v_1 and v_2 for the *perturbed* dynamic model (1) and (20).

Pick two constants k_3 and k_4 such that

$$A_1 = \begin{bmatrix} 0 & 1 \\ k_3 & k_4 \end{bmatrix} \quad \text{is stable.}$$

With the help of linear systems theory, it is easy to prove that the observer-based tracking controller

$$v_1 = \dot{u}_{1d} + k_3\eta_{n+1} + k_4\eta_{n+2} \tag{27}$$

drives the states η_{n+1} and η_{n+2} to zero at an *exponential* rate.

Now, we proceed with the design of v_2. As above, introduce the following new variables

$$\bar{\eta}_i = \eta_i - \bar{\alpha}_{i-1}(\eta_1, \ldots, \eta_{i-1}, \phi_1, \ldots, \phi_{i-1}) \tag{28}$$

where $\bar{\alpha}_0 = 0$ and, for $1 \leq i \leq n-1$,

$$\begin{aligned} \bar{\alpha}_i = & -\bar{c}_i\bar{\eta}_i\phi_1(u_{1d}) - \bar{\eta}_{i-1} + \sum_{j=1}^{i-1} \frac{\partial\bar{\alpha}_{i-1}}{\partial\eta_j}\eta_{j+1} \\ & + \sum_{j=1}^{i-1} \frac{\partial\bar{\alpha}_{i-1}}{\partial\phi_j}\phi_{j+1} \quad \text{for some } \bar{c}_i > 0 \end{aligned} \tag{29}$$

In new $\bar{\eta}$-coordinates, the $(\eta_1, \ldots, \eta_n)$-subsystem of (25) is rewritten as

$$\begin{aligned} \dot{\bar{\eta}}_i &= -u_{1d}\bar{\eta}_{i-1} - \bar{c}_i u_{1d}^{2\ell+2}\bar{\eta}_i + u_{1d}\bar{\eta}_{i+1} \\ & \quad + \bar{\varphi}_i(\eta_{n+2} + e_1^u), \quad 1 \leq i \leq n-2 \\ \dot{\bar{\eta}}_{n-1} &= -u_{1d}\bar{\eta}_{n-2} - \bar{c}_{n-1}\bar{\eta}_{n-1} + \bar{\eta}_n \\ & \quad + e_2^u + \bar{\varphi}_{n-1}(\eta_{n+2} + e_1^u) \\ \dot{\bar{\eta}}_n &= v_2 + \bar{\varphi}_{n+1} \\ & \quad + k_2 e_2^u - \frac{\partial\bar{\alpha}_{n-1}}{\partial\eta_{n-1}} e_2^u + \bar{\varphi}_n(\eta_{n+2} + e_1^u) \end{aligned} \tag{30}$$

where all $\bar{\varphi}_i$ are known functions Now, consider the quadratic Lyapunov function for system (30) and (24)

$$\bar{V} = \frac{1}{2}\bar{\eta}_1^2 + \cdots + \frac{1}{2}\bar{\eta}_n^2 + \frac{1}{2}|e_2^u|^2 \tag{31}$$

When we choose the tracking control law v_2 as

$$v_2 = -\bar{c}_n\bar{\eta}_n - \bar{\eta}_{n-1} - \bar{\varphi}_{n+1} \tag{32}$$

with $\bar{c}_n > 0$, the time derivative of $\bar{V}$ along the solutions of (30) and (24) satisfies

$$\begin{aligned} \dot{\bar{V}} = & -\sum_{i=1}^{n-2} \bar{c}_i u_{1d}^{2\ell+2}\bar{\eta}_i^2 - \bar{c}_{n-1}\bar{\eta}_{n-1}^2 - \bar{c}_n\bar{\eta}_n^2 - k_2|e_2^u|^2 \\ & + (\kappa_1(u_1) - \kappa_1(\widehat{u}_1))\,\kappa_2(x)e_2^u + \bar{\eta}_{n-1}e_2^u \\ & + \sum_{i=1}^{n} \bar{\eta}_i\bar{\varphi}_i(\eta_{n+2} + e_1^u) + (k_2 - \frac{\partial\bar{\alpha}_{n-1}}{\partial\eta_{n-1}})\bar{\eta}_n e_2^u \end{aligned} \tag{33}$$

Under the conditions of the following result, the tracking control laws (32) and (27) solve the tracking problem via partial-state feedback.

Proposition 3. Under Assumptions 1 and 2, if $u_{1d}^{(n-1)}$ and $\dot{u}_{2d}$ are also bounded on $[0, \infty)$, then the tracking errors $x_e(t)$ associated with the resulting closed-loop system *globally exponentially* converge to zero.

4. AN ILLUSTRATIVE EXAMPLE

We illustrate the presented methodology with the help of a simple benchmark nonholonomic knife-edge example. This mechanical system has often served as an elementary illustration for theoretical studies on nonholonomic control systems. We refer the reader to Kolmanovsky and McClamroch (1995); Jiang and Nijmeijer (1999b) for the details on this system.

The dynamics of a knife-edge moving on the plane satisfy the following differential equations Kolmanovsky and McClamroch (1995):

$$\begin{aligned} \ddot{x}_c &= \frac{\gamma}{m}\sin\phi + \frac{\tau_1}{m}\cos\phi \\ \ddot{y}_c &= -\frac{\gamma}{m}\cos\phi + \frac{\tau_1}{m}\sin\phi \\ \ddot{\phi} &= \frac{\tau_2}{I_c} \\ \dot{x}_c\sin\phi &= \dot{y}_c\cos\phi \end{aligned} \tag{34}$$

where (x_c, y_c) denotes the coordinates for the center of mass of the knife-edge, ϕ denotes the heading angle measured from the x-axis, and τ_1 is the pushing force in the direction of the heading angle, τ_2 the steering torque about the vertical axis through the center of mass. The constants (m, I_c) are the mass and the moment of inertia of the knife-edge respectively, γ is the scalar constraint multiplier. Note that the fourth-equation in (34) represents the nonholonomic constraint on the linear velocity of the knife-edge system.

It was shown in Jiang and Nijmeijer (1999b) that (34) can be brought into a chained form (1) with two integrators, via *state*-feedback and coordinate changes. However, here, we assume that (x_c, y_c, ϕ)-only is available to the designer, but the velocities $(\dot{x}_c, \dot{y}_c, \dot{\phi})$. System (34) will be transformed only to a *perturbed* dynamic model

$$\dot{x}_1 = x_4, \; \dot{x}_2 = x_5, \; \dot{x}_3 = x_2 x_4 \tag{35}$$

$$\dot{x}_4 = v_1, \quad \dot{x}_5 = v_2 - x_4^2 x_2 \tag{36}$$

provided that we apply

$$x_1 = \phi, \; x_2 = x_c\cos\phi + y_c\sin\phi$$
$$x_3 = x_c\sin\phi - y_c\cos\phi, \; x_4 = \dot{\phi}$$

$$x_5 = \dot{x}_c \cos\phi + \dot{y}_c \sin\phi + \dot{\phi}(-x_c \sin\phi + y_c \cos\phi)$$
$$v_1 = \frac{\tau_2}{I_c},\ v_2 = \frac{\tau_1}{m} + \frac{\tau_2}{I_c}(-x_c \sin\phi + y_c \cos\phi)$$

Note that the new state variables x_4 and x_5 are unmeasured. As in Jiang and Nijmeijer (1999b), consider the following reference trajectory

$$\phi^{\mathrm{r}}(t) = t\,, \quad x_c^{\mathrm{r}}(t) = \sin t\,, \quad y_c^{\mathrm{r}}(t) = -\cos t$$

which corresponds to the center of mass of the knife-edge moving along the circle centered at the origin of unit radius with uniform angular rate.

A direct application of Proposition 3 yields a global observer-based partial-state feedback tracker that forces the knif-edge to follow the reference trajectory. (In this case, $\eta_1 = x_3 - 1$, $\eta_2 = x_2$, $\eta_3 = \widehat{\widehat{u}}_2 + k_2 x_2$, $\eta_4 = x_1 - t$, $\eta_5 = \widehat{\widehat{u}}_1 + k_1 x_1 - 1$.) For simulations and comparison, we choose the same parameters and initial conditions as in the state-feedback case Jiang and Nijmeijer (1999b). The time histories of the tracking errors and torques (τ_1, τ_2) are given in Figure 5. As seen from Figure 5, the output-feedback law performs as well as the state-feedback case after a short transient period.

5. CONCLUSION

Global exponential tracking solutions were obtained for a class of nonholonomic control systems using output feedback. The constructive control design procedure has been inspired from our recent tracking approaches proposed in Jiang and Nijmeijer (1997, 1999b,a) and yields tracking controllers with stronger stability properties. In contrast to earlier approaches to the tracking problem for nonholonomic systems in Kanayama et al. (1990); Walsh et al. (1994); Fierro and Lewis (1995); Kolmanovsky and McClamroch (1995); Fliess et al. (1995); Escobar et al. (1998), our recursive approach does dot exhibit restrictive features such as the singularity issue, the smallness of region of feasibility, and the discontinuity in the control laws. It is under current investigation to explore the underlying flexibility of our constructive tracking scheme and see how the tracking problem can be addressed for other types of nonlinear mechanical systems with nonholonomic constraints.

References

G. Escobar, R. Ortega, and M. Reyhanoglu. Regulation and tracking of the nonholonomic double integrator: A field-oriented control approach. *Automatica*, 34:125–131, 1998.

R. Fierro and F. L. Lewis. Control of a nonholonomic mobile robot: backstepping kinematics into dynamics. *Proc. 34th IEEE Conf. Dec. Control*, pages 3805–3810, 1995.

M. Fliess, J. Levine, P. Martin, and P. Rouchon. Design of trajectory stabilizing feedback for driftless flat systems. *Proc. 3rd European Contr. Conf.*, pages 1882–1887, 1995.

Z. P. Jiang and H. Nijmeijer. Tracking control of mobile robots: a case study in backstepping. *Automatica*, 33:1393–1399, 1997.

Z. P. Jiang and H. Nijmeijer. Observer-controller design for global tracking of nonholonomic systems. In H. Nijmeijer and T. Fossen, editors, *New Directions in Nonlinear Observer Design*, pages 205–228. Springer, London, 1999a.

Z. P. Jiang and H. Nijmeijer. A recursive technique for tracking control of nonholonomic systems in chained form. *IEEE Trans. Automat. Control*, 44:265–279, 1999b.

Z. P. Jiang and J.-B. Pomet. Global stabilization of parametric chained systems by time-varying dynamic feedback. *Int. J. Adaptive Contr. Signal Processing*, 10:47–59, 1996.

Y. Kanayama, Y. Kimura, F. Miyazaki, and T. Noguchi. A stable tracking control scheme for an autonomous mobile robot. *Proc. IEEE 1990 Int. Conf. on Robotics and Automation*, pages 384–389, 1990.

P. V. Kokotović. The joy of feedback: nonlinear and adaptive. *IEEE Control Systems Magazine*, 12:7–17, 1992.

I. Kolmanovsky and N. H. McClamroch. Developments in nonholonomic control problems. *IEEE Control Systems Magazine*, 15:6:20–36, 1995.

C. Samson and K. Ait-Abderrahim. Feedback control of a nonholonomic wheeled cart in cartesian space. *Proc. 1991 IEEE Int. Conf. Robotics Automation*, pages 1136–1141, 1991.

G. Walsh, D. Tilbury, S. Sastry, R. Murray, and J. P. Laumond. Stabilization of trajectories for systems with nonholonomic constraints. *IEEE Trans. Automat. Contr.*, 39:216–222, 1994.

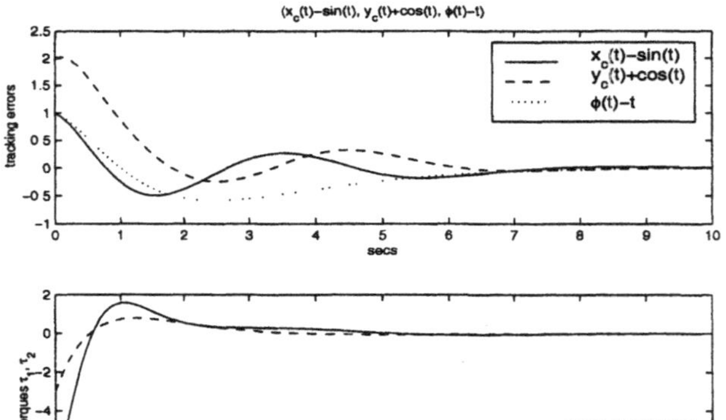

Fig. 1. Global output-feedback tracking of the knife-edge.

MATCHING CONTROL LAWS FOR A BALL AND BEAM SYSTEM

F. Andreev[‡], **D. Auckly**[‡], **L. Kapitanski**[§], **A. Kelkar**[¶] and **W. White** [§]
Kansas State University, Manhattan, KS 66506

Abstract: This note describes a method for generating an infinite-dimensional family of nonlinear control laws for underactuated systems. For a ball and beam system, the entire family is found explicitly. *Copyright © 2000 IFAC*

Keywords: Nonlinear control, mechanical systems

1. THE MATCHING CONDITION

This note presents an application of the method developed by Auckly, *et al.* (2000), to stabilization of a ball and beam system. The results are fully described in (Andreev, *et al.*, (2000), Auckly, Kapitanski (2000), and Auckly, *et al.* (2000)). An experimental comparison of a linear control law versus the nonlinear control laws described here will be given in the full paper, (Andreev, *et al.*, (2000)).

Let Q denote a configuration space. Let $g \in \Gamma(T^*Q \otimes T^*Q)$ be a metric. Let $c, f : TQ \to TQ$ be fiber-preserving maps. We assume that $c(-X) = -c(X)$. Let $V : Q \to \mathbf{R}$. The differential equation that we consider is

$$\nabla_{\dot\gamma}\dot\gamma + c(\dot\gamma) + grad_\gamma V = f(\dot\gamma). \tag{1}$$

Let $P \in \Gamma(T^*Q \otimes TQ)$ be a g-orthogonal projection. We consider the situation where a constraint $P(f) = 0$ is imposed. A system is called underactuated if $P \neq 0$.

Several recent papers propose to find control inputs so that the closed-loop system (1) would have a natural candidate for a Lyapunov function (Bloch, *et al.* (1998), Hamberg (1999), and van der Shaft (1986)). Auckly, *et al.* (2000) introduced the following matching condition and characterization of matching in terms of linear partial differential equations. A control input, f, satisfies the matching condition if there are functions $\widehat{g}$, $\widehat{c}$, and $\widehat{V}$ so that the closed loop equations take the form:

$$\widehat{\nabla}_{\dot\gamma}\dot\gamma + \widehat{c}(\dot\gamma) + \widehat{grad_\gamma}\widehat{V} = 0. \tag{2}$$

The motivation for this method is that $\widehat{H} = \frac{1}{2}\widehat{g}(\dot\gamma,\dot\gamma) + \widehat{V}(\gamma)$ is a natural candidate for a Lyapunov function because $d\widehat{H}/dt = -\widehat{g}(\widehat{c}(\dot\gamma),\dot\gamma)$. A straightforward computation shows that, the matching condition is satisfied if and only if

$$P(\nabla_X X - \widehat{\nabla}_X X) = 0, \tag{3}$$

$$\begin{aligned} P(grad_\gamma V - \widehat{grad_\gamma}\widehat{V}) &= 0, \\ P(c(X) - \widehat{c}(X)) &= 0. \end{aligned} \tag{4}$$

Equation (3) is a system of non-linear first order PDE's for $\widehat{g}$. It is perhaps surprising and pleasing that all of the solutions to (3), (4) may be obtained by first solving one first order linear system of PDE's and then solving a second set of linear PDE's. This is accomplished by introducing a new variable, λ, by $g(X,Y) = \widehat{g}(\lambda X, Y)$.

Theorem 1 *The metric, $\widehat{g}$, satisfies (3) if and only if λ and $\widehat{g}$ satisfy*

$$\nabla g\lambda\big|_{Im\ P^{\otimes 2}} = 0, \quad L_{\lambda PX}\widehat{g} = L_{PX}g. \tag{5}$$

In the special case of a system with two degrees of freedom, it is possible to write out the general solution to this set of differential equations. Following Auckly, Kapitanski (2000), express the underactuaded subspace as the span of a unit length vectorfield, PX. Choose coordinates x^1, x^2 so that $PX = \frac{\partial}{\partial x^1}$, and write $\lambda PX = \sigma\frac{\partial}{\partial x^1} + \mu\frac{\partial}{\partial x^2}$. For the λ-equation, (5), to be consistent the following compatibility condition must hold: $\partial([11,2]\,\mu)/\partial x^2 = \partial([12,2]\,\mu)/\partial x^1$. Starting with this equation and working backwards, all of the equations may be solved via the method of characteristics.

2. THE BALL AND BEAM SYSTEM

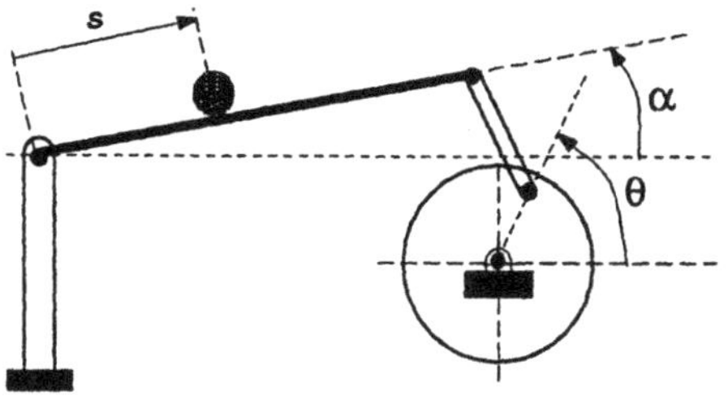

Fig.1. Nonlinear mechanical system.

As an application of our method consider the stabilization problem for the ball and beam system

[‡] Department of Mathematics, on leave from Steklov Institute of Mathematics, St.-Petersburg, Russia

[§] Department of Mathematics

[¶] Department of Mechanical and Nuclear Engineering

[1] This work was partially supported by NSF Grant No. CMS-9813182.

described schematically in figure 1. One can express α as an explicit function of θ. After rescaling, the kinetic energy of the system is given by:

$$T = \frac{1}{2}\dot{s}^2 + \alpha'\dot{s}\dot{\theta} + \frac{1}{2}\left(a_4 + \left(a_3 + \frac{5}{2}s^2\right)(\alpha')^2\right)\dot{\theta}^2$$

and $V = a_5 \sin(\theta) + (s + a_6)\sin(\alpha)$, where the a_k are dimentionless parameters. The projection, $P = (ds + \alpha' d\theta) \otimes \partial/\partial s$, so the control input u is related to f in (1) by $f = (u d\theta)^\sharp$. The resulting equations of motion are

$$\ddot{s} + \alpha'\ddot{\theta} + (\alpha'' - \frac{5}{2}s\alpha'^2)\dot{\theta}^2 + \sin(\alpha) = 0$$

$$\begin{aligned} &\alpha'\ddot{s} + [a_4 + (a_3 + \tfrac{5}{2}s^2)\alpha'^2]\ddot{\theta} + 5\alpha'^2 s\dot{s}\dot{\theta} \\ &+(a_3 + \tfrac{5}{2}s^2)\alpha'\alpha''\dot{\theta}^2 + a_5\cos\theta \\ &\quad +(a_6 + s)\cos(\alpha)\,\alpha' + a_7\dot{\theta} = u\,, \end{aligned}$$

where a_7 corresponds to inherent dissipation.

The general solution to the matching equations is

$$\widehat{g}_{11}(s,\theta) = \psi^2(\alpha)\,(h(y(s,\theta)) + 10\int_0^\alpha \frac{d\varphi}{\mu_1'(\varphi)\psi^2(\varphi)})$$

$$\widehat{g}_{12} = \frac{1}{\mu}(g_{11} - \sigma\widehat{g}_{11}), \quad \widehat{g}_{22} = \frac{1}{\mu}(g_{12} - \sigma\widehat{g}_{12}),$$

$$\widehat{V}(s,\theta) = w(y) + 5\,(y + s_0)\int_0^\alpha \frac{\sin(\varphi)}{\mu_1'(\varphi)\psi(\varphi)}\,d\varphi$$

$$-5\int_0^\alpha \frac{\sin(\varphi)}{\mu_1'(\varphi)\psi(\varphi)}\int_0^\varphi \psi(\tau)\,d\tau\,d\varphi,$$

where $y = \psi(\alpha)s - s_0 + \int_0^\alpha \psi(\tau)\,d\tau$, $\psi(\alpha) = \exp\{-5\int_0^\alpha \frac{\mu_1(\kappa)}{\mu_1'(\kappa)}\,d\kappa\}$, $\mu(s,\theta) = \frac{\mu_1'(\alpha)}{5s\,\alpha'}$, $\sigma(s,\theta) = \mu_1(\alpha) - \frac{1}{5s}\mu_1'(\alpha)$ and μ_1, h, and w are arbitrary functions. Also, $\widehat{c}^1 = -\alpha'\widehat{c}^2$, where $\widehat{c}^2(s,\theta,\dot{s},\dot{\theta})$ is an arbitrary function which is odd in $\dot{s}$ and $\dot{\theta}$. The final nonlinear control law is $u = u_g + u_V + u_c$, where $u_g = g(\nabla_{\dot{\gamma}}\dot{\gamma} - \widehat{\nabla}_{\dot{\gamma}}\dot{\gamma}, \frac{\partial}{\partial\theta})$, $u_V = \frac{\partial V}{\partial\theta} - g(\widehat{grad}_\gamma \widehat{V}, \frac{\partial}{\partial\theta})$, and $u_c = a_7\,\dot{\theta} - g(\widehat{c}(\dot{\gamma}), \frac{\partial}{\partial\theta})$. Using $\widehat{H}$ as a Lyapunov function, we obtain the following conditions that guarantee local asymptotic stability of the equilibrium: $\det(\widehat{g}(0)) > 0$, $\mathrm{tr}(\widehat{g}(0)) > 0$, $\det(\widehat{g}\widehat{c}(0)) > 0$, $\mathrm{tr}(\widehat{g}\widehat{c}(0)) > 0$, $\det(D^2\widehat{V}(0)) > 0$, and $\mathrm{tr}(D^2\widehat{V}(0)) > 0$,

Another way to check local asymptotic stability is to find the poles of the linearized closed-loop system. It is a theorem (Andreev, *et al.* (2000), Auckly, Kapitanski (2000)) that any linear full state feedback control law can be obtained as a linearization of some control law in our family.

A good stabilizing control law will produce a large basin of attraction, send solutions to the equilibrium in a short period of time, and will require little control effort. It is, unfortunately, not clear how to quantify these goals.

We have done some numerical simulation of various control laws in our family. We always pick the arbitrary functions in our nonlinear control law in such a way that the linearization at the desired equilibrium, $u_{lin} = a_8 + K_{bp}(s - s_0) + K_{ap}\theta + K_{bd}\dot{s} + K_{ad}\dot{\theta}$, is exactly the linear state feedback control law provided by the manufacturer of a commercially available system (Apkarian, (1994)). The numerical and experimental response of the system to various initial conditions will be recorded in the full version of the paper.

3. CONCLUSION

We believe that nonlinear control laws have the potential to achieve better performance than linear control laws. There are, however, several subtle questions which must be resolved before nonlinear control laws may be fully exploited in practice. The first question is how to quantify performance. The second question is how to pick a control law which will come close to optimizing performance. One interesting idea is to restrict attention to a class of control laws which generate a closed loop system of a special form. The hope is then that it will be easier to quantify the performance of such systems. We have shown that, in many situations it is possible to find all control laws which will result in a closed loop system of the form (2).

REFERENCES

Andreev, F., D. Auckly, L. Kapitanski, A. Kelkar, and W. White (2000). Matching, linear systems, and the ball and beam. *Preprint.*

Apkarian, J. (1994). Control System Laboratory, Quanser Consulting, Hamilton, Ontario, Canada L8R 3K8.

Auckly, D., L. Kapitanski, and W. White (2000). Control of nonlinear underactuated systems. To appear in *Commun. Pure Appl. Math.*

Auckly, D., L. Kapitanski (2000). Mathematical Problems in the Control of Underactuated Systems. *Preprint.*

Bloch, A., N. Leonard and J. Marsden (1998). Matching and stabilization by the method of controlled Lagrangians. *Proc. IEEE Conf. on Decision and Control*, Tampa, FL, pp. 1446-1451.

Bloch, A., N. Leonard and J. Marsden (1999). Stabilization of the pendulum on a rotor arm by the method of controlled Lagrangians. *Proc. IEEE Int. Conf. on Robotics and Automation*, Detroit, MI, pp. 500-505.

Hamberg, J.(1999). General matching conditions in the theory of controlled Lagrangians. *Proceedings of the 38th Conference on Decision and Control*, Phoenix, AZ.

van der Schaft, A. J. (1986). Stabilization of Hamiltonian systems. *Nonlinear Analysis, Theory, Methods & Applications*, **10**, 1021-1035.

ON PERTURBATION METHODS FOR MECHANICAL CONTROL SYSTEMS

Francesco Bullo *

* *General Engineering and Coordinated Science Laboratory*
1308 W. Main St, Urbana, IL 61801
University of Illinois at Urbana-Champaign
Url: `http://motion.csl.uiuc.edu`

Abstract: In this note we investigate open-loop control of underactuated mechanical systems and draw connections between averaging and controllability theory. Two sets of results are presented: averaging under high-amplitude high-frequency forcing, and series expansions for the evolution of a forced mechanical system starting at rest.

Keywords: mechanical control systems, averaging, controllability theory

1. INTRODUCTION

Perturbation methods for mechanical systems are a classic topic at the center of the attention of numerous mathematicians as well as practitioners. This note reviews two sets of results recently obtained on mechanical systems subject to time-varying forcing. The results build on the contributions on vibrational stabilization via the *averaged potential* in (Baillieul, 1993) and on configuration controllability via the *symmetric product* operation in (Lewis and Murray, 1997).

First, we study the behavior of mechanical systems forced by high amplitude and highly oscillatory inputs. The averaged system is shown to be again a mechanical system subject to an appropriate forcing. By investigating the class subclass of simple systems, i.e., systems with "Hamiltonian equal to kinetic plus potential energy," we precisely characterizes how the averaged potential is related to the symmetric products of certain vector fields. We refer to (Bullo, 1999*b*) for the application of these results to vibrational stabilization problems.

Next, we present a series expansion that describes the evolution of a mechanical system starting at rest and subject to a time-varying external force. We provide a first order description to the solutions of a second order initial value problem. Simplified expressions can be written for simple mechanical systems or systems defined on Lie groups, see (Bullo, 1999*a*).

2. MODELING MECHANICAL SYSTEMS VIA AFFINE CONNECTIONS

The notion of affine connection provides a coordinate-free mean of describing various types of mechanical systems, see (Lewis and Murray, 1997). We write the Euler-Lagrange equations for a system subject to a time-varying force as:

$$\nabla_{\dot{q}}\dot{q} = Y(q,t). \tag{1}$$

Alternatively, if m input forces, potential and damping forces are present, we write

$$\nabla_{\dot{q}}\dot{q} = Y_0(q) - D(q)\dot{q} + \sum_{a=1}^{m} Y_a(q)u^a(t). \tag{2}$$

We assume $q(0) = q_0$, and $\dot{q}(0) = v_0$. We assume the affine connection, the input fields and the input forcing are smooth functions of their respective arguments.

Affine connections are instrumental in defining a key operation, the symmetric product of two vector fields, that is: $\langle Y_1 : Y_2\rangle = \nabla_{Y_1}Y_2 + \nabla_{Y_2}Y_1$.

3. AVERAGING UNDER HIGH AMPLITUDE HIGHLY OSCILLATORY FORCING

Introduce an $\epsilon > 0$, and let $u^a(t) = v^a(t/\epsilon)/\epsilon$, where the v^a are T-periodic functions that satisfy

$$\int_0^T v^a(s_1)ds_1 = 0, \quad \int_0^T \int_0^{s_1} v^a(s_2)ds_1 ds_2 = 0.$$

Let $v(t) = [v^1(t), \ldots, v^m(t)]'$ and define the matrix V according to:

$$V = \frac{1}{2T}\int_0^T \left(\int_0^{s_1} v(s_2)ds_2\right)\left(\int_0^{s_1} v(s_2)ds_2\right)' ds_1.$$

Theorem 1. Consider the initial value problem

$$\nabla_{\dot{r}}\dot{r} = Y_0(r) - D(r)\dot{r} - \sum_{a,b=1}^{m} V_{ab}\langle Y_a : Y_b\rangle(r),$$

with initial conditions $r(0) = q_0$, and $\dot{r}(0) = v_0$. Then $q(t) - r(t) = O(\epsilon)$ as $\epsilon \to 0$ on the time scale 1, and $q(t) - r(t) = O(\delta(\epsilon))$ as $\epsilon \to 0$ for all t, if $(r, \dot{r}) = (0, 0)$ is an asymptotically stable critical point.

Remark 2. Classic averaging in Hamiltonian systems typically relies on the assumption that the system is integrable and that the force is of size ϵ. Here instead it is the Hamiltonian dynamics that is negligible in the first approximation.

Next, we focus on simple systems with integrable inputs, and to expedite the treatment, we assume the configuration space to be $\mathbb{R}^n$. Such systems are completely characterized by their Hamiltonian:

$$H(q,p,u) = V(q) + \frac{1}{2}p'M(q)^{-1}p - \sum_{a=1}^{m}\varphi_a(q)u^a,$$

where M is the inertia matrix, V the potential energy and φ_a are m arbitrary smooth functions. Neglecting the dissipation term, Hamilton's equation are equivalent to the formulation in equation (2).

Gradient vector fields play a natural key role in this setting. Let φ_1, φ_2 be two functions and define a symmetric product between functions via

$$\langle\varphi_i : \varphi_j\rangle \triangleq \frac{\partial\varphi_i}{\partial q}M^{-1}\frac{\partial\varphi_j}{\partial q}.$$

Remarkably, $\langle \operatorname{grad}\varphi_1 : \operatorname{grad}\varphi_2\rangle = \operatorname{grad}\langle\varphi_1 : \varphi_2\rangle$.

Theorem 3. Consider a simple mechanical control system with Hamiltonian defined above. Then the averaged system is a simple mechanical system subject to no input forces and with Hamiltonian

$$H_{\text{averaged}}(q,p) = V_{\text{averaged}}(q) + \frac{1}{2}p'M(q)^{-1}p,$$

where the ***averaged potential*** is defined as

$$V_{\text{averaged}}(q) \triangleq V(q) + \sum_{a,b=1}^{m} V_{ab}\langle\varphi_a : \varphi_b\rangle(q).$$

4. A SERIES EXPANSION FOR THE FORCED EVOLUTION FROM REST

The procedure underlying the averaging results in the previous section can be iterated. Assuming zero initial velocity and dropping the damping force), the evolution of the second order initial value problem in equation (1) can be described via a first order differential equation. Precise statements and proof are available in (Bullo, 1999*a*).

Theorem 4. Define recursively the time-varying vector fields V_k:

$$V_1(q,t) = \int_0^t Y(q,s)ds$$

$$V_k(q,t) = -\frac{1}{2}\sum_{j=1}^{k-1}\int_0^t \Big\langle V_j(q,s) : V_{k-j}(q,s)\Big\rangle ds.$$

The solution $t \to q(t)$ to equation (1) satisfies

$$\dot{q}(t) = \sum_{k=1}^{+\infty} V_k(q(t),t),$$

where the series $(q,t) \mapsto \sum_{k=1}^{\infty} V_k(q,t)$ converges absolutely and uniformly in a neighborhood of q_0 and for $t \in [0,T]$.

5. CONCLUSION

This brief note brings together a number of exciting recent results. Point stabilization, analysis of locomotion gaits, and motion planning for underactuated systems will provide plenty of challenges.

This research was supported by the Campus Research Board of the University of Illinois.

6. REFERENCES

Baillieul, J. (1993). Stable average motions of mechanical systems subject to periodic forcing. In: *Dynamics and Control of Mechanical Systems: The Falling Cat and Related Problems* (M. J. Enos, Ed.). Vol. 1. pp. 1–23. Field Institute Communications.

Bullo, F. (1999*a*). Series expansions for the evolution of mechanical control systems. *SIAM Journal of Control and Optimization*. Submitted.

Bullo, F. (1999*b*). Vibrational control of mechanical systems. *SIAM Journal of Control and Optimization*. Submitted.

Lewis, A. D. and R. M. Murray (1997). Configuration controllability of simple mechanical control systems. *SIAM Journal of Control and Optimization* **35**(3), 766–790.

THE ROLE OF MODEL VALIDATION FOR CHOOSING THE ORDER OF AN IDENTIFIED MODEL: APPLICATION TO CONTROL IN MECHANICAL ENGINEERING

J. C. Carmona * V. M. Alvarado **,[1]

* *Laboratoire d'Automatique, Ecole Supérieure de Mécanique de Marseille, France*
** *Laboratoire d'Automatique de Grenoble, France*

Abstract: Model validation plays a basic role in system identification as it allows to select the model structure as well as the order of the model. Moreover robust control design focuses on the interest of having reliable model error bounds, for linear models preferably described as bounds on frequency functions. A new approach for validating the order of a model is proposed. The principle of this attempt is to calculate the error bound in frequency plan of two identified models difference. This paper describes the theoretical basis and presents some interesting experimental results and their first coments. *Copyright © 2000 IFAC*

Keywords: Idenfication, Model order, Error bound.

1. INTRODUCTION

Very important approaches have been recently developed for model validation. See for example (Goodwin *et al.*, July 1992). In our contribution we shall point to the problem of the identification of systems with an infinity of oscillatory modes, weakly damped, as we can encountered in mechanical engineering. The classical problem is to choose the order of the troncated model for a given data set, a given model structure and a given estimation method. The solution is generally based on the contribution of energy of each mode frequency. Iterative approaches determine the highest mode to be considered as well as the modes to be neglected in the bandwith thus obtained.

We place ourselves in a slightly different situation. For a given data set and for a given model structure, could a given identification method lead to a significant extra mode and in this case what is the extra cost? In order to bring a solution to this problem, its seems interesting to give a clear and straightforward assessment of the distance between the two identified models. In recent literature, the distance between an estimated or a given model and the "true" description has been evaluated giving bounds to a weighted integral of this distance in the frequency domain. See (Ljung and Guo, Sept. 1997). Scalar quantities expressing the correlation between the past imputs and the model residuals on the one hand, and the disturbance on the other hand has been cleverly used as well as a discussion about the tail of the impulse response of the model distance. But it is always delicate to use the "true" system. This implies to make some doubtful assumptions. On the other hand considering the distance between two estimated models allows to make strong assumptions and hence lead to more reliable results.

2. THE MAIN RESULT

Before we give our main technical result, let us make some main assumptions:

[1] Partially supported by Conacyt-Seit/Dgit

A0: We shall consider the input-output data set: $Z^N = \{y(1), u(1), \ldots, y(N), u(N)\}$ as well as the past inputs vector on the past horizon M $\phi(t) = [u(t), u(t-1), \ldots, u(t-M+1)]^T$. Then we can define the matrix: $R_N = \frac{1}{N}\sum_{t=1}^{N}\phi(t)\phi(t)^T$. We assume that R_N is invertible, i.e. the vectors set $\{\phi(t), t = 1, \ldots, N\}$ generates $\Re^N$.

A1: There exists a system, called the "true" system, generated from Z^N such that:

$$y(t) = G_0(q)u(t) + v(t) \qquad \forall t = 1, 2..., N \quad (1)$$

where $v(t)$ is called the disturbance of the system.

A2: $\widehat{G}_1(q)$ et $\widehat{G}_2(q)$ are two models belonging to the same structure as G_0, estimated from the same data set Z^N, such that $\widehat{G_2}$ estimates one high frequency mode more than $\widehat{G_1}$. We assume that the tails of the impulse response of these models are approximately equals.

Theorem :

Under the assumptions **A0**, **A1** and **A2**, the distance between the two estimated models $\widehat{G}_{12}(q) = \widehat{G}_1(q) - \widehat{G}_2(q)$ is such that :

$$\left[\frac{1}{2\pi}\int_{-\pi}^{+\pi}|\widehat{G}_{12}(e^{i\omega})|^2|L(e^{i\omega})|^2|U_N(\omega)|^2 d\omega\right]^{\frac{1}{2}} \le (1+\eta)\left[\frac{1}{N}\theta_N^M\right]^{\frac{1}{2}} + (2+\eta)\tilde{\rho}_M C_u \quad (2)$$

where $L(e^{i\omega})$ is a given linear stable filter, and:

$$\theta_N^M = \frac{1}{N}\left\|\sum_{t=1}^{N}\phi(t)\nu(t)\right\|^2_{R_N^{-1}} \quad (3)$$

evaluates the correlation between the past inputs and $\nu(t) = L(q)(\varepsilon_2(t) - \varepsilon_1(t))$ the difference between the filtered models residuals. Here we use the input periodogram $|U_N(\omega)|^2$, δ the smallest (positive) eigenvalue of R_N and $\{\rho_k\}$ the impulse response of $L(q)\widehat{G}_{12}(q)$.

$$|U_N(\omega)|^2 = \frac{1}{N}\left|\sum_{t=1}^{N}u(t)e^{-i\omega t}\right|^2 \quad (4)$$

$$C_u = Max_{1\le t\le N}|u(t)| \quad (5)$$

$$\eta = \frac{C_u M}{\sqrt{N\delta}} \quad (6)$$

$$\tilde{\rho}_M = \sum_{k=M+1}^{+\infty}|\rho_k| \approx 0 \quad (7)$$

3. THE FIRST EXPERIMENTAL RESULTS

In addition to the theoretical interest of this result, it is our concern to apply it to an actual process in order to show that the uncertainty bounds thus obtained leads to practically exploitable outcomes, from which relevant conclusions can be deduced. The process used is a semi-finite acoustical duct through which the noise generated by an air conditioner propagates. Firstly output error models are estimated using classical parametric identification methods. The time delay and the zeros of the models mainly due to the propagative phenomenum are maintained. Only the model order mainly due to the acoustical modes is investigated. Our first results are given in the folowing table:

order	Z_1^{date1}	Z_2^{date2}	Z_3^{date3}	M
4-3	74.39	74.39	74.39	114
5-4	40.98	40.98	40.98	129
6-5	55.99	55.99	55.99	151
7-6	31.61	31.61	31.61	114
8-7	99.97	99.97	99.97	198
9-8	96.62	96.62	96.62	200
7-5	51.38	51.38	51.38	167
9-7	88.33	88.33	88.33	200

Here there are the results of three identification campaigns exploiting three data sets ($N = 1024$) done at different times, using the same estimation methodology. A difference of one week exists between Z_1^{date1} and Z_2^{date2} and several months between Z_2^{date2} and Z_3^{date3}. M is the value which started from leading to a difference less than 1% between the two tails of the impulse response. One can already notice in the one hand, in each period, an outstanding repetability of the results, and in the other hand a good sensitivity to the detection of an extra mode. In particular one can notice that the value obtained for the order change between 5 and 7 is quite different than the values corresponding to the transitions 5 to 6 and 6 to 7.

4. CONCLUSION

In conclusion we note that the appealing theoretical result we present here has every chance that very relevant outcomes can be carried out. A first fundamental step has just been done. It is now our concern to investigate all the opportunities that this new tool can provide.

5. REFERENCES

Goodwin, G.C., M. Gevers and B. Ninnes (July 1992). Quantifying the error in estimated transfer functions with applications to model order selection. *IEEE Trans. on Autom. Contr.* **37**, 913–929.

Ljung, L. and L. Guo (Sept. 1997). The role of model validation for assessing the size of unmodeled dynamics. *IEEE Trans. on Autom. Contr.* **42-9**, 1230–1239.

Copyright © IFAC Lagrangian and Hamiltonian Methods for
Nonlinear Control, Princeton, New Jersey, USA, 2000

ROBUST CONTROL OF FLAT NONLINEAR SYSTEM

F.CAZAURANG, B.BERGEON, and S.YGORRA

Laboratoire d'Automatique et Productique
Université Bordeaux I, 351 cours de la Libération,
33405 Talence cedex, France
Tel. 33-56842416 - Fax: 33-56846644
e-mail: cazaurang@lap.u-bordeaux.fr

Abstract: This paper proposes a systematic approach for robust tracking control design of a class of nonlinear systems, refered to dynamic flat systems. First, a nonlinear dynamic feedback is designed to ensure nominal path tracking performance. Next, a compact set of models is elaborated in the vicinity of the nominal path, taking into account both state space disturbances and parametric uncertainties. Finally, a robust linear controller is designed which guarantees the path tracking perfomance objectives for the above so-obtained compact set. Simulations results obtained from speed control of a synchronous actuator demonstrate the potential of the proposed approach. *Copyright © 2000 IFAC*

Key words : Nonlinear control, path planning, path tracking, robust control, H-infinity control, synchronous actuator.

1. PROBLEM SETTING

This paper presents a methodology for robust tracking control design of dynamic flat systems. The problem of reference trajectory generation is easy to solve for such systems, as there exists a simple parametrization of these trajectories which depends on state variables and a finite number of input derivatives (Fliess *et al.*, 1995). However, this parametrization is based only on a nominal model, without considering any unknown inputs affecting the system. This paper suggests a method for designing a control system to solve the problem of path tracking by optimisation of an appropriate cost function. The approach is based on a compact set of all possible models and takes into account the presence of external disturbances. The compact set of models is elaborated in the vicinity of the nominal path, taking into account both state space disturbances and parametric uncertainties.

The synthesis procedure is based on a generic two degree of freedom structure which ensures a clear division between nominal tracking objectives and robust tracking and disturbance rejection ones (Prempain *et al.*, 1998). The design procedure is based on a Linear Fractional Transformation Representation of an augmented model including a model error model as well as the desired performances specifications.

The methodology is applied to speed control of a synchronous actuator.

The main contributions of the paper are considered to be the generation of nonlinear compact set of models, and its application to linear tracking control design based on a μ-synthesis approach.

2. FLAT SYSTEMS

Flat systems correspond to a class of nonlinear systems, which are equivalent to linear ones via a special type of dynamic feedback. This dynamic feedback, (called endogenous feedback), is defined as a real-analytic function of state, input and a finite number of its derivatives. For a flat system there are m scalar functions y_i of state x, input u and a finite number of its derivatives such that the dynamic behaviour with input u and outputs $y_i,\ldots,y_m$ can be linearized from an input to state point of view. Outputs $y_i,\ldots,y_m$, which might be regarded as a fictitious output, are called linearizing or flat outputs. The major property of a flat system is that the state x and input u variables can be directly expressed, without integrating a differential equation, in terms of flat output y and a finite number of its derivatives. This property is useful when dealing with trajectories. From y trajectories, x and u trajectories are immediately deduced.

3. COMPACT SET OF MODELS

When we use an endogenous feedback, the input u_{ref} is determinated for a trajectory of the output y assuming the disturbance-free case and perfectly known dynamic behaviour of the system. Here we take into account the state disturbances as well as parametric uncertainties in order to construct a compact set of possible models. This compact set is determined by a differential calculus on the disturbance system in a vicinity of nominal trajectories. This approach is based on differential geometric theory of jets and prolongation of infinite order developed in particular by Vinogradov (1989). This approach is also used by Fliess *et al.* (1999) to generalize the differentially flat non linear systems to orbitally flat systems.

Briefly, if we consider a flat system, then it is possible to make a linearization by endogenous feedback. Then, the flat output y_m is governed by the new input v_m and $y_m^{(\alpha+1)} = v_m$. The compact set of model is given by:

$$y^{(\alpha+1)} = v_m + \delta\left(L_{F_e}^{(\alpha+1)} h_m'\right)(x_m, \xi_m, \bar{v})$$
$$+\left(DL_{F_{,me}}^{(\alpha+1)} h_m' + \delta\left(DL_{F_e}^{(\alpha+1)} h_m'\right)\right)(x_m, \xi_m, \bar{v}) . \begin{pmatrix} \delta x \\ \delta \xi \\ \delta \bar{v} \end{pmatrix} \cdots$$
$$+o\left(|\delta x|^2 + |\delta \xi|^2\right)$$

The input v includes the reference input v_m and the linear regulator output δv. The effect due to the distance between the nominal extended field $F_{e,m}$ and disturbance extended field F_e (second term on the right hand side) is reduced by the output regulator action. It contributes also to reducing the effect of the gap between the nominal path and the disturb path. (third term on the right hand side).

4. LINEAR FRACTIONAL REPRESENTATION AND μ-SYNTHESIS

The two degree-of-freedom design methodology is based on a decoupling scheme, according to the Youla parametrization. A two step methodology has been developed initially for the synthesis of robust tracking control in the case of linear multivariable systems. Here, a similar scheme (*Fig. 1*) is proposed for flat systems in which x is the state space vector, u the input vector and y is the so-called flat output.

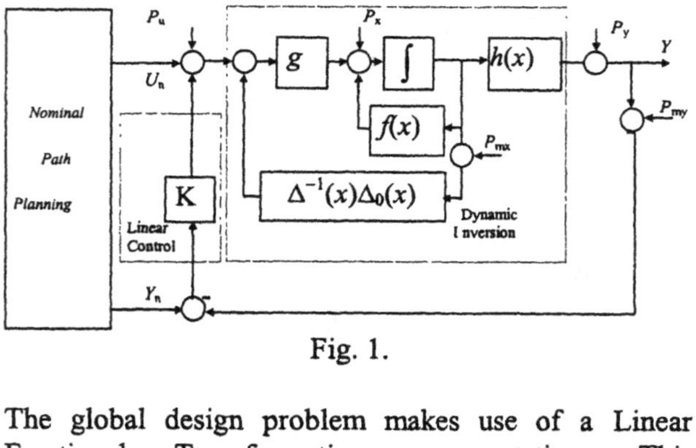

Fig. 1.

The global design problem makes use of a Linear Fractional Transformation representation. This description allows an augmented model including a compact set of models, and performance specifications. To take into account the structure of the representation and in order to be less conservative we use the μ-synthesis technique instead of the H Infiniti approach.

5. SPEED CONTROL OF SYNCHRONOUS ACTUATOR:

Finally, we investigate the application of the proposed methodology to robust control design of a synchronous actuator. We show that rotor position and one component of the current vector in the dq frame are the flat outputs. We suggested flat output trajectories with respect to classical approach (Leonhard (1996)). In the vicinity of these trajectories we determine a compact set of models due to state space disturbances and parametric variation. Following this, the compact set of models and the performance tracking specifications are included in an augmented system. This augmented system uses Linear Fractional Transformation representation. We then designed a linear controller with a μ-synthesis. To conclude we implemented this control on an actuator simulator. (Cazaurang, 1997).

6. REFERENCES:

Cazaurang F., (1997), *Commande robuste des systemes plats, Application a la commande d'une machine synchrone,* PhD Thesis, Université Bordeaux I.

Fliess M., J. Lévine, Ph. Martin and P. Rouchon (1995), Flatness and defect of non-linear systems: introduction theory and examples. *Int. Journal of Control*, **61**, pp. 1327-1361.

Fliess M., J. Lévine, Ph. Martin and P. Rouchon (1999), A Lïe-Bäcklund approach to equivalence and flatness of nonlinear systems, *IEEE Trans. Automat. Control*, **44** N°5, pp 922-937.

Prempain E. and B. Bergeon (1998). Multivariable two-degree of freedom control methodology, *Automatica*, **34**, pp. 1601-1606.

Leonhard W. (1996), *Control of Electrical Drives*, Springer-Verlag 2nd ed.

Vinogradov A. M. (1989), Ed. *Symetries of partial Differential Equations,* Kluwer, Dordrecht.

CONSTRAINED JOINT PD+ CONTROLLER FOR FLEXIBLE LINK ROBOTS

Leonid Freidovich [1]

Department of Mathematics, Michigan State University, MI 48824, USA

Abstract: A class of globally asymptotically stable regulators for a finite-dimensional model of robot arm with flexible links under gravity is presented. The control law is formed as the sum of static compensation of gravity at the desired position and constrained state feedback. Only some of generalized coordinates (joint positions and velocities) are assumed available for measurement and saturation in amplifier characteristic curves is taken into account.

Keywords: Mechanical systems; Robotic manipulators; Flexible arms; Set-point control; Constrained feedback stabilization; Lyapunov method; Global stability.

1. INTRODUCTION

The point-to-point problem for underactuated mechanical systems (with both, free and controllable degrees of freedom) is one of the basic in engineering practice. In particular, many different approaches were proposed to solve this problem (Takegaki and Arimoto, 1981; Spong, 1987; Dunskaya and Pyatnitskii, 1988; Tomei, 1991; De Luca and Siciliano, 1992; Loria *et al.*, 1996; Burkov and Freidovich, 1997; Ortega *et al.*, 1999) for robotic manipulators with dynamics described by different mathematical models which are particular cases of the model studied here.

Nevertheless, the simplest for implementation, joint proportional-differential feedback regulators with constant gravity and static deflection compensation (PD+), are still very popular in practical robot control. In this paper, it is shown that saturation effects present in all real systems can be taken into account, more precisely, that constrained PD+ controllers ensure global asymptotic stability (GAS) of the desired equilibrium under the quite natural (and mild) assumption that control forces dominate gravity.

The article extends our previous results for flexible joint manipulators (Burkov and Freidovich, 1997) that were inspired by well-known regulation results (Dunskaya andPyatnitskii, 1988) for rigid robots (i.e. for fully actuated mechanical systems).

The proof of the main theorem is given with help of the direct Lyapunov approach, more precisely, using Barbashin - Krasovski theorem (Barbashin and Krasovski, 1952; Rouche *et al.*, 1977) with the mechanical energy of the closed-loop system as the Lyapunov function candidate (Salvadori, 1966; Rouche *et al.*, 1977).

The main technical difficulty is to prove that the shaped pseudopotential energy of the closed-loop system is globally positive definite and radially unbounded, despite the fact that corresponding Hessian is not positive definite.

2. FINITE-DIMENSIONAL MODEL FOR A FLEXIBLE ROBOTIC MANIPULATOR

The finite-dimensional model for a robotic manipulator with several flexible links (De Luca and Siciliano, 1991; Arteaga, 1998) can be written as the following system of $n + m$ second order nonlinear differential equations:

[1] e-mail: leonid@math.msu.edu, http://www.math.msu.edu/~leonid/

$$\frac{d}{dt}\frac{\partial L}{\partial \dot{q}} - \frac{\partial L}{\partial q} + \frac{\partial R}{\partial \dot{q}} = \begin{bmatrix} u \\ 0 \end{bmatrix} \qquad (1)$$

Here $q = [\theta^T, \delta^T]^T$ is the vector describing the full arm configuration; $\theta \in \mathbf{R}^n$ is the vector of joint coordinates; $\delta \in \mathbf{R}^m$ is the vector of link deflection coordinates; $L = T - U$ is the system Lagrangian; $T = \frac{1}{2}\dot{q}^T B(q)\dot{q}$ is the kinetic energy; $U = U_g(q) + \frac{1}{2}\delta^T K \delta$ is the potential energy; $R = \frac{1}{2}\dot{\delta}^T D \dot{\delta}$ is the Reyleigh's dissipation function; $B(q) \in \mathbf{R}^{(n+m)\times(n+m)}$ is the positive definite inertia matrix; $K \in \mathbf{R}^{m\times m}$ and $D \in \mathbf{R}^{m\times m}$ are the positive definite diagonal matrices of stiffness and modal damping coefficients; $u \in \mathbf{R}^n$ is the vector of generalized control forces.

Let us denote by $g(q) = [g_\theta(q)^T, g_\delta(\theta)^T]^T$ the vector of gravitational forces, that is the gradient of the potential function $U_g(q)$. One may assume (De Luca and Siciliano, 1992; Arteaga, 1998) that there exists $\alpha > 0$ such that

$$\|g(q_1) - g(q_2)\| \leq \alpha\|q_1 - q_2\|, \quad \forall q_1, q_2 \in \mathbf{R}^{n+m},$$

and $A > 0$ such that $\|g(q)\| \leq A, \ \forall q \in \mathbf{R}^{n+m}$.

3. CONSTRAINED STATIC FEEDBACK

It was shown in (De Luca and Siciliano, 1992) that under some pretty natural assumptions the simple linear feedback

$$u = K_p(\theta_d - \theta) - K_d\dot{\theta} + g_\theta(q_d), \qquad (2)$$

where θ_d is the vector of the desired positions of joint variables and $q_d = [\theta_d^T, (-K^{-1}g_\delta(\theta_d))^T]^T$, ensures GAS of the closed-loop system (1), (2).

It is easy to see, that although GAS is guaranteed, the larger the difference between the initial and the desired values, the larger the magnitudes of control forces that are required. In "real" systems there are always some restrictions on control and all practically realizable regulators include amplifiers with saturated characteristic curves. To take into account such nonlinear effects, one may consider the following control law

$$u = -F(\theta - \theta_d + \Lambda\dot{\theta}) + g_\theta(q_d). \qquad (3)$$

Here Λ is a positive definite diagonal matrix and the components of the vector-function $F(p) = [f_1(p_1), \ldots, f_n(p_n)]^T$ are continuous strictly increasing functions vanishing at zero and such that for some $\gamma > 0$ and $\beta > 0$:

$$|f_i(t)| \geq \gamma \min\{|t|, \beta\}, \quad \forall i = 1, \ldots, n. \quad (4)$$

Note that (Burkov and Freidovich, 1997) these functions can be linear as in (2) or bounded (g.e. $\gamma\frac{t}{1+|t|}$, $\gamma\tanh(t)$, $\gamma\tan^{-1}(t)$).

4. MAIN RESULT

Theorem. *The following conditions:*

$$\min\{\gamma, \|K^{-1}\|^{-1}\}\beta > (A + \|g(q_d)\|),$$
$$\min\{\gamma, \|K^{-1}\|^{-1}\} > \alpha$$

are necessary for GAS of the system (1), (3).

The proof is based on the use of the energy type Lyapunov function candidate $V(q, \dot{q}) = T(q, \dot{q}) + U(q) - U(q_d) + (\theta - \theta_d)^T g_\theta(q_d) + \int_0^{\theta-\theta_d} F(\xi)^T\, d\xi$. This function was shown to be positive definite and radially unbounded by estimating its directional derivatives along rays starting at the desired position, and using the following result.

Lemma. *Let $V : \mathbf{R}^N \to \mathbf{R}$ be a smooth function, $V(0) = 0$, $\frac{\partial V}{\partial p} = 0$ iff $p = 0$, and let $\Omega \subset \mathbf{R}^N$ be a compact convex set such that $0 \in int\Omega$, V is positive definite on Ω, and $\forall \tilde{p} \in \partial\Omega$ and $t \geq 1$: $\frac{d}{dt}(V(t\tilde{p})) \geq c > 0$. Then V is globally positive definite and radially unbounded.*

5. REFERENCES

Arteaga M.A. (1998). On properties of a dynamic model of flexible robot manipulators, *Trans. ASME. J. of Dyn. Syst., Meas. and Contr.*, **120**, 8-14.

Barbashin E.A. and N.N. Krasovskii (1952). On stability of motion in the large, *Dokl. Akad. Nauk SSSR (in Russian)*, **86**, 453-456.

Burkov I.V. and L.B. Freidovich (1997). Stabilization of Lagrangian systems with elastic elements under the restrictions on control with and without velocity measurements, *J. Appl. Math. Mechan.*, **61**, 97-106.

Dunskaya N.V. and E.S. Pyatnitskii (1988). Stabilization of controlled mechanical and electromechanical systems, *Automat. and Remote Control*, No. **12**, 40-51.

Loria A., R. Kelly, R. Ortega, and V. Santibañez (1996). On output feedback control of Euler-Lagrange systems with bounded inputs, *IEEE Trans. on Autom. Control*, **42**, 1138-1142.

De Luca A. and B. Siciliano (1991). Closed form dynamics of planar multilink lightweight robots, *IEEE Trans. on Syst., Man, and Cybern.*, **21**, 826-839.

De Luca A. and B. Siciliano (1992). An asymptotically stable joint PD controller for robot arms with flexible links under gravity, in: *Proc. 31st IEEE Conf. on Dec. and Control*, 325-326, Tucson.

Ortega R., A. Loria, H. Sira-Ramirez, and P.J. Nicklasson (1999). *Passivity-based Control of Euler-Lagrange Systems: Application to Mechanical, Electromechanical and Power systems.* Communication and Control Engineering Series. Springer Verlag.

Rouche M., P. Habets and M. Laloy (1977). *Stability theory by Liapunov's direct method.* Springer-Verlag, N.Y.

Salvadori L. (1966). Sull' estensione ai sistemi dissipativi del criterio di stabilità del Routh, *Richershe Mat.*, **15**, 162-167.

Spong M.W. (1987). Modeling and control of robots with elastic joints, *Trans. ASME. J. Dyn. Syst. Meas. Contr.*, **109**, 310-319.

Takegaki M. and S. Arimoto (1981). A new feedback method for dynamic control of manipulators, *Trans. ASME. J. of Dyn. Syst. Meas. Contr.*, **103**, 119-125.

Tomei P. (1991). A simple PD controller for robots with elastic joints, *IEEE Trans. on Autom. Control*, **36**, 1208 -1212.

AN AUTONOMOUS LOCOMOTION CONTROL OF A MULTI-JOINT SNAKE-LIKE ROBOT WITH CONSIDERATION OF THE DYNAMIC MANIPULABILITY

Yoshikatsu Hoshi * Mitsuji Sampei * Masanobu Koga *

* *Tokyo Institute of Technology, Tokyo*

Abstract: This paper discusses a locomotion control of a snake-like robot. This research utilizes a notion of manipulability to evaluate the locomotability. A proposed controller can make the robot autonomously locomote in a desired direction without decreasing the manipulability. This method realizes that the robot spontaneously generates its gait, thus any design of gaits are needed. *Copyright ©2000 IFAC*

Keywords: Snake-like Robot, Dynamic Manipulability, Autonomous Locomotion, Singular Posture, Hyper-redundant Systems

1. INTRODUCTION

This research discusses an autonomous locomotion control of snake-like robot which is modeled by some links with passive wheels and active joints. Such a robot has nonholonomic constraint just like trailers. However, the robot is unlike trailers because it can not actuate wheels directly. In order to locomote, it actuates its joints and uses side force of wheels (Hirose, 1987). Hence it should always keep the high locomotable shape, and should avoid the shape in which it can not move (singular posture) (Prautsch *et al.*, 1999). In some previous researches, suitable gaits must be given to avoid singular posture.

In this paper, a notion of dynamic manipulability (Yoshikawa, 1983) is utilized to evaluate the locomotability and a new controller is proposed to make the robot autonomously locomote in the desired direction. In this method, the robot spontaneously generates the gait which avoids singular posture. Some simulations show that the proposed method realizes the ceaseless winding locomotion.

2. MODELING OF SNAKE-LIKE ROBOT

In this research, a model of snake-like robot consists of n rigid links($n \geq 4$), like Fig. 1 (Prautsch *et al.*, 1999). Each link has free passive wheels at the center, and it is assumed that they never slip to the perpendicular direction of each link. The input of this model is joint torque. The moving force arises from constraining force between wheels and ground. The robot moves only on a horizontal plane, thus the effect of the gravity is ignored.

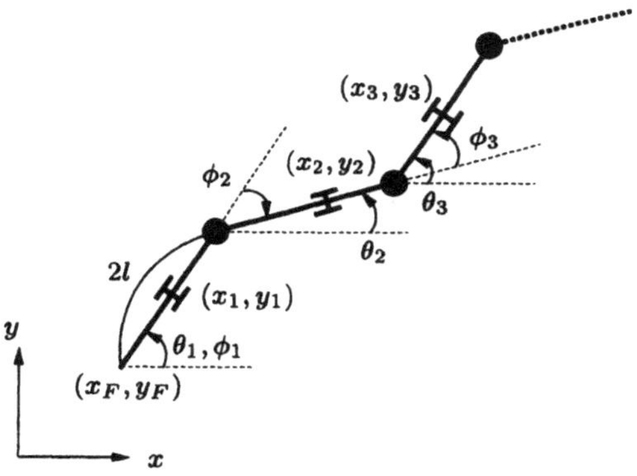

Fig. 1. Model of snake-like robot.

If slipless condition of wheels is neglected, the dynamic property of the model is given by Lagureange's equation of motion; that is

$$M(\boldsymbol{\theta})\ddot{\boldsymbol{q}} + C(\dot{\boldsymbol{\theta}}, \boldsymbol{\theta})\dot{\boldsymbol{\theta}} + D(\boldsymbol{\theta})\dot{\boldsymbol{q}} = \begin{bmatrix} \tilde{E}_n \boldsymbol{\tau} \\ 0 \end{bmatrix}, \quad (1)$$

where $\boldsymbol{q} = [\theta_1, \ldots, \theta_n, x_F, y_F]^T$ is the generalized coordinates and $\boldsymbol{\tau} = [\tau_2, \ldots, \tau_n]^T$ is the input torque. On the other hand, slipless condition of passive wheels is given by the form of Pfaffian constraint (Murray *et al.*, 1993)

$$\left[\, I_n \;\; -F_\theta(\boldsymbol{\theta}) \,\right] \begin{bmatrix} \dot{\boldsymbol{\theta}} \\ \dot{\boldsymbol{r}} \end{bmatrix} = A(\boldsymbol{q})\dot{\boldsymbol{q}} = 0. \qquad (2)$$

When such constraint is imposed, the constraint force is needed to realize the constraint motion. This force is expressed in terms of Lagrange's multiplier $\boldsymbol{\lambda} \in \mathbf{R}^n$ and the term $A^T\boldsymbol{\lambda}$ must be added to left-hand member of the Eq. (1).

By eliminating $\boldsymbol{\lambda}$ term, the equation of motion of snake's head, whose coordinate is $\boldsymbol{r} = [x_F, y_F]^T$, is given by

$$\tilde{M}\ddot{\boldsymbol{r}} + \tilde{C}\dot{\boldsymbol{r}} + \tilde{D}\dot{\boldsymbol{r}} = F_\theta^T \tilde{E}_n \boldsymbol{\tau}. \qquad (3)$$

3. CONTROL WITH MANIPULABILITY

From the Eq. (3), the acceleration of snake's head $\ddot{\boldsymbol{r}}$ is given by

$$\ddot{\boldsymbol{r}} = \tilde{M}^{-1}F_\theta^T \tilde{E}_n \boldsymbol{\tau} - \tilde{M}^{-1}\left[\tilde{C} + \tilde{D}\right]\dot{\boldsymbol{r}}. \qquad (4)$$

According to the notion of dynamic manipulability (Yoshikawa, 1983), the property of dynamic manipulability ellipsoid of the Eq. (4) is characterized by the singular values ρ_1 and ρ_2 of the matrix $\tilde{M}^{-1}F_\theta^T \tilde{E}_n$. In this research, the dynamic manipulability of a snake-like robot ω_d is defined as the ratio of two singular values $\omega_d = \rho_2/\rho_1$, and the locomotability of a snake-like robot is evaluated by ω_d.

Our control strategy is only control the position of the head of a snake-like robot and leaving other dynamics as zero dynamics, which is stable. Our control objective is to make it follow the desired line trajectory.

At first, let $\boldsymbol{\alpha}_t$ be an acceleration vector of the head which is desirable for tracking trajectory. Next, for the present shape of a model, let $\boldsymbol{\alpha}_m$ be an acceleration vector which improves manipulability (see Fig. 2).

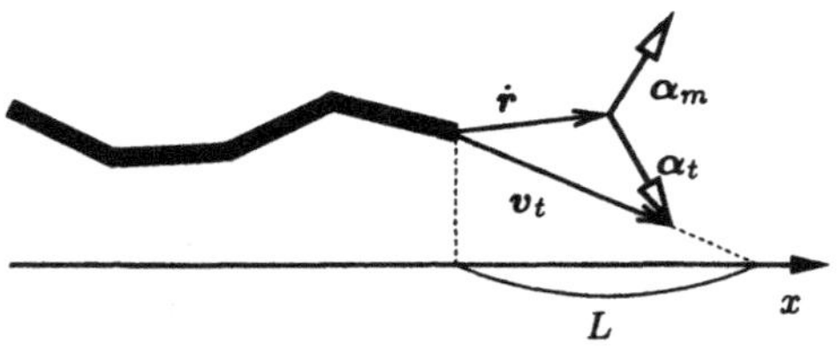

Fig. 2. Two kind of accelaration vector.

Then, let $\boldsymbol{\alpha}_{\text{next}}$, which is a suitable weighted average between $\boldsymbol{\alpha}_t$ and $\boldsymbol{\alpha}_m$, be a desired acceleration of next step. Since the robot is given the acceleration $\boldsymbol{\alpha}_{\text{next}}$ to its head, it locomotes getting trade-off between tracking and manipulability.

When the number of links are large(almost more than 7 or 8), the robot often locomotes with very little winding. Such winding increases loss of energy because the actuators have viscosity friction. To avoid such behavior, let $\boldsymbol{\alpha}_e$ be an acceleration vector which minimizes loss of energy. Then let $\boldsymbol{\alpha}_{\text{next}}$ be a weighted average between $\boldsymbol{\alpha}_t$, $\boldsymbol{\alpha}_m$ and $\boldsymbol{\alpha}_e$.

The input $\boldsymbol{\tau}$ which generates the desired acceleration $\boldsymbol{\alpha}_{\text{next}}$ is not unique. Then the input $\boldsymbol{\tau}_{\text{next}}$ is computed from the Eq. (3) by using the pseudo-inverse matrix of $F_\theta^T \tilde{E}_n$ as

$$\boldsymbol{\tau}_{\text{next}} = \left[F_\theta^T \tilde{E}_n\right]^+ \left[\tilde{M}\boldsymbol{\alpha}_{\text{next}} + \tilde{C}\dot{\boldsymbol{r}} + \tilde{D}\dot{\boldsymbol{r}}\right]. \quad (5)$$

The Euclidian norm of such input $\|\boldsymbol{\tau}_{\text{next}}\|_2$ is the minimum of the inputs which generate $\boldsymbol{\alpha}_{\text{next}}$.

4. SIMULATION RESULTS

Some simulation results using proposed controller are shown as follows. The positive direction of x-axis is the desired trajectory. Whole length of the model is 0.4[m] and whole weight is 1.2[kg]. As shown in them, the robot tracks the desired trajectory with ceaseless winding locomotion and without making singular posture.

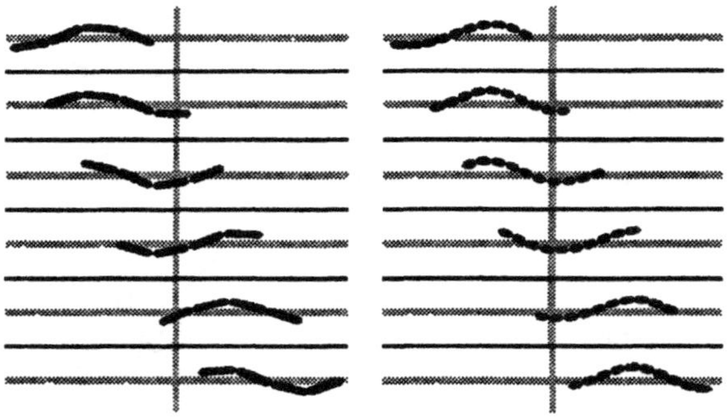

Fig. 3. Simulation results.(L.: 4 links, R.: 10 links)

5. REFERENCES

Hirose, S. (1987). *Biomechanical Engineering.* Kougyou Chousa Kai. Tokyo.

Murray, R.M., Z. Li and S.S. Sastry (1993). *A Mathmatical Introduction to Robotic Manipulation.* CRC Press. Florida.

Prautsch, P., T. Mita and T. Iwasaki (1999). Snake-like robot: Control of the gait. *28th SICE Symposium on Control Theory* pp. 161–164.

Yoshikawa, T. (1983). Analysis and control of robot manipulators with redundancy. In: *Proc. of the 1st International Symposium of Robotics Recearch.* pp. 735–747. MIT Press.

LAGRANGIAN DAES: A STARTING POINT FOR APPLICATIONS IN CONTROL

Richard A. Layton and Marwan U. Bikdash

North Carolina A&T State University
Greensboro, NC

Abstract: This paper presents the status of our effort to reformulate the basic tools and algorithms of dynamic systems and control in terms of energy, work, and constraints using Lagrangian differential-algebraic equations (DAEs). We review the physical basis of Lagrangian DAEs and illustrate their application in modeling, simulation, and linear analysis. Our purpose is to show that this approach provides a promising starting point for applications in control in general. *Copyright © 2000 IFAC*

Keywords: dynamic systems, modelling, simulation, linear analysis.

1. MODELLING USING LAGRANGIAN DAES

We adopt the well-known effort-flow variable classification for modelling multidisciplinary systems (Paynter, 1961). Effort $e(t)$ and flow $f(t)$ are the pair of variables in each discipline whose product is power, $P(t) = e(t)f(t)$, where $e(t)$ is the effort an element exerts on a system and t is time. Momentum $p(t)$ and displacement $q(t)$ are defined as the time-integrals of effort and flow.

The work-energy principle underlying the Lagrangian differential-algebraic equations (DAEs) is given by

$$\delta(T + V) = \delta W, \qquad (1)$$

where $T(p, q, t)$ is kinetic energy, $V(q, t)$ is potential energy, δT and δV are variations in stored energy consistent with the constraints, and δW is the virtual work of nonpotential efforts (Layton, 1998). From this work-energy principle is obtained the virtual-work form of Lagrange's equation, given by

$$\sum_j \left(\frac{d}{dt}\frac{\partial T^*}{\partial f_j} - \frac{\partial T^*}{\partial q_j} + \frac{\partial V}{\partial q_j} + \frac{\partial D}{\partial f_j} - Q_j\right) \delta q_j = 0, \qquad (2)$$

where $T^*(f, q, t)$ is kinetic coenergy, $D(f, q, t)$ is content, $Q(e, t)$ is the vector of applied efforts, and the δq_j are virtual displacements. Displacement constraints $\Phi(q, t) = 0$ and flow constraints $\Psi(f, q, t) = B(q, t)f + b(t) = 0$ are appended to (2) using Lagrange multipliers. New constraint classifications—effort constraints $\Gamma(s, e, f, q, t) = 0$ and dynamic constraints $\dot{s} - \Lambda(s, e, f, q, t) = 0$, where s is a "dynamic" variable not part of the energy manipulation described by (2)—are defined in (Fabien and Layton, 1997; Layton and Fabien, 1996). The resulting Lagrangian DAEs are given by

$$\begin{aligned} \dot{q} &= f, \\ M\dot{f} + \Phi_q^T \kappa + \Psi_f^T \mu &= \Upsilon, \\ \Phi &= 0, \\ \Psi &= 0, \\ \Gamma &= 0, \\ \dot{s} &= \Lambda, \end{aligned} \qquad (3)$$

where

$$
\begin{aligned}
M &= \nabla_f^2 T^* = \partial^2 T^*/\partial f^2, \\
\Phi_q &= \nabla_q \Phi = \partial\Phi/\partial q, \qquad (4)\\
\Psi_f &= \nabla_f \Psi = \partial\Psi/\partial f, \\
\Upsilon &= Q - (\nabla_f T^*)_q f + (\nabla_f T^*)_t - \nabla_q V - \nabla_f D.
\end{aligned}
$$

2. NUMERICAL INTEGRATION OF DAES

Algorithms for solving ODEs are generally unsuitable for solving DAEs. Computer packages for numerical solution of initial-value problems in DAEs include *DASSL* (Brenan *et al.*, 1996), *DYNAST* (Mann, 1996), *GODESS* (Soederlind and Olsson, 1996), and *LDAE* (Fabien and Layton, 1996). The *LDAE* package is specialized for the equation structure given by (3)-(4). This is an area of active research and we recommend that the interested reader investigate these packages.

3. LINEAR ANALYSIS

Guidelines toward an energy-based, linear analysis of discrete systems are presented in (Bikdash and Layton, 1999). The analysis accommodates nonholonomic constraints and explicit inputs. An equilibrium postulate is proposed, stating "The necessary conditions for equilibrium of a physical system are constant total energy and kinematic constraint compliance." Thus constant total energy $T + V$ characterizes both static equilibrium and steady-state operating conditions.

Lagrangian DAEs are linearized using a local, indirect approach. The resulting DAEs have the linear singular form $E\dot{x} = Ax + Bu$ given by

$$
\begin{bmatrix} I & 0 & 0 & 0 \\ 0 & M & 0 & 0 \\ 0 & 0 & 0 & 0 \\ 0 & 0 & 0 & 0 \end{bmatrix}_o \Delta\dot{x} = \begin{bmatrix} 0 & I & 0 & 0 \\ -K & -C & -\Phi_q^{\mathrm{T}} & -\Psi_f^{\mathrm{T}} \\ \Phi_q & 0 & 0 & 0 \\ \Psi_q & \Psi_f & 0 & 0 \end{bmatrix}_o \Delta x + \begin{bmatrix} 0 \\ L \\ \Phi_u \\ \Psi_u \end{bmatrix}_o \Delta u, \qquad (5)
$$

where the symbol $[A]_o$ indicates that A is evaluated at an equilibrium condition, where

$$
\begin{aligned}
M &:= \left[T_{ff}^*\right]_o, \\
C &:= \left[T_{fq}^* - T_{qf}^* - Q_f + D_{ff}\right]_o, \qquad (6)\\
K &:= \left[(\Phi_q^{\mathrm{T}}\kappa)_q + (\Psi_f^{\mathrm{T}}\mu)_q - T_{qq}^* + V_{qq} - Q_q\right]_o, \\
L &:= \left[Q_u - (\Phi_q^{\mathrm{T}}\kappa)_u - (\Psi_f^{\mathrm{T}}\mu)_u\right]_o
\end{aligned}
$$

and where $\Delta x^{\mathrm{T}} := \left[\, \Delta q^{\mathrm{T}} \ \Delta f^{\mathrm{T}} \ \Delta\kappa^{\mathrm{T}} \ \Delta\mu^{\mathrm{T}} \,\right]$ is the set of descriptors.

The symmetric matrix E admits a well-behaved eigenvalue decomposition, enabling a ready transformation to first equivalent form (EF1). Because EF1 is well understood, this form is a good basis for further developing the tools of linear analysis for Lagrangian DAEs.

4. CONCLUSION

A brief overview is given of an approach to modelling, simulation, and linear analysis of discrete systems using Lagrangian DAEs. The unifying concept is to formulate a given dynamic systems and control problem in terms of energy, work, and constraints. The approach is general and systematic—a starting point for applications in control in general.

5. REFERENCES

Bikdash, M.U. and R.A. Layton (1999). Towards an energy-based linear analysis of nonholonomic systems. In: *ASME Dyn. Sys. and Control Div., IMECE'99.* Nashville.

Brenan, K.E., S.L. Campbell and L.R. Petzold (1996). *Numerical Solution of Initial-Value Problems in Differential-Algebraic Equations.* SIAM.

Fabien, B.C. and R.A. Layton (1996). Modeling and simulation of physical systems II: An approach to solving Lagrangian DAEs. In: *4th IASTED Int. Conf., Robotics and Manufacturing.* Honolulu.

Fabien, B.C. and R.A. Layton (1997). Modeling and simulation of physical systems III: An approach for modeling dynamic constraints. In: *IASTED Int. Conf., Applied Modeling and Simulation.* Banff.

Layton, R.A. (1998). *Principles of Analytical System Dynamics.* Springer-Verlag. New York.

Layton, R.A. and B.C. Fabien (1996). Modeling and simulation of physical systems I: An introduction to Lagrangian DAEs. In: *4th IASTED Int. Conf., Robotics and Manufacturing.* Honolulu.

Mann, H. (1996). A versatile modeling and simulation tool for mechatronics control system development. In: *IEEE Symposium on Computer-Aided Control System Design.* Dearborn. pp. 524–529. See also http://icosym-nt.cvut.cz/dyn/.

Paynter, H.M. (1961). *Analysis and Design of Engineering Systems.* MIT Press.

Soederlind, G. and H. Olsson (1996). Godess—a generic ODE solving system. In: *IFIP/TC2/WG Conf., Quality of Numerical Software: Assessment and Enhancement.* Oxford. See also http://w1.461.telia.com/~u46108092/Godess/ index.html.

NONLINEAR CONTROL FOR MANIPULATOR WITH FLEXIBLE LINK BASED ON A BACKSTEPPING APPROACH

Kazuya Ogata, Masunori Shibata, Yoshikazu Hayakawa

Department of Electronic-Mechanical Engineering
Nagoya University, Japan

Abstract: A systematic approach to construct a lyapunov function for closed loop system of flexible manipulator is considered. Backstepping approach and nature of the mechanical system are used for the design of the control system. However it is found that the controller becomes complicated. *Copyright ©2000 IFAC*

Keywords: Flexible arms, Nonlinear control systems, Lyapunov function, Backstepping

1. INTRODUCTION

In order to make the energy consumption small, it is preferable to make a manipulator arm as light as possible. However such a light robot arm tends to bend and vibrate because of its elastic movement. In this case the arm needs to be modeled as a flexible arm considering the elasticity.

To design a nonlinear controller, lyapunov function is often used to guarantee the closed loop stability. A controller for a mechanical system which has enough control inputs correspondent to the generalized coordicates can be easily designed using a lyapunov function, that is the sum of the kinematic energy and potential energy. However flexible arm has a bending mode which is driven just by the motor movement. Therefore it is not easy to find a lyapunov function which guarantees desirable attenuation of the link oscillation.

Using backstepping approach(Kristić *et al.*, 1995), a lyapunov function can be found systematically. In this case a good property for the manipulator as a mechanical system can not be used to design the control system. In this paper, using a systematic approach in the backstepping strategy and a good property of the mechanical system, to construct a lyapunov function is attempted.

2. MODELING

One link flexible manipulator is considered as an example to design the controll system. Using the modal analysis and modeling several modes of elasticity, the dynamical equation between the control torque τ, motor angle θ and modal coordinates q for elasticity can be written as

$$M(x)\ddot{x} + \{D(x,\dot{x}) + D_0\}\dot{x} + Kx = b\tau \quad (1)$$

where $x = [\theta^T \; q^T]^T$, $b = [I \; 0]^T$ Because of the cross term for $\ddot{\theta}$ and $\ddot{q}$ in the inertia matrix, iterative approach to construct the lyapunov function, which will be described later, can not be used. Hence new coordinates are introduced to make the inertia matrix block diagonal,

$$p = Eq + C\theta$$

where C, E are matrices appeared in $M(x)$. The dynamical equation turns into the following expression.

$$\bar{M}(x)\ddot{\bar{x}} + \{\bar{D}(x,\dot{x}) + \bar{D}_0\}\dot{\bar{x}} + \bar{K}\bar{x} = b\tau \quad (2)$$

where

$$\bar{x} = \begin{bmatrix} \theta \\ p \end{bmatrix} \qquad \bar{M}(x) = \begin{bmatrix} \bar{A} + q^T E q & 0 \\ 0 & E^{-1} \end{bmatrix}$$

$$\bar{D}(x,\dot{x}) = \begin{bmatrix} q^T E\dot{q} & q^T\dot{\theta} \\ -\dot{\theta}q & 0 \end{bmatrix} \quad \bar{K} = \begin{bmatrix} K_1 & K_2^T \\ K_2 & K_3 \end{bmatrix}$$

3. CONTROL DESIGN BASED ON THE BACKSTEPPING APPROACH

Here a regulation problem is considered. Control system will be designed to eliminate the initial deviation of the motor angle or link oscillation.

First the bending mode and motor angle are chosen as a state variable $x_b = [\theta\ p^T\ \dot{p}^T]^T$ for the dynamics of the link oscillation.

$$\dot{x}_b = A_b(\dot{\theta})x_b + B_b\dot{\theta} \tag{3}$$

$$A_b(\dot{\theta}) = \begin{bmatrix} 0 & 0 & 0 \\ 0 & 0 & I \\ EK_2 - C\dot{\theta}^2 & -EK_3 + I\dot{\theta}^2 & 0 \end{bmatrix} \quad B_b = \begin{bmatrix} I \\ 0 \\ 0 \end{bmatrix}$$

This is a nonlinear equation but not so complicated as original equation (1). The property of the oscillation seems to be affected by the motor angle velocity $\dot{\theta}$.

Suppose virtual control input u_b can be designed instead of $\dot{\theta}$ in the second term of the right side of (3). Then the subsystem

$$\dot{x}_b = A_b(\dot{\theta})x_b + B_b u_b \tag{4}$$

will be stabilized by state feedback $u_b = K_b x_b$. If the maximum range of the motor velocity v_{max} is known in advance, it is sufficient for the following inequality to be satisfied to stabilize the link oscillation.

$$\exists P > 0, Q > 0 \ \ s.t.$$
$$(A_b(\dot{\theta}) + B_bK_b)^T P + P(A_b(\dot{\theta}) + B_bK_b) \leq -Q$$
$$for\ \forall\dot{\theta} \in \{\dot{\theta} \mid |\dot{\theta}| \leq v_{max}\} \tag{5}$$

Because $\dot{\theta}$ can't be driven equal to u_b, the error $z = \dot{\theta} - u_b$ must be considered. In this case the closed loop for the link oscillation becomes

$$\dot{x}_b = (A_b(\dot{\theta}) + B_bK_b)x_b + B_b z \tag{6}$$

Next, using the motor movement dynamics in (2),

$$(\bar{A} + q^T Eq)\ddot{\theta} + q^T E\dot{q}\dot{\theta} + q^T\dot{\theta}\dot{p}$$
$$+ D_m\dot{\theta} + K_1\theta - K_2^T p = \tau \tag{7}$$

asymptotic stability of the overall system will be achieved. The candidate of the lyapunov function for overall closed loop system is made to be sum of the lyapunov function for the closed loop of the link oscillation (6) and kinematic and potential energy for motor dynamics.

$$V = x_b^T P x_b + \dot{\theta}^T(\bar{A} + q^T Eq)\dot{\theta} + \theta^T K_1\theta \tag{8}$$

Using (6)(7), time derivative of V becomes

$$\dot{V} \leq -x_b^T Q x_b + z^T B_b^T P x_b + x_b^T P B_b z$$
$$+(\tau + \phi)^T(z + K_b x_b) + (z + K_b x_b)^T(\tau + \phi)$$
$$-2\dot{\theta}^T D_m\dot{\theta}$$

$$\phi = -q^T\dot{\theta}\dot{p} + K_2^T p$$

Therefore if control input for τ is designed as

$$\tau = -\phi - K_z z \tag{9}$$

then

$$\dot{V} \leq [\,x_b^T\ z^T\,]\,\bar{Q}\begin{bmatrix} x_b \\ z \end{bmatrix} - 2\dot{\theta}^T D_m\dot{\theta}$$

$$\bar{Q} = \begin{bmatrix} (A_b + B_bK_b)^T P + P(A_b + B_bK_b) & PB_b - K_b^T K_z \\ B_b^T P - K_z^T K_b & -K_z^T - K_z \end{bmatrix}$$

If the feedback gain K_b, K_z could satisfy $\bar{Q} < 0$, $x_b \to 0, z \to 0$ and $\dot{\theta}_a \to 0$ would be achieved. However, it is easily found that there exists an especial value $z = -K_b x$ which makes

$$\dot{V} \leq x_b^T(A_b^T P + PA_b)x_b - 2\dot{\theta}^T D_m\dot{\theta}$$

$\dot{V}$ is found not to be negative definite, so asymptotic stability can not be guaranteed.

4. CONCLUDING REMARK

In this paper, a controller with simple structure comparing the one using standard backstepping technique is attempted to construct using the nature of the mechanical system. However, desirable attenuation of the link oscillation can not be achived.

Using standard bachstepping technique, the candidate of the lyapunov function becomes as follows.

$$V = x_b^T P x_b + z^T(\bar{A} + q^T Eq)z$$

In this case, it is verified that $\dot{V}$ can be negative definite function. However the resulting controller becomes very complicated.

$$\tau = -\phi - B_b^T P x_b - K_z z$$

$$\phi = -(\bar{A} + q^T Eq)K_b(A_b(\dot{\theta})x_b + B_b\dot{\theta})$$
$$-q^T E\dot{q}K_b x_b - q^T\dot{\theta}\dot{p} - D_m\dot{\theta} + K_2^T p - K_1\theta$$

5. REFERENCES

Kristić, M., I. Kanellakopoulos and P. Kokotović (1995). *Nonlinear and Adaptive Control Design.* John Wiley & Sons, Inc.

BIFURCATION ANALYSIS OF AN INVERTED PENDULUM WITH SATURATED HAMILTONIAN CONTROL LAWS

Enrique Ponce *,1 Javier Aracil**,1 Francisco Salas **,1
Daniel Juan Pagano ***,2

* *Dep. Matemática Aplicada II, e-mail:enrique@matinc.us.es*
** *Dep. Ingeniería de Sistemas y Automática, Universidad de Sevilla, Camino de los Descubrimientos, 41092-Sevilla, Spain*
*** *Automação e Sistemas, Universidade Federal de Santa Catarina, 88040-900 Florianópolis, Brazil*

Abstract: The saturation effects in the global state space structure of an inverted simple pendulum with a Hamiltonian control law are studied via bifurcation analysis.

Keywords: Qualitative analysis, Saturation, Nonlinear systems

1. INTRODUCTION

Bifurcation theory constitutes a very interesting tool for understanding the complex behavior of nonlinear systems, see (Kuznetsov, 1998). Dealing with control systems, one is sometimes enforced to consider actual nonlinearities, for instance when saturations in the actuators have to be taken into account. In this context, bifurcation theory allows to obtain a global picture of the state space structure and how this structure changes depending on the values of parameters, see (Pagano *et al.*, 1999). In this paper, a bifurcation analysis is made for an inverted pendulum with saturated Hamiltonian control laws. This simple but nontrivial example illustrates the kind of results that can be obtained from the methodology proposed when studying other more complex problems.

The equation of a simple inverted pendulum of length $2l$ is

$$J_a\ddot{\theta} + \rho\dot{\theta} - mgl\sin\theta = \tau,$$

where θ is the angle (measured with respect to the upper position), J_a stands for the motor and arm inertia, m is the arm mass, g is the gravity, ρ is the damping coefficient due to the friction, and τ represents the motor torque.

In state variables, after normalizing time and rescaling angular velocity, one gets

$$\dot{x}_1 = x_2,$$
$$\dot{x}_2 = \sin x_1 - bx_2 + au,$$

where $x_1 = \theta$, $b = \frac{\rho}{\omega_0 J_a}$, $a = \frac{1}{\omega_0^2 J_a}$ and $\omega_0^2 = \frac{mgl}{J_a}$.

In controlling this generalized Hamiltonian control system example, see (van der Schaft, 1996), one achieves the following equation

$$\dot{\mathbf{x}} = [J - R]\frac{\partial H}{\partial \mathbf{x}} + gu = [J - R]\frac{\partial H_d}{\partial \mathbf{x}}, \quad (1)$$

where

$$J = \begin{bmatrix} 0 & 1 \\ -1 & 0 \end{bmatrix}, R = \begin{bmatrix} 0 & 0 \\ 0 & b \end{bmatrix}, g = \begin{bmatrix} 0 \\ a \end{bmatrix},$$

H_d stands for the desired closed loop generalized Hamiltonian which stabilizes the system, and

$$H(x_1, x_2) = \frac{1}{2}x_2^2 + \cos x_1 - 1.$$

---

[1] Partially supported by the Spanish Ministry of Education and Science grant TAP97-0553
[2] Partially supported by the Brazilian Ministry of Education (CAPES) grant BEX0439/95-6

In the sequel, it will be assumed

$$H_d(x_1, x_2) = \frac{1}{2}x_2^2 + 1 - \cos x_1$$

in order to preserve the configuration space $S^1 \times \mathbb{R}$ (breaking this configuration space, another common choice would be $H_d(x_1, x_2) = \frac{1}{2}x_2^2 + \frac{1}{2}x_1^2$, to be analyzed elsewhere).

Thus, global asymptotic stability for the upper position is obtained. From (1), the corresponding control law is $u = -\frac{2}{a}\sin x_1$.

If friction is to be modified, the more general control law $u = -l_1 \sin x_1 - l_2 x_2$ should be considered. When a normalized saturation is included, the system becomes

$$\dot{x}_1 = x_2,$$
$$\dot{x}_2 = \sin x_1 - bx_2 + a\,\mathrm{sat}(-l_1 \sin x_1 - l_2 x_2).$$

2. BIFURCATION ANALYSIS

The first natural step for qualitative analysis is the determination of the number and stability character of possible equilibrium points. All these points have $x_2 = 0$, and the corresponding x_1-value can be obtained by solving the equation

$$\sin x_1 = a\,\mathrm{sat}(l_1 \sin x_1). \qquad (2)$$

From the symmetry and periodicity, the points $(0,0)$ and $(\pi, 0)$ are always equilibrium points. A careful analysis of equation (2) leads to new *saturation-induced* equilibria (see the bifurcation set of fig. 1). Stability of the origin is preserved for $al_1 > 1$, but its global stability is lost for $a < 1$, due to the double saddle-node bifurcation at $a = 1$, which gives rise to four new equilibria (two of them are stable). The line $al_1 = 1$ is made up by degenerate pitchfork bifurcation points for $a < 1$ (transcritical bifurcation points for $a > 1$), and it is another boundary for the origin local stability (in fact, for that points there appear a continuum of equilibria).

Letting l_2 to be negative (reducing friction), more complex bifurcations appear organized around a degenerate Bogdanov-Takens bifurcation point at $l_1 = 1/a$, $l_2 = -b/a$. Thus, new bifurcation lines of degenerate Hopf and saddle-connection points arise, stating the boundaries of parameter regions where limit cycles can be found. The degenerate Hopf points appear for $l_2 = -b/a$ (the system is then conservative) and they resemble the bifurcation analyzed in (Freire *et al.*, 1999). Detailed computations justifying the above assertions for the considered control law will be reported elsewhere.

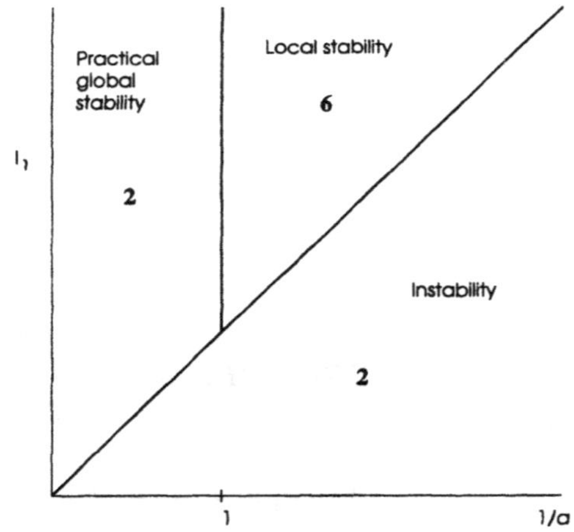

Fig. 1. Bifurcation set in the plane $(1/a, l_1)$. Zone digits indicate the number of equilibria.

As a consequence, local stability for the upper position of the pendulum is guaranteed when $l_1 > 1/a$ and $l_2 > -b/a$, but practical global stability also requires $a > 1$. Despite the seeming simplicity of above results, what is important is that the above analysis allows one to predict the global behaviour of the system for any parameter values.

3. CONCLUSIONS

Although the problem here considered is straightforward to analyze, the methodology is general enough to be taken into account for more complex problems. It is claimed that, in controlling actual plants and to have a global perspective on the actual behaviour modes that the system can display, the use of Hamiltonian methods should be complemented by additional analysis as the one proposed in the paper. Thus, it is then possible to be warned about the appearance of undesired phenomena and, in particular, to guarantee global stability for the target operating point.

4. REFERENCES

Freire, E., E. Ponce and J. Ros (1999). Limit cycle bifurcation from center in symmetric piecewise-linear systems. *International J. Bifurcation and Chaos* **9**, 895–907.

Kuznetsov, Y.A. (1998). *Elements of Applied Bifurcation Theory.* 2nd ed.. Springer. New York.

Pagano, D.J., E. Ponce and J. Aracil (1999). Bifurcation analysis of time-delay control systems with saturation. *International J. Bifurcation and Chaos* **9**, 1089–1109.

van der Schaft, A. (1996). *L_2-Gain and Passivity Techniques in Nonlinear Control.* Springer. London.

MODIFICATION OF HAMILTONIAN STRUCTURE TO STABILIZE AN UNDERWATER VEHICLE

Craig A. Woolsey *,1,2 **Naomi Ehrich Leonard** *,1

* *Department of Mechanical and Aerospace Engineering, Princeton University, Princeton, NJ 08544*

Abstract: This paper presents new results on stabilization of an underwater vehicle using internal rotors. In previous work, a stabilizing control law was derived which preserves the open-loop Hamiltonian structure in the closed-loop system but which modifies the Hamiltonian. The results presented here illustrate the utility of feedback control that not only shapes the energy but also modifies the Hamiltonian structure. *Copyright ©2000 IFAC*

Keywords: Lyapunov methods, feedback stabilization, rotors, underwater vehicle

1. INTRODUCTION

Internal rotors can provide energy shaping that stabilizes steady forward motion of an underwater vehicle with dynamics described by Kirchhoff's equations (Leonard and Woolsey, 1998; Woolsey and Leonard, 1999*b*; Bloch *et al.*, 2000). Kirchhoff's equations, which are Hamiltonian (Lie-Poisson), are stabilized with a feedback law that preserves the Lie-Poisson structure in the closed loop but which modifies the Hamiltonian. Stability requires choosing control gains so that the equilibrium is a maximum of a Lyapunov function constructed from the modified Hamiltonian. Physical dissipation (in this case, viscous drag due to the body's motion through the fluid) decreases the energy and tends to destabilize the equilibrium. The control law proposed here provides stabilization in such a way that physical dissipation enhances stability. The control law provides a Hamiltonian closed-loop system with a modified Hamiltonian structure, as well as a modified Hamiltonian. Modification of Hamiltonian structure is inspired in part by Krishnaprasad (1985).

[1] Research partially supported by NSF grant BES-9502477 and ONR grant N00014-98-1-0649
[2] Research partially supported by NDSEG Fellowship

2. FEEDBACK STABILIZATION

Consider an ellipsoidal vehicle with three internal rotors and let the ellipsoid principal axes define a body-fixed coordinate frame. Each rotor is axisymmetric and spins about its symmetry axis under the influence of a control torque. The rotors are mounted orthogonally within the vehicle so that each rotor's symmetry axis is aligned with a body coordinate axis. Assume that the vehicle mass is uniformly distributed so that the center of gravity (CG) is also the center of buoyancy (CB).

Let the diagonal matrix $\boldsymbol{I} = \mathrm{diag}(I_1, I_2, I_3)$ represent the inertia of the vehicle without rotors plus the added inertia of the fluid. Similarly, let the diagonal matrix $\boldsymbol{M} = \mathrm{diag}(m_1, m_2, m_3)$ represent the mass of the vehicle multiplied by the identity matrix plus the added mass matrix of the fluid. We assume that the vehicle's 1-axis is longest and that its 3-axis is shortest. Then, $m_1 < m_2 < m_3$.

Let $\mathrm{diag}(J_1^i, J_2^i, J_3^i)$ be the inertia matrix of the rotor which spins about the ith coordinate axis ($i = 1, 2$, or 3). The total inertia, with the rotors locked in place, is $\boldsymbol{\Lambda} = \mathrm{diag}(\lambda_1, \lambda_2, \lambda_3)$ where

$$\lambda_j = I_j + J_j^1 + J_j^2 + J_j^3, \quad j = 1, 2, 3.$$

We also define the matrix $\boldsymbol{J_r} = \mathrm{diag}(J_1^1, J_2^2, J_3^3)$.

Let $\boldsymbol{\Omega}$ and $\boldsymbol{v}$ represent the body angular and linear velocity, respectively, in body coordinates. Also, define $\boldsymbol{\Omega_r} = (\Omega_{r_1}, \Omega_{r_2}, \Omega_{r_3})^T$, where Ω_{r_i} is the angular rate of the ith rotor relative to the vehicle. The body momenta are given by

$$\begin{pmatrix} \boldsymbol{\Pi} \\ \boldsymbol{P} \\ \boldsymbol{l} \end{pmatrix} = \begin{pmatrix} \boldsymbol{\Lambda} & 0 & \boldsymbol{J_r} \\ 0 & \boldsymbol{M} & 0 \\ \boldsymbol{J_r} & 0 & \boldsymbol{J_r} \end{pmatrix} \begin{pmatrix} \boldsymbol{\Omega} \\ \boldsymbol{v} \\ \boldsymbol{\Omega_r} \end{pmatrix}.$$

The Hamiltonian is the total kinetic energy,

$$H = \frac{1}{2} \begin{pmatrix} \boldsymbol{\Pi} \\ \boldsymbol{P} \\ \boldsymbol{l} \end{pmatrix} \cdot \begin{pmatrix} \boldsymbol{\Lambda} & 0 & \boldsymbol{J_r} \\ 0 & \boldsymbol{M} & 0 \\ \boldsymbol{J_r} & 0 & \boldsymbol{J_r} \end{pmatrix}^{-1} \begin{pmatrix} \boldsymbol{\Pi} \\ \boldsymbol{P} \\ \boldsymbol{l} \end{pmatrix}.$$

The equations of motion are

$$\begin{pmatrix} \dot{\boldsymbol{\Pi}} \\ \dot{\boldsymbol{P}} \\ \dot{\boldsymbol{l}} \end{pmatrix} = \begin{pmatrix} \hat{\boldsymbol{\Pi}} & \hat{\boldsymbol{P}} & 0 \\ \hat{\boldsymbol{P}} & 0 & 0 \\ 0 & 0 & 0 \end{pmatrix} \nabla H + \begin{pmatrix} 0 \\ 0 \\ \boldsymbol{u} \end{pmatrix},$$

where $\boldsymbol{u} = (u_1, u_2, u_3)^T$ is the control and u_i is the torque applied to the ith internal rotor about its spin axis. With $\boldsymbol{u} = \boldsymbol{0}$, it can be verified that steady translation of the vehicle along its long axis is an unstable relative equilibrium (Lamb, 1932).

The control law developed in our earlier work is

$$\boldsymbol{u} = k\dot{\boldsymbol{\Pi}} = k(\boldsymbol{\Pi} \times \boldsymbol{\Omega} + \boldsymbol{P} \times \boldsymbol{v}), \qquad (1)$$

where k is a control gain. Let

$$\boldsymbol{\zeta} = \frac{1}{1-k}(\boldsymbol{l} - k\boldsymbol{\Pi})$$

and change coordinates from $(\boldsymbol{\Pi}, \boldsymbol{P}, \boldsymbol{l})$ to $(\boldsymbol{\Pi}, \boldsymbol{P}, \boldsymbol{\zeta})$. Note that $\boldsymbol{\zeta}$ is conserved, by construction. The closed-loop equations of motion are Lie-Poisson with respect to the energy $H_R =$

$$\frac{1}{2}(\boldsymbol{\Pi} - \boldsymbol{\zeta}) \cdot \boldsymbol{I_C}^{-1}(\boldsymbol{\Pi} - \boldsymbol{\zeta}) + \frac{1}{2}\boldsymbol{P} \cdot \boldsymbol{M}^{-1}\boldsymbol{P},$$

where $\boldsymbol{I_C} = \frac{1}{1-k}\bar{\boldsymbol{I}}$. The equations are

$$\begin{pmatrix} \dot{\boldsymbol{\Pi}} \\ \dot{\boldsymbol{P}} \\ \dot{\boldsymbol{\zeta}} \end{pmatrix} = \begin{pmatrix} \hat{\boldsymbol{\Pi}} & \hat{\boldsymbol{P}} & 0 \\ \hat{\boldsymbol{P}} & 0 & 0 \\ 0 & 0 & 0 \end{pmatrix} \nabla H_R.$$

Conditions for stability may be determined by applying the energy-Casimir method. The method proves stability of steady long-axis translation if we choose $k > 1$ ($\boldsymbol{I_C} < \boldsymbol{0}$). The method also provides a Lyapunov function, H_Φ, constructed from the energy H_R and other conserved quantities. The desired equilibrium is a maximum of H_Φ. Fluid drag, which is not included in this model, tends to destabilize the stabilized equilibrium by decreasing H_Φ. Asymptotic stability thus requires additional feedback dissipation to dominate drag (Woolsey and Leonard, 1999*a*).

The new control law that we propose is

$$\boldsymbol{u} = k(\boldsymbol{P} \times \boldsymbol{v}). \qquad (2)$$

This control law is a modification of the original control law (1) and was formulated from physical intuition (see also Leonard (1996)). The closed-loop dynamics are

$$\begin{pmatrix} \dot{\boldsymbol{\Pi}} \\ \dot{\boldsymbol{P}} \\ \dot{\boldsymbol{\zeta}} \end{pmatrix} = \begin{pmatrix} \hat{\boldsymbol{\Pi}} & \hat{\boldsymbol{P}} & 0 \\ \hat{\boldsymbol{P}} & 0 & 0 \\ 0 & 0 & \frac{k}{1-k}\hat{\boldsymbol{\Pi}} \end{pmatrix} \nabla H_R.$$

This system is almost Lie-Poisson (an implicit generalized Hamiltonian system in the sense of van der Schaft (1998)); that is, the corresponding Poisson bracket does not satisfy the Jacobi identity. Asymptotic stability of steady long-axis translation can be proven by choosing $k > 1$, using H_R to construct a negative semidefinite Lyapunov function, and applying feedback dissipation. In this case, however, fluid drag tends to increase the Lyapunov function, enhancing stability of the desired equilibrium. In addition, when extending these ideas to the case of noncoincident CG and CB, a generalization of the control law (1) requires that the CG be above the CB, whereas a generalization of (2) requires the more practical low CG for stability.

3. REFERENCES

Bloch, A. M., N. E. Leonard and J. E. Marsden (2000). Controlled Lagrangians and the stabilization of Euler-Poincaré mechanical systems. Preprint.

Krishnaprasad, P. S. (1985). Lie-Poisson structures, dual-spin spacecraft and asymptotic stability. *Nonlinear Anal., Theory, Meth. & App.* **9**(10), 1011–1035.

Lamb, H. (1932). *Hydrodynamics*. 6th ed.. Dover.

Leonard, N. E. (1996). Stability and stabilization of underwater vehicle dynamics. In: *Proc. CISS*. Princeton, NJ. pp. 771–775.

Leonard, N. E. and C. Woolsey (1998). Internal actuation for intelligent underwater vehicle control. In: *Proc. 10th Yale Workshop on Adaptive and Learning Sys.*. pp. 295–300.

van der Schaft, A.J. (1998). Implicit Hamiltonian systems with symmetry. *Reports on Mathematical Physics* **41**(2), 203–221.

Woolsey, C. and N. E. Leonard (1999*a*). Global asymptotic stabilization of an underwater vehicle using internal rotors. In: *Proc. IEEE CDC*. Vol. 38. pp. 2527–2532.

Woolsey, C. and N. E. Leonard (1999*b*). Underwater vehicle stabilization by internal rotors. In: *Proc. ACC*. pp. 3417–3421.

MATCHING AND STABILIZATION OF THE UNICYCLE WITH RIDER

Dmitry V. Zenkov *,1,2 Anthony M. Bloch *,2
Naomi E. Leonard **,3 Jerrold E. Marsden ***,4

* *Department of Mathematics, University of Michigan, Ann Arbor, MI 48109*
** *Department of Mechanical and Aerospace Engineering, Princeton University, Princeton, NJ 08544*
*** *Control and Dynamical Systems, California Institute of Technology 107-81, Pasadena, CA 91125*

Abstract: In this paper we apply matching techniques for controlled Lagrangians to the stabilization problem of a nonholonomic system consisting of a unicycle with rider. We show how generalized matching results may be applied to the Routhian associated with this nonholonomic system. *Copyright © 2000 IFAC*

Keywords: Feedback stabilization, Lyapunov methods, Nonlinear control

1. INTRODUCTION

In this paper we apply the method of controlled Lagrangians to the stabilization of slow vertical steady state motions of the unicycle with rider. The controlled Lagrangian approach for stabilization was introduced in Bloch, *et al.* (1997) for underactuated holonomic systems with the control force acting along the symmetry directions. Later on this approach was extended to handle certain systems with broken symmetry, see (Bloch, *et al.*, 1999; Auckly, *et al.*, 1998; Hamberg, 1999).

This method requires that specific *matching conditions* are satisfied. These conditions allow one to introduce a *controlled Lagrangian* and to rewrite the equations for the controlled (forced) system as the Euler-Lagrange equations for this controlled Lagrangian.

The method proposed here extends the technique of controlled Lagrangians to a class of nonholonomic systems and gives a systematic procedure for control design in both the linear and the nonlinear settings.

[1] Research partially supported by a University of Michigan Rackham Fellowship
[2] Research partially supported by NSF grant DMS-9803181, AFOSR grant F49620-96-1-0100, and an NSF group infrastructure grant at the University of Michigan
[3] Research partially supported by NSF grant BES-9502477 and ONR grant N00014-98-1-0649
[4] Research partially supported by AFOSR grant F49620-95-1-0419

2. NONHOLONOMIC MATCHING

The system considered here, the unicycle with rider, is modeled by a homogeneous disk that moves on a horizontal plane without slipping and has a mass and a pendulum attached. The pendulum is free to move in the plane orthogonal to the disk, while the attached mass stays in the disk's plane. In this system only the sideways motion of the rider (such as the rider's limbs) is modeled and not any pedaling control. The configuration space is $Q = S^1 \times S^1 \times S^1 \times SE(2)$, which we parametrize with coordinates $(r^1, r^2, \psi, \phi, x, y)$, as in Figure 1. This mechanical system is $SO(2) \times SE(2)$-invariant; the group $SO(2)$ represents the symmetry of the wheel,

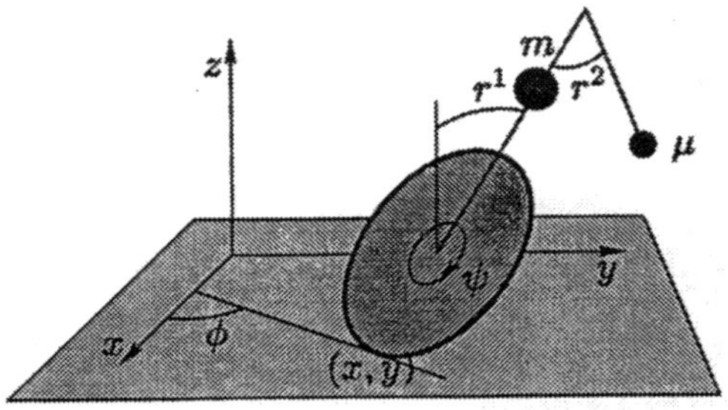

Fig. 1. The configuration variables for the unicycle with rider.

that is in the ψ variable, while the group $SE(2)$ represents the Euclidean symmetry of the overall system.

The equations of motion with a control torque u on the pendulum are those derived in the standard way from the Lagrange-d'Alembert principle:

$$\frac{d}{dt}\frac{\partial \mathcal{R}}{\partial \dot{r}^1} = \nabla_1 \mathcal{R}, \qquad \frac{d}{dt}\frac{\partial \mathcal{R}}{\partial \dot{r}^2} = \nabla_2 \mathcal{R} + u, \qquad (1)$$

$$\frac{dp_1}{dt} = \mathcal{D}^b_{1\alpha} p_b \dot{r}^\alpha, \qquad \frac{dp_2}{dt} = \mathcal{D}^b_{2\alpha} p_b \dot{r}^\alpha, \qquad (2)$$

where $\mathcal{R} = \frac{1}{2} g_{\alpha\beta} \dot{r}^\alpha \dot{r}^\beta - U(r,p)$ is the *Routhian*, U is the *amended potential*, (p_1, p_2) is the nonholonomic momentum where p_1 is conjugate to ϕ, p_2 is conjugate to ψ, and the covariant derivatives in the shape equations (1) are defined by

$$\nabla_\alpha = \partial_{r^\alpha} + \mathcal{D}^b_{a\alpha} p_b \partial_{p_a}.$$

See (Zenkov, *et al.*, 1998, 1999) and references therein. The full dynamics is governed by equations (1) and (2) coupled with the *reconstruction equation* for the group variables ψ, ϕ, x, y. This reconstruction equation is not needed here as it does not affect the evolution of the shape and the momentum variables, and thus is not used in our stabilization analysis.

Our key observation is that the vertical steady state motions of the unicycle

$$r^1 = 0, \quad r^2 = 0, \quad p_1 = 0, \quad p_2 = c \qquad (3)$$

for each value of c are dynamically equivalent to the equilibria of an auxiliary holonomic system, the inverted double pendulum. The Lagrangian for this auxiliary system is the Routhian of the original system restricted to the level set $p = (0, c)$ of the nonholonomic momentum. Applying holonomic matching techniques, we obtain the *controlled metric* $\widetilde{g}_{\alpha\beta}$ and the *controlled amended potential* $\widetilde{U}$ and form the *controlled Routhian* $\widetilde{\mathcal{R}} = \frac{1}{2}\widetilde{g}_{\alpha\beta}\dot{r}^\alpha \dot{r}^\beta - \widetilde{U}$. The *controlled energy* corresponding to this controlled Routhian can be chosen, with appropriate choices of gains, to be positive definite at equilibrium (3). We then show that there exist *controlled covariant derivatives* $\widetilde{\nabla}_\alpha$, such that the equations

$$\frac{d}{dt}\frac{\partial \widetilde{\mathcal{R}}}{\partial \dot{r}^\alpha} = \widetilde{\nabla}_\alpha \widetilde{\mathcal{R}} \qquad (4)$$

coupled with (2) are equivalent to the original equations (1) and (2). Equations (2) and (4) linearized at (3) have two zero and four pure imaginary eigenvalues. By adding appropriate dissipative terms to the control input, we force the four nonzero eigenvalues to the left half plane. By the Lyapunov-Malkin theorem, the slow vertical motion of the unicycle (3) (with small c) becomes orbitally stable. See (Zenkov, *et al.*, 1998, 1999) for the details on the Lyapunov-Malkin theorem and on the dissipative terms.

The explicit formula for the control u is

$$u = \nabla_2 U - g_{2\beta}\widetilde{g}^{\alpha\beta}\widetilde{\nabla}_\alpha \widetilde{U} - g_{2\gamma}(\widetilde{\Gamma}^\gamma_{\alpha\beta} - \Gamma^\gamma_{\alpha\beta})\dot{r}^\alpha \dot{r}^\beta + \{\text{dissipative terms}\},$$

where $\Gamma^\gamma_{\alpha\beta}$ and $\widetilde{\Gamma}^\gamma_{\alpha\beta}$ are the Christoffel symbols of the metrics $g_{\alpha\beta}$ and $\widetilde{g}_{\alpha\beta}$.

3. CONCLUSION

The method developed here extends the matching technique to the class of nonholonomic systems with no curvature terms in the shape equation and the momentum equation in the form of a parallel transport equation. We intend in a future publication to consider more general nonholonomic systems, in particular with control inputs acting along some of the symmetry directions as well.

REFERENCES

Auckly, D., L. Kapitanski and W. White (1998) Control of Nonlinear Underactuated Systems. *Preprint.*

Bloch, A.M., N.E. Leonard and J.E. Marsden (1997) Stabilization of Mechanical Systems Using Controlled Lagrangians. *Proc. CDC*, **36**, 2356–2361.

Bloch, A.M., N.E. Leonard and J.E. Marsden (1999) Potential Shaping and the Method of Controlled Lagrangians *Proc. CDC*, **38**, 1652–1657.

Hamberg, J. (1999) General Matching Conditions in the Theory of Controlled Lagrangians. *Proc. CDC*, **38**, 2519–2523.

Zenkov, D.V., A.M. Bloch and J.E. Marsden (1998) The Energy-Momentum Method for Stability of Nonholonomic Systems. *Dynamics and Stability of Systems*, **13**, 123–165.

Zenkov, D.V., A.M. Bloch and J.E. Marsden (1999) Stabilization of the Unicycle with Rider. *Proc. CDC*, **38**, 3470–3471.

AUTHOR INDEX

www.ingramcontent.com/pod-product-compliance
Lightning Source LLC
Chambersburg PA
CBHW072058220326
41598CB00068BA/4455
9780080436586